Frontiers in Urban Water Management

Frontiers in Urban Water Management

Deadlock or hope

Edited by

Čedo Maksimović
Urban-Water Research Group
Department of Civil and Environmental Engineering,
Imperial College of Science, Technology and Medicine,
London, UK
and
Director, Centre for Urban Water, CUW-UK, London, UK

and

José Alberto Tejada-Guibert
Secretariat of the International Hydrological Programme,
Division of Water Sciences, UNESCO, Paris, FRANCE

IWA Publishing

UNESCO

Published by IWA Publishing, Alliance House, 12 Caxton Street, London SW1H 0QS, UK
Telephone: +44 (0) 20 7654 5500; Fax: +44 (0) 20 7654 5555; Email: publications@iwap.co.uk
www.iwapublishing.com

First published 2001
Reprinted 2005

Original cover design by Ivana Maksimović.

British Library Cataloguing in Publication Data
A CIP catalogue record for this book is available from the British Library

Library of Congress Cataloging- in-Publication Data
A catalog record for this book is available from the Library of Congress

ISBN: 1900222760

Contents

Preface

Urban water problems are mounting all over the globe. Widespread mismanagement of water resources, growing competition for the use of freshwater, degraded sources – sometimes by pollutants of unpredictable effects – only heighten the acuteness of the problems. The explosive growth of urbanization, particularly in the developing world, exhibiting as its most visible expression the emergence of megacities – often obeying a massive internal migration toward the cities that feeds an uncontrolled, unplanned expansion – exacerbates the situation in those locations.

Facing this grim state of affairs represents is a formidable challenge to society at large and to urban water planners, managers and decision makers, in particular, in the next generations. It is now commonly acknowledged that "business as usual" attitudes will not work, and that solutions cannot simply be transferred from one reality to the other – whether we are talking about technological components, institutional structures, or expected responses of the people. Thus, new approaches have to be developed, the range of applicable technologies must be broadened, and awareness and communication between stakeholders must be enhanced. But how?

This book endeavours to address these issues in a realistic and constructive way. It is forward looking and investigates the frontiers of knowledge and experience in urban water management. It analyses the current circumstances from multiple points of views both in the thematic and disciplinary realms and points to promising directions. For this purpose, it has gathered a distinguished group of international authors widely recognized for their expertise in their respective fields and representing different corners of the world.

The book is organized in a sequence of chapters, aiming first to define the nature of the problem, then continues examining relevant planning, environmental and infrastructural issues, it explores the emerging technologies, and finally delves into economic, social and institutional aspects. The book successively deals with:

The challenge or urban water management: A global view of developmental and environmental issues conditioning future urban water management and quality of life.

Urban water as a part of integrated catchment management: The effects of urbanization and urban water planning are considered from the catchment perspective, considering the physical, planning and management processes.

Interactions with the environment: The effect of withdrawals of freshwater and disposal of wastewater, as well as the interactions between urban water systems are addressed.

Infrastructure integration issues: The relationship and the integration of the different components of the urban water services are examined within an urban planning context.

Emerging paradigms in water supply and sanitation: New approaches and alternative technologies that can alter traditional ways of solving age-old problems of water supply and sanitation are brought forward.

Problems of developing countries: The water problems of cities in developing countries are reviewed critically: pattern of growth (including periurban areas), the nature of water services, technological and social factors, etc.

Economic and financial aspects: The economic nature of water, financing of urban water projects in developing countries, sustainable financing of urban water management institutions and the role of international funding agencies are discussed.

Social, institutional and regulatory issues: The shift of the management paradigm, the relevance of the institutional framework and of the role of the stakeholders and the legal background are examined.

Outlook for the 21st century: Based on the findings in the above aspects, perspectives for the 21st century on urban water issues and promising approaches are presented.

The chapters may be considered a sequence of stand-alone essays with the distinctive style and points of view of the authors of each, yet with sufficient overlap to provide continuity and cover integrally the major theme of the book.

This book represents one of the culminating activities on the theme of *Integrated Urban Water Management* of the Fifth Phase (1996–2001) of the International Hydrological Programme (IHP) of UNESCO. It was prepared as a contribution to the International Symposium "Frontiers of Urban Water Management: Deadlock Or Hope?" (Marseille, 18–20 June 2001), which was organized by UNESCO and the Académie de l'Eau de France, with the support of the City of Marseille and of the Secretariat of the World Water Council.

The above symposium, in addition to the institutions mentioned, had the backing of numerous partners, who collectively gave decisive help to make this book feasible. Among them the Agence de l'Eau Seine Normandie deserves a special mention, having been responsible for producing an extended chapter on infrastructure and for publishing the French version of this book. The other partners included (in alphabetical order): Agence de l'Eau Rhône-Mediterranée-Corse; Chambre de Commerce et d'Industrie Marseille Provence; Conseil Général des Bouches-du-Rhône; Conseil Régional Provence-Alpes-Côte d'Azur; Groupe des Eaux de Marseille; International Association of Hydraulic Engineering and Research; International Association of Hydrological Sciences; International Water Association; International Water Resources Association; Ministère de l'équipement, des transports et du logement; Office International de l'Eau; Préfecture des Bouches-du-Rhône; UNCHS – Habitat, the United Nation Environment Programme; and the World Health Organization.

Our deep thanks go to all contributing authors, whose sustained effort over the past year and a half has made this book a reality. Likewise, we thank the numerous individuals and institutions who authorised us to use their material as illustrations in the book.

It is the fervent hope of the editors that this book may not only constitute a valuable contribution to the above symposium, but that it will also be widely read and be considered a helpful reference in setting the course for tomorrow in urban water management practices.

Čedo Maksimović
Urban Water Research Group
Department of Civil and Environmental Engineering
Imperial College of Science, Technology & Medicine
and Centre for Urban Water (CUW-UK)
London, UK

José Alberto Tejada-Guibert
International Hydrological Programme (IHP)
Division of Water Sciences
UN Educational, Scientific and Cultural
Organisation (UNESCO)
Paris, FRANCE

April 2001

List of contributors

The ideas and opinions expressed in this book are those of the individual authors and do not necessarily represent the views of their employers.

Chapter 1: The Challenge of Urban Water Management

Prof. Evan VLACHOS
Sociology & Civil Engineering, Colorado State University, Fort Collins, Colorado 80523-1784, USA
Tel: +1 970/491-6089; Fax: +1 970491-2191;
E-mail: evlachos@engr.colostate.edu

Prof. Benedito BRAGA
Director, Agencia Nacional de Aguas – ANA, SCN quadra 1 Bloco A - 5 andar 70710-500 Brasilia, DF, Brazil
Tel: (+55) 61-425-2233; Fax: (+55) 61-327-3427,
E-mail: benbraga@uol.com.br <mailto:benbraga@uol.com.br>

Chapter 2: Urban Water as a Part of Integrated Catchment Management

Dr. Jiri MARSALEK
National Water Research Institute, 867 Lakeshore Road, Burlington, Ontario L7R 4A6, CANADA
Tel: +1-905-336-4899; Fax: +1-905-336-4420;
E-mail: jiri.marsalek@cciw.ca

Mr. Quintin ROCHFORT
National Water Research Institute, 867 Lakeshore Road, Burlington, Ontario L7R 4A6, CANADA
Tel: +1-905-336-4434; Fax: +1-905-336-4420;
Email: quintin.rochfort@cciw.ca

Prof. Dragan SAVIĆ
School of Engineering and Computer Science, Department of Engineering, University of Exeter, Exeter EX4 4QF, UK
Tel: +44 1392 263637; Fax: +44 1392 217965;
E-mail: d.savic@ex.ac.uk

Chapter 3: Interactions with the Environment

Dr. David BUTLER
Department of Civil and Environmental Engineering, Imperial College of Science, Technology and Medicine, London, UK
Tel: + 44 207 594 60 99; Fax: +44 207 225 2716;
E-mail: d.butler@ic.ac.uk

Prof. Čedo MAKSIMOVIĆ
Department of Civil and Environmental Engineering, Imperial College of Science, Technology and Medicine, London SW7 2BU, UK
Tel: + 44 207 594 6013; Fax: +44 207 225 2716 / +44 207 7068105;
E-mail: c.maksimovic@ic.ac.uk
(also Director, Centre for Urban Water, CUW-UK, London, UK)

Chapter 4: Infrastructure Integration Issues

Mr. Pierre-Alain ROCHE
Directeur Général, Agence de l'Eau Seine-Normandie, 51,
rue Salvador Allende, 92027 Nanterre Cedex , FRANCE
Tel: +33 (0) 1 41 20 17 21; Fax: +33 (0) 1 41 20 16 09;
E-mail: dirgene@aesn.fr

Mr. François VALIRON
Académie de l'eau, 51 rue Salvador Allende, 92027 NANTERRE cedex,, France
Tel: +33 (0) 1 41 20 18 56; Fax: +33 (0) 1 41 20 16 09;
E-mail: academie@oieau.fr

Mr. Rene COULOMB
Vice-president, World Water Council, Special Advisor to the President,
Suez Lyonnaise des Eaux, 1 rue d'Astorg, 75383 Paris Cedex 08, FRANCE
Tel: +33 (0) 1 40 06 66 31; Fax: +33 (0) 1 40 06 66 36;
E-mail: rene_coulomb@mail.suez-lyonnaise.eaux.fr

Mr. Daniel VILLESSOT
Technical Director, France Technology and Research Division,
18 square Edouard VII, 75316 Paris Cedex 09, FRANCE
Tel +33 1 46 95 52 44, Fax +33 1 46 95 46 14;
E-mail: daniel.villessot@lyonnaise-des-eaux.fr

Chapter 5: Emerging Paradigms in Water Supply and Sanitation

Prof. Saburo MATSUI
Research Center for Environmental Quality Control, Kyoto University,
1-2 Yumihama, Otsu City, Shiga, JAPAN 520-0811
Tel: +81-77-527-6620; Fax: +81-77-524-9869;
E-mail: matsui@biwa.eqc.kyoto-u.ac.jp

Prof. Mogens HENZE
Department of Environmental Science and Engineering, Building 115,
Danish Technical University, DK-2800 Lyngby, DENMARK
Tel: +4525 1600 / +4525 1477 (direct); Fax: + 4593 2850;
E-mail: mh@imt.dtu.dk

Prof. Ralf OTTERPOHL
TUHH Technical University Hamburg-Harburg,
Director, Dept. for Municipal and Industrial Wastewater Management
Eissendorfer Str.42, D-21073 Hamburg, GERMANY
Tel: +49 40 42878 3207; Fax: +49 40 42878 2684:
E-mail: otterpohl@tu-harburg.de

Prof. Gohen HO
Murdoch University, Environmental Engineering, South Street, Murdoch, WESTERN AUSTRALIA
Tel: + 61 8 9360 2167; Fax: +61 8 9310 4997;
E-mail: ho@essun1.murdoch.edu.au

Chapter 6: Problems of Developing Countries

Prof. David STEPHENSON
Department of Civil Engineering, Director of Water Systems Research Group, University of the Witwatersrand, Johannesburg, PO Box 277, Wits, 2050 SOUTH AFRICA
Tel: +27 11 7162560; Fax: +27 11 4032062;
E-mail: steph@civil1.civil.wits.ac.za

Chapter 7: Economic and Financial Issues

Mr. Terence Richard LEE
Consultant (formerly in Economic Commission for Latin America and the Caribbean)
Ismael Valdes Vergara 360, dpto.71, Santiago, CHILE
Tel: (56-2) 639-1240; Fax: (56-2) 664-3806;
E-mail: tlee@netline.cl

Mr. Jean-Louis OLIVER
Ingenieur en Chef des Ponts et Chausées,
Ministère de l'equipement, des transports et du logement, Conseil général des ponts et Chausées, Tour Pascal B, 92055 La Défense Cedex, FRANCE
Tel: +33 1 4081 2772; Fax: +33 1 4081 2770;
E-mail: kalitka@daei.equipement.gouv.fr

Mr. Pierre-Frédéric TENIÈRE-BUCHOT
UNEP/GPA, The Hague, THE NETHERLANDS
Tel: +31 70 311 44 71 or +33 (0)6 09 69 2992 (France); Fax: +31 70 345 6648;
E-mail: pf.teniere-buchot@unep.nl

Mr. Lee TRAVERS
1818 H St. NW, Washington, DC 20433
Tel: 202 458-4076; Fax: 202 522-3228:
E-mail: stravers@worldbank.org

Mr. François VALIRON
(See Chapter 4 for details)

Chapter 8: Social, Institutional and Regulatory Issues

Prof. Jan LUNDQVIST
Tema Institute, Faculty of Arts and Sciences, Linkoping University,
581 83 Linkoping, SWEDEN
Tel: +46 13 282272; Fax: +46 13 133630;
E-mail: janlu@tema.liu.se

Ms. Sunita NARAIN
Director, Centre for Science and Environment, 41, Tughlakabad Institutional Area, New Delhi-110062, India
Tel: +91 11 6081110; Fax: +91 11 6085879;
E-mail: sunita@cseindia.org

Mr. Anthony TURTON
Head, African Water Issues Research, University of Pretoria, Pretoria 0002, Republic of South Africa
Tel: +27 12 4204486; Fax: + 27 12 4203886;
E-mail: awiru@postino.up.ac.za

Chapter 9: Outlook for the 21st Century

Mr. José Alberto TEJADA-GUIBERT
Division of Water Sciences, UNESCO - International Hydrological Programme,
1, rue Miollis, 75732 Paris Cedex 15, FRANCE
Tel: +33 1 45 68 40 96; Fax: +33 1 45 68 58 11;
E-mail: ja.tejada-guibert@unesco.org

Prof. Čedo MAKSIMOVIĆ
(See Chapter 3 for details)

1

The challenge of urban water management

Evan Vlachos and Benedito Braga

1.1 INTRODUCTION

While other parts of this volume examine in considerable detail various aspects of urban water management, the present overview addresses broader concerns with water and human settlements, as well as transformations of a fast changing environment. The introductory argument attempts to set the stage for understanding the opportunities and challenges involved in planning and managing water in increasingly expanding urban settlements all over the planet.

The organising concept in the following pages revolves around the need for a forward-looking, participatory, integrated water management. Four interrelated

© 2001 IWA Publishing. *Frontiers in Urban Water Management: Deadlock or Hope*
Edited by Čedo Maksimović and José Alberto Tejada-Guibert. ISBN: 1 900222 76 0

parts relate past practices, present efforts and potential futures. Beginning with a brief overview of the *urban context,* larger forces of change, patterns of growth and shifting paradigms attempt to outline the critical connection of population, water and cities. A second part outlines *the search for liveable cities* in terms of what sustainable growth implies and the linkages between urban ecosystems and sustainable cities. The third part turns attention to efforts for *implementing sustainable urban water management.* Here, a series of new configurations of urban water systems lead to the need for capacity building, integrated management, and new planning horizons. Finally, the last part allows for *speculating about urban water futures,* including alternative urban water scenarios, a vision of eco-cities, and some broad guidelines linking theoretical premises with actual practices in a variety of cultural settings. Such a sequential presentation is organised around the central concept of a search for sustainable growth and development that underscores conceptual and methodological challenges in a shifting paradigm of integrated water management.

1.2 THE URBAN CONTEXT

1.2.1 The larger forces of change and urbanisation

A constant theme at the end of the second millennium has been both the rapidity of social change as well as increasing complexity, both united in the composite term of 'raplexity' as emphasising both rate and extent of transformation. Increasing size, density and heterogeneity call for massive efforts for managing water supply, waste water (quality), and storm water systems. They also call for escalating investment costs of spatially diffused entities and reinforce cumulative and interactive effects of various interrelated aspects of urbanisation and comprehensive land use planning.

If one looks at the cities of both developed and developing countries it becomes apparent that rather urgent decisions must be made with regard to such overriding freshwater problems as increasing demands for water for various users; changes in the physical environment altering the water balance; and the disposal of wastes that may contaminate streams and groundwater. An initial list of trends and concerns that contribute to a whole series of problems characterising, especially larger metropolitan complexes, would include among others: the presence of large numbers of people in limited space; increasing costs of public services and inadequate tax structures; decay and obsolete facilities, especially in the core city; pronounced disparities between core and suburbs and the attendant problems of social differentiation and inequitable distribution of opportunities; fractionalised governmental units and overlapping jurisdictions; and, haphazard urban sprawl and piecemeal, ad hoc sectoral planning.

Although the above concerns relate to specific problems of urban settlements, they are at the same time part of the backdrop of a more generalised concern for developing new approaches to planning and natural resources management that would:

- place emphasis on stronger interdisciplinary approaches integrating a variety of "environments" in any planning effort;
- underscore the search for a "social calculus" of project effects and the establishment of a wider range of activities;
- expand the time horizon from a narrow enumeration of immediate impacts, to a long-range forecasting of higher-order consequences;
- combine the historical experience of the past, the constraints of the present, and the visions of both probable and desirable futures.

Such observations can become even more complex, if they are seen as part of a series of "strains" that contemporary society is facing. In the literature, a whole litany of such "strains" and "stresses" has been presented, usually in terms of a fast approaching clash between expanding demands and limited resources. Specifically three "strains" are seen as central in any discussion of the physical–social environment interfaces: a) *ecological strains*, part of exhaustion of strategic natural resources and of spillovers of the technical apparatus of society; b) *economic strains*, resulting from large economic paralysis, global interdependence, and crisis of confidence; and c) *cultural strains*, result of a lag between technology and non-material culture, changing values, and the collapse of expectations vis-à-vis limited achievements in many countries.

All in all, rapid population growth and sprawling urbanisation (Figure 1.1), coupled with increasing environmental impacts and consequences, contribute to increasing societal complexity and interdependence through magnification of effects on the surrounding environment, in terms of intensity, severity and duration; give rise to fundamental questions as to the distribution of effects to various segments of population; and, require strong efforts to cope in a flexible and responsive manner to both short and long-term problems of rapid urbanisation. At the same time, the rise of metropolitan and megalopolitan[1] agglomerations give rise to renewed concerns as to the control of the natural

1 coming together of a lot of metropolitan areas (e.g. Boswas the Boston-Washington Megalopolis or SanSan the San Francisco-San Diego Megalopolis)

metabolism of infrastructure networks; as to the continuous demand for public services and facilities; as to the need for new institutional arrangements for efficient and equitable water supplies; and as to the quest for comprehensive management that incorporates also public involvement and participation.

It is at this junction that past historical practices, limited if not obsolete water systems, and expanding or new demands pose a tremendous challenge to all efforts of comprehensive planning and construction of larger scale, integrated systems able to respond to changing socio-economic circumstances. It is also important to notice that most urban activities are intricately related to the development and use of the land. In the case of water resources, traditionally, the rural or low-density suburban area outside a city initially develops its water supply and sewage disposal facilities on a private, house-by-house basis. As the area becomes more developed and density increases, small water systems emerge. Finally, as an urban constellation takes place and population increases rapidly, the limited water system becomes strained and proves inadequate, forcing communities to consider much larger, permanent facilities.

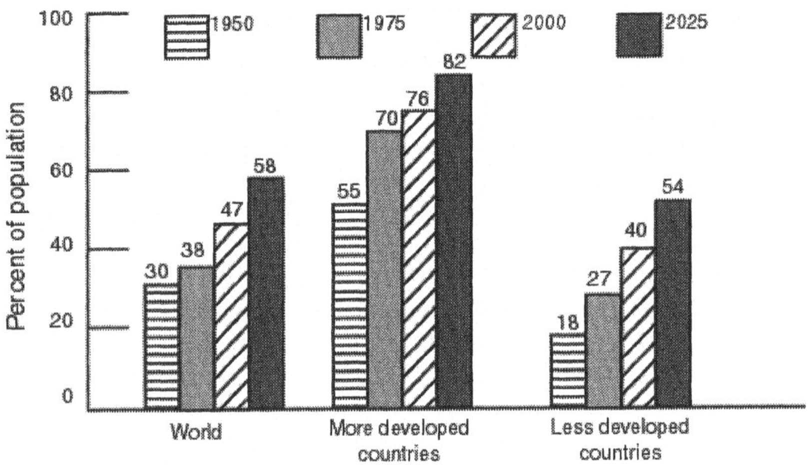

Figure 1.1 Percentage of population living in urban areas in major world regions in the years 1950, 1975, 2000 and 2025 (estimated)

The previous short remarks as to increasing complexity, the accelerating rate of change, and the anticipated as well as unanticipated effects of technology and development become more meaningful in the context of rapid urbanisation and enlargement of the spatial envelope of human settlements. As related in a variety of official and unofficial documents, most notably United Nations projections, in Asia, Africa and Latin America, the unprecedented population growth that characterised much of the 20th century has evolved into unparalleled urban growth. By 2020, a majority of population in LDC (less-developed countries) will be living in urban areas.

The concentration of population in large cities of 1 million or more is increasingly associated with developing countries. There were an estimated 292 such "million-plus" cities in LDCs in 2000. Megacities and megalopolitan conurbations with 10 million or more residents are also becoming more numerous and are to be found predominantly in developing countries. Dhaka, for example, is expected to add about 9 million people between 2000 and 2015 while New York will expand in the same period by about 800,000 residents. In most developing countries, cities are growing between two to three times faster than the national population (Hinrichsen and Roby, 2000).

The unprecedented magnitude of urban growth has added fuel to long-standing debates about overpopulation and carrying capacity, by extending the concern as to both the liveability in such large urban concentrations, and as to the capability to govern and provide good urban management. The argument here has to do not so much with the economic viability and competitiveness of such cities, but more with the notion that unplanned development is depleting non-renewable resources, that cities become sources of fast spreading epidemics, and that even contribute to climate change. Similarly, in developed countries, sprawl and "exurbs" contribute to the disappearance of surrounding rural lands and test the ecological fragility of the environment, by both urban growth and a consumption-intensive life style.

Thus, the urban population changes related to water resources, must be seen in the context of the levels of urbanisation, the rate of urban growth, and changing life styles that demand more consumption of water and higher degrees of water quality. Again, because more developed countries are already highly urbanised, their urban share is not expected to increase substantially. On the other hand, the less developed world is urbanising as quickly as developed areas at the first half of the 20th century. As a total the urban population in less developed countries is expected to nearly double in size between 2000 and 2025, from less than 2 billion to more than 3.5 billion, concentrating, once again, in very large cities. The latest 1999 revision of the UN World Population Prospects shows that virtually all the population growth expected between 2000–2030 will be concentrated in the urban areas of the world. In terms of major regions, by 2030 Asia is likely to be 55% urban and Africa 53%, whereas 83% of the population of Latin America and the

Caribbean will be urban. In Europe and Northern America the proportion urban is expected to reach 84% in 2030, whereas Oceania is likely to increase from 70% in 2000 to 74% in 2030.

The basic characteristics of ongoing demographic changes and projected population transformations point out that urban areas of less developed countries will incur almost all of the world's population growth. Urban areas of more developed countries will experience population aging, centrifugal movement to the periphery and in-migration. All such trends will impose, especially in LDCs, more pressure on the space infrastructure and resources of the cities, with heavy demands for already inadequate housing, growth of squatter settlements, and persistent problems with running water and waste disposal. Even in cases of economic development, manufacturing plants move to increasingly distant places (e.g., in Mexico City and Sao Paulo) with the potential for further geographical dispersion, encroachment of agricultural land, and the need for elongated and expensive urban water services.

A word of caution is needed at this point. The "frightening" images of a transition to a predominant urbanised world, reflect both long-existing economic and political changes as well as recent transformations of a more global economy. Also, despite all the above changes, there are elements of continuity since the average size of the world's largest cities may have changed, but their location has not changed as much. Finally, the rates of growth seem to be decreasing, especially as definitions of what is an "urban" settlement differ widely between countries. This is why some recent writing (as e.g. E. Brennan, 1999) not that contrary to the alarmist prediction about "exploding cities," the growth of most of the world's megacities has been slowing down. Part of the explanation appears to be a deceleration in rates of national population growth. But even with this "optimism" the fact remains that today and in the immediate future, megacities are being inundated by their own waste, suffer from inadequate sewer systems, and face tremendous difficulties in supplying running water to sprawling conurbations, illegal settlements, and decaying central city water systems.

Coming back to the relevance of all the above for the urban water manager it becomes obvious that we are facing an almost paradoxical combination of growth and decay. On the one hand, we are facing on an unprecedented global scale centrifugal forces of rapid, haphazard metropolitan growth. On the other, older cities are facing the natural decay of their inner core, result of both neglect and of the life expectancy of facilities built for a rather limited span of time. Thus, looking at the patterns of future urban growth, a number of developments will be central in any understanding and management of the urban water system, such as: megalopolitanisation and increased scattering; recognition of limits to growth and the necessity to conserve resources; diffusion of urban functions to outlying subcentres and, therefore, a more diversified internal structure of the city of the future; increased complexity and the need for more centralised (but deconcentrated) control; the use of potential technological breakthroughs associated with the "post-industrial city" altering the form, structure and operation of urban aggregations and offering solutions to urban water problems; and alterations in life styles with emphasis on recycling, reuse and reclamation, which ultimately may be conducive to diminishing demands rather than continuously increasing supply. In the Figure 1.2 the world's 100 largest cities and selected other cities are presented.

Figure 1.2 The world's 100 largest cities in 2000 and selected other cities

It is the last remark that points out to a basic debate today in all discussions of natural resources management, i.e., a probable, some would say imperative, move from "technical fixes" to "non-structural" solutions (or a combination of both) to many of our contemporary problems or challenges. Two observations, then, become particularly relevant at this point. First, that in general, more technology is available than is being effectively used. And, second, that the design of community water systems is not fundamentally limited by available materials and technologies; rather it centres around the knowledge of socio-economic requirements and the policy alternatives related to sustainable development rather than simple, but inequitable, economic growth.

1.2.2 Population, water and cities

All previous remarks as to the global urban predicament point out to the twin phenomena of rapid population growth and rapid urbanisation. Growing cities have been the engines of economic growth, vital centres of human creativity, and the requisite infrastructure for industrial activity. Cities owe also their early existence and development to available good quality water. But their fast growth, contaminants released by their large populations, and industrial activities have created new problems of both quantity and quality. Urban pressures on water and sanitation are result of overload on water sources, improper waste disposal, contamination of stream and rivers, unwise extraction of water from depleting aquifers, and last, but not least, inefficient service management and absence of appropriate institutional support.

There is no need to elaborate the interdependence of water and urbanisation. The historical roots of settlements are to be found in the riverine régimes of our world, and the evolution and welfare of cities have been intricately connected to both quantity and quality of their water supplies. At the same time, urbanisation has a variety of impacts on rivers. As small towns grow into cities, urban runoff increases and reaches water bodies, faster and with greater force. Paved surfaces cannot absorb and store rainwater, with heavier sediment loads that destroy habitat. While urban rivers inherit certain problems from upstream, they also add new pollutants including bacteria, heavy metals, a large variety of toxic substances as well as trash and debris. More broadly, however, environmental problems that face cities can be grouped into four overlapping and interacting categories: (a) problems of access to environmental infrastructure and services (notably water and sanitation systems); (b) problems of pollution from urban wastes; (c) problems of resource degradation; and (d) problems of environmental hazards. In developed countries (a) and (d) categories are either under control or less prevalent. Less-developing countries and lower-income urban areas often face all four sets of problems (intensified in many cases as they tend to occur simultaneously and with higher intensity over longer periods of time).

Many authors have identified throughout the developing world a whole series of environmental and social stresses and strains, including the sheer magnitude of population growth; economic development and the quantity and quality (of often untreated) urban waste; contamination of water supplies and poor sanitation systems; high economic costs for both new water supply sources and high economic costs for both new water supply sources and wastewater treatment; financial management constraints; and, inefficient and inadequate public provisions and overall management of water.

The "World Water Vision" Hague declaration of March 2000 most emphatically underlined that there is undeniable evidence that freshwater ecosystems have been declining in some parts of the world for hundreds of years. The problem has been getting more serious and it threatens the economic, social and environmental security of human society. Furthermore, despite the revolutions in science, industry and economy, inadequate management has contributed to a deterioration of water quality at an increasing rate throughout the world. Water quality has become a central water problem, especially in the industrial world.

1.2.3 Patterns of urban growth and shifting paradigms

In the first part of this chapter the case was made as to recent dramatic changes in population growth, over-urbanisation and megacities' growth patterns, and as to the complex character of interlocking urban water problems of both quantity and quality.

The past 20 to 30 years in particular are characterised by significant changes in planning, design, and management of urban water resources all over the planet. Mounting concerns about the environmental impacts of human activities, potential climatic shifts, expanding populations and the use of megacities as well as new knowledge are part of the pressing need to develop alternative institutional schemes for managing in an integrated manner scarce natural resources. Many nations and regions have increasingly been turning attention to both streamlining existing administrative mechanisms and to introducing innovative institutional arrangements with regard to quantitative and qualitative aspects of their water resources. In the context of conceptual, methodological and administrative developments particular emphasis is placed on the need for integrated approaches; the mobilisation of resources, personnel and facilities; and in incorporating new or emerging professional practices and technological innovations.

A series of trends and developments are viewed as the backdrop of an imminent crisis regarding water supplies and their utilisation. Factors underlying this context of urgency include: a) the high variance of water supply in many parts of the planet, resulting in dramatic fluctuations, exacerbated by periodic droughts or floods; b) rapid population growth and significant consumptive demands, especially as a result of dramatic shifts from rural to urban areas; c) expanding agricultural uses and intensive irrigation developments; d) deterioration of water quality, the result of both intensive agricultural practices and of urban and industrial uses; e) decreasing groundwater availability coupled with contamination of many aquifers; f) increasing environmental concerns and ecosystemic considerations, including natural changes and the entire gamut of anthropogenic disturbances in the surrounding environment; and g) trans-frontier water dependencies, and challenging questions of overlapping and shifting political and administrative boundaries affecting shared water bodies.

But what has further complicated the search for integrated, coordinated and long-range approaches in planning and management have been rapid growth and attendant social changes and their institutional consequences. The sources of these rapid changes can be tracked to fundamental shifts in values, to morphological changes in the structure and functioning of many societies, to technological developments and to such notable exogenous changes as climatic anomalies and socio-economic interdependencies. Further aggravating the nature and rate of all such transformations are also the confluence of popular and scientific interest in environmental protection; data and information explosion; new understandings as to the structure and functioning of environmental systems; and, an increased emphasis on globalisation and interdependence. The end result of all such considerations can be summarized in three interrelated conditions: complexity (if not complexification) or an entwining process of causes and effects resulting in the near inability to cope with any problems at all; conflict and the presence of increasing numbers of competing stakeholder groups and constituencies; and control or the necessity to cope with and adjust to complexity, interdependence, turbulence, uncertainty and large systems' vulnerability.

One may ask at this point, how different are the patterns of present urbanisation with those of previous past periods, especially those during the industrial revolution. This question is of particular importance for the proposed theme and challenge of integrated urban water management, especially in complex systems such presented in Figure 1.3. Extending remarks made throughout the volume of Uitto and Biswas (1999) we should note:

- The explosive rate of population growth and cities' spatial expansion.
- The lag between economic development and urban population growth or the investment capacity for appropriate infrastructure. This is a case

of growth exceeding the capacity of national and local governments to plan and manage requisite water services.

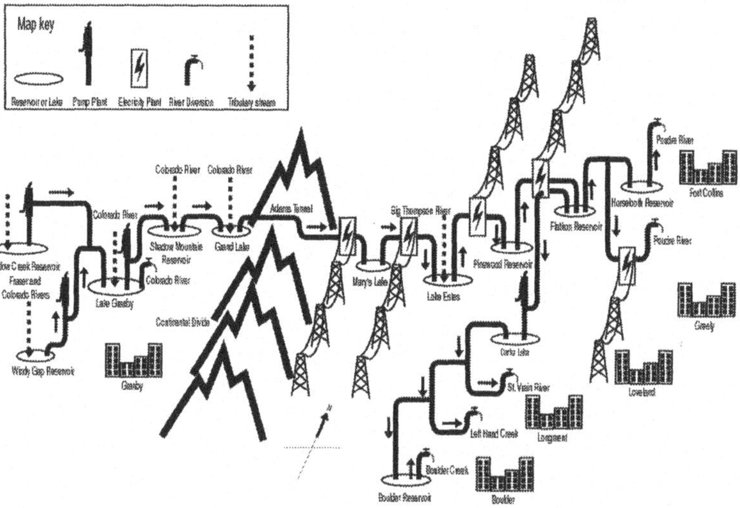

Figure 1.3 Colorado Big Thompson project. An example of complex urban development through water transfer from the western side of the Rockies to the northern part of the Eastern Colorado megapolis

- Unique spatial patterns of city growth including the matching of water and sanitation and unusual urban densities in the periphery (such as favelas shanty towns and squatter settlements).
- Overall patterns of development and provision of water. On top of water scarcity, there are excessive economic costs; the notion of water as a free good contrasted to commodity; extensive management problems and difficulties in institutional mobilisation; lack of national water policy or an extended planning time horizon, etc.
- The increasing importance of environmental concerns and the ideological emphasis on sustainability. Here one finds the concerns with conservation, a shift from structural to non-structural solutions, new concerns with public health, and water-intensive lifestyle and its consequences. In addition, extensive environmental legislation and regulatory provision, including the National Environmental Policy Act (NEPA) in the U.S., a whole series of similar environmental protection acts as the Clean Water Act in the U.S., and the recently adopted Water Framework Directive (WFD, 2000) in the

European Union. The last has far-reaching provisions for the protection of quantity and quality of surface and ground water, introduction of water pricing policies, integrated river basin management, and the strengthening of public participation. The common approach and objectives of the WFD emphasise not only quantitative and qualitative aspects of water resources, but contribute to the central concern of EU in ensuring sustainable water consumption and use.
- Exogenous factors such as climatic changes (and new epidemiological apprehensions), globalisation, and new management principles, most notably the emphasis on non-hierarchical administrative focus and the search for anticipatory, participatory integrated forms of organisational structures and managerial mind sets.

Combined with the increased emphasis on sustainability, the response to the forces of social changes and the cry for integrated water management, imply first of all an alternative vision of the means by which ecologically sustainable and socially just communities will emerge. Second as to how can urban settlements and the space they occupy be transformed into balanced eco-cities. And third, how can we conceptualise and institute water management mechanisms across different socio-cultural settings, or introduce diversified experiences of responding to the forces of rapid social changes in increasingly expanding megacities. In essence, we are being asked to confront growth and development differently. Urban authorities are asked to move from a commitment to protecting and regulating environmental quality to the more complex commitment to sustainable development where urban management strategies should incorporate the central notion of the finite nature of water ecosystems; the need for an ethos of conservation; changing technologies (for both innovation and renovation); a predominant political institutional context; and, changing goals that recognise multi-purpose, multi-objective and multi-disciplinary approaches.

This paradigm shift, responsive to complexity, intergenerational equity, balanced ecosystem principles, and economic efficiency permeates current writing and practices of urban water management. Water resources systems development has an obligation to see that those systems provide sufficient quantities and qualities, at acceptable and fair prices, while at the same time protect the environment and preserve the health of ecosystems for future generations (see here in particular the extended discussion in the ASCE/UNESCO monograph 1998, pp. xxii ff). The emerging sustainability paradigm has been variably called integrated, holistic, total, or interdisciplinary and, thus, calls for a far-reaching re-examination of current approaches to water resources planning and management. In this context, the remaining two parts of the present introductory chapter, attempt to outline aspects of both the search for sustainable cities and for implementing sustainable urban water management.

1.3 THE SEARCH FOR LIVEABLE CITIES

The demands for integrated water resources management imply also the application of sustainability principles to cities. Today's sustainable communities efforts are in many respects a reaction to the inability to constrain and effectively manage urban sprawl, to guarantee adequate infrastructures, and to enhance both Quality of Life (QOL) and Social Well-Being (SWB). The backdrop for building sustainable cities and efficient and effective water systems are part of or a combination of capital theory, urban design, ecosystem management, and metropolitan/regional governance.

Thus, viewing the city as an ecosystem with its goal being maximisation of sustainability one can discern an extended metabolic model of human settlements. Such an approach becomes even more demanding if one combines city size with sustainability and the consideration of a variety of systems and subsystems over an extended spatial context. In addition, not only system and subsystem elements are numerous and related in complex ways; but, the necessity for integrated/comprehensive planning and management requires a skilful merging of physical/environmental planning with socio-economic conditions, as well as with innovative technological elements. Urban sustainability implies policy integration, ecosystems thinking, cooperation and partnership, and sharing of goals through participatory envisioning and long-range commitment.

1.3.1 What sustainable growth and development imply

Almost 14 years ago, the seminal report of the World Commission on Environment and Development, *Our Common Future* (WCED, 1987) set up the tone and summarised the effort to articulate the transformation in theory and practice toward the environment. Its basic definition of sustainability as "...development that meets the needs of the present without compromising the ability of future generations to meet their own needs" became the rallying cry for articulating the premises and components of such a strategic thrust.

Thus, the debates that followed crystallised earlier efforts appearing under different terms of "sustainability" beginning with the UN Stockholm Conference on the Human Environment (1972) and the sequence of efforts in the 1980s by IUCN (World Conservation Union), UNEP and the World Wildlife

Fund (WWF), and the formal milestone of the 1992 United Nations Conference on Environment and Development (UNCED or Earth Summit) in Rio de Janeiro. A context, then, began to emerge for a growing number of national sustainable development strategies or plans. Such a context was coupled with political and regulatory pressures, exemplified with methodological efforts of enforcing environmental impact assessments, and as part of the confluence of popular and scientific interest for integrated environmental approaches.

As Munasinghe and Shearer indicated (1995), the expression "sustainable development" was coined to demonstrate that economic growth and environmental protection can be compatible. The wide variety of definitions, however, point out not only the difficulties of articulating this broad concept in the "Brundtland Report" as well as in Agenda 21; but, also, the various epistemological perspectives of the social and natural sciences, as well as schools of thought and ideological camps. Whatever the interpretation, sustainability became the development paradigm and the reigning slogan of the 1990s. Yet, when concentrating on the underlying thrust, sustainability acquired a normal bend emphasising justice to people within a generation ("intragenerational"); justice to people between generations ("intergenerational"); and, justice to "nature" as part of the effort to maintain ecosystemic integrity.

The emerging vocabulary associated with sustainable development points out toward the search for more meaningful planning and for comprehensive, integrated, holistic approaches. In this context a sustainable ecological base has always been essential for the long-term survival of humanity. Thus, in any discussion as to efforts for sustainable development in increasingly complex, turbulent and uncertain socio-political environments, future-oriented and balanced approaches became the all-encompassing concepts for a wise use of renewable resources. When all is said and done, three terms became part of the strategy for sustainability, i.e., *efficiency* (benefit–cost appropriateness); *equity* (fair distribution of resources to all segments of population); and *ecosystem integrity* (the maintenance of sustainable ecological processes).

A virtual torrent of academic writings, key forums, declarations, symposia and meetings attempted throughout the 1980s and early 1990s to define and outline means for resolving long-term environmental, social and economic issues associated with balanced growth and development. Converging concepts such as "carrying capacity," "ecodevelopment," "limits to growth," "tradeoffs," "multi-objective planning," etc., all contributed to an idealised relationship between dynamic and fast changing human systems and larger, but slower changing ecological systems. The interest, then, for sustainability has remained high, as larger philosophical questions of appropriate growth and of global change vis-à-vis the passage to the 21st century have created not only an international agenda of developmental goals, but also national and sectoral economic and environmental policies and plans.

The development and tracking of environmental indicators became early a centrepiece in the formulation of sustainable natural resources strategies. As a UNEP (1994) document summarised it, the search for indicators took three different directions:

- by use, in terms of early warning or general policy;
- by subject or theme, such as water, air, or acidification;
- by position in causality chains, in terms of status of the environment, stress, or societal responses.

Similarly, OECD in a series of documents approached indicators of sustainability by recognising such criteria as policy relevance and utility for users, analytical soundness, and measurability. Other organisations and countries, such as "Environment Canada," "United States EPA," or "The National Environmental Policy Plan Plus," (Netherlands), have developed operational definitions, criteria and categories of indicators, often culminating to "models' sometimes linked to "sub-models" attempting to show integration and interaction, especially with regard to urban water management. An interesting current example for comparability is the search for an Environmental Sustainability Index (ESI) developed by a research team of Yale University with support from Columbia University and the World Economic Forum (WEF). This multivariate index attempts to measure environmental health and sustainable development. The researchers have been sorting 67 variables through clusters of five broad areas, namely environmental systems, reduction of environmental stresses, reduction of human vulnerability, social and institutional capacity, and global stewardship. (For an interesting journalistic commentary and the pitfalls of the ensuing quantitative quagmire see the story in *The Economist*, 27 January 2001, pp. 74 and 77.)

What is important at this point is the repeated emphasis on *justice* and *equity* and the normative character of all such efforts attempting to describe "fairness", duty to posterity, changed life-styles and similar notions of broad environmental policies. This implies the centrality of goals and objectives, an extended time horizon, institutional mobilisation or, in other words, policies and expressed goals measured eventually by relevant indicators. Furthermore, a more current view of sustainability adds up such dimensions as the possibility of global disturbances; a new ethical dimension in the linkage between the rights of nature and rights of cultures; and, the recognition of mechanisms for environmental accounting and allocation based on collective and long-term analytical approaches. At the end, sustainable development (and this is particularly true for water resources) includes two key components: the concept of *need* and the idea of *limitations*. With regard to water (and with reference to the U.S. context following the President's Council on Sustainable Development—PCSD) the component and themes of such a broad policy identify three overarching areas, namely economic efficiency, environmental health, and social equity. Furthermore, the Council's report emphasises that the three areas are interdependent and must be pursued "simultaneously" and in a "balanced" manner if sustainable development goals are to be achieved. Especially for the third theme (social equity) it is also interesting to notice how the previously quoted monograph of ASCE/UNESCO, "Sustainable Criteria for Water Resources Systems" emphasise such social objectives as empowerment, participation, social mobility, social cohesion, cultural identity and institutional development. It is truly a "revolutionary" vocabulary not usually encountered (certainly in the past) in technical reports.

By now, it becomes apparent that sustainable development of water resources requires not only comprehensive policies but also valid and reliable indicators, and hopefully, coherent models. Here belongs also the discussion of the shifting paradigms in the relationship between humans, their cultures and their surrounding environment; our new understanding of "development" models; the ideological differentiations between ecocentric and technocentric perspectives and, the new "practising ethos" in environmental thinking, planning and implementation. When combined with "appropriate technology" and "global change" sustainable development becomes part of a much more complex worldview and new ways of thinking about social change and society's relationship to the environment.

To recapitulate, then, the quest for comprehensive or integrated urban water management shifted from simpler hierarchical, top-down, decompositional and projective management, to more heterarchical[2], participatory, holistic and normative models of complex management schemes. For pragmatists and realists of current urban water management, this idealistic search for a "utopian" sustainability common thread is an ambiguous and frustrating quest full of conceptual cul-de-sacs void of concrete action, and a context-intensive term whole ultimate meaning defies final interpretation. Essentially, we know more what is "unsustainable" and, thus, we have to settle for some relative agreements (as various guidelines and formal documents have attempted so far) as to what sustainability and integrated management imply. This debate is exacerbated by the combination of the two terms, sustainable development, or sustainability for short, implying both stability and change. The concept has become the "guiding

---

[2] non-linear, non-hierarchical complex systems

principle" (or implicit paradigm) for efforts to reconcile economic and ecological imperatives (see here the wealth of definitions in Mazmanian, 1999, p. 47).

The focus of this concept of sustainability and its relationship to cities, became apparent not only in the political agenda of the 1972 UN Conference on the Human Environment in Stockholm (linking global environmental problems to the economic development of poor, especially those in the Third World); but also to the principles elaborate in the 1987 Brundlandt Report, articulating the notion that sustainability can occur predominantly with community-based approaches. The sustainability paradigm links these community developments with both economic and ecological development.

1.3.2 Sustainability and water resources systems

Given the multitude of definitions regarding the key concepts of sustainability, it is obvious that there are also disagreements as to what such a paradigm may imply for integrated urban water management. The report of ASCE/UNESCO (1998) opens with the following statement: "Water resource professionals have an obligation to design and manage water resource systems so that they can fully contribute to an improved quality of life for all humans." And it continues "... Sustainability is a unifying concept that emphasises the need to consider the long-term future as well as the present. This includes the future economic, environmental, ecological, physical and social impact that will result from decisions and actions taken today. Although we cannot know with certainty what all these impacts will be, or what future generations of individuals or societies will want or value, we can attempt to predict what we think might happen and what future generations may want or value as we develop our current plans, designs and management policies. Admittedly we can only guess at what future generations would like us to do now in our generation for them in their generations. We must take these guesses into account as we make our decisions or take actions to satisfy our immediate demands and desires." And, then, the monograph introduces the following definition: "Sustainable water resource systems are those designed and managed to fully contribute to the objectives of society, now and in the future, while maintaining their ecological, environmental, and hydrological integrity."

As far as water resources are concerned, the topic is vast and a lot of competent authors have repeatedly articulated the context of change and sources of uncertainty regarding water and the need for more integrated, holistic and ecosystem sensitive practices in water resources planning and management. One way of summarising long lists of concerns, conflicts and crises linking sustainable development and integrated water management is through a cross-cutting list of interacting issues and crises:

- A water supply and demand crisis that represents a predominantly engineering dimension. Here one should include management and population growth issues such as the promotion of more desirable levels and patterns of use; augmenting fresh water supplies through e.g. conservation, reuse, desalination, transport of water from other areas; conjunctive water use with withdrawals from aquifers, reservoir storage, etc.
- A deteriorating water quality crisis that can be translated into an ecological dimension of urban water problems. Here one encounters a variety of health issues, poor water quality, waterborne diseases, lack of adequate safe drinking water supplies and sanitation qualities, groundwater contamination, as well as interference of water resources systems with the proper functioning of natural life cycles.
- A transboundary dependencies crisis representing in many cases a geopolitical dimension not only in terms of international boundaries, but also intra-national transfers of water across administrative boundaries (as e.g. in such volatile situations as in Cyprus or the Western Bank cities).
- An organisational crisis exemplified in a management dimension, i.e., the appropriate mix of competent personnel, facilities and procedures, as well as legal procedures and administrative guidelines. The literature refers here to capacity building and to the fact that the people who can best manage a system effectively and efficiently are those who know what an urban water system is supposed to accomplish and understand how it functions.
- A data and information crisis, not only in terms of availability, validity, reliability or comparability but also as part of combining data and judgement, modelling, and the building of useful Decision Support Systems.

Related to the notion of sustainability and such expressions as carrying capacity, equity, integrated, etc., is the new and controversial concept of urban areas as ecosystems. The simplest and most straight-forward definition of this concept may be "a biological community where humans represent the dominant or keystone species and the built environment is the dominant element controlling the physical structure of the ecosystem" (World Resources 2000–2001, 2000, p. 142). This could be an interesting expansion of the emerging complex understanding of cites and their management, as urbanisation can change also the ecological structure and composition of a given territory, producing new sources of environmental stress. Many urban streams are so polluted, or canalised that only certain vegetation can survive and, thus, losing aquatic diversity. Together with other forms of environmental degradation cities lose the capacity of their natural areas to provide environmental goods and services. Such areas contain vital green spaces and aquatic amenities. The rapid growth in urban population adds to the mounting stress on urban ecosystems.

The last discussion adds a new and demanding dimension to urban water management when treating cities as natural systems. Students of cities and planners recognise early the relationship between cities and nature as well as an organic approach to planning and management that stresses the harmonious relationship between city and region (read perhaps here "watershed"), between city and environment, and of land uses within cities. In other words, integrated urban water planning and management. This certainly underlies the spirit of the Brundlandt Report not only in terms of the critical interaction between cities and the environment, but also as an expression of the problems of environmental infrastructure and health in urban areas of developing countries (Cities and Sustainability, 2000:32). Finally this ecosystemic urban dimension was incorporated in the Settlement Section of Agenda 21, especially in terms of "integrated provisions of environmental infrastructure, policies to minimise environmental damage, promotion of policies to recover infrastructure costs and the seeking of joint solutions where issues cross localities" (Agenda 21, 1993, Section 7 "Sustainable Human Settlements").

One final remark here. This new vocabulary and concept on sustainability reflects evolving frameworks in the planning and management of cities and urban water. The historical backdrop of the Garden City movement, Bioregional planning, New Town movements all over the planet, Grassroots communitarian movements, Urban ecology and Eco-city movements, "Limits to Growth", New Urbanism and Neotraditional Towns, the coupling of environment and development efforts, social capital debates, environmental audits, sustainability indicators, and so on are expressions of the challenge subsumed under Sustainable Communities (with the variations of Green Cities, Eco-cities, eco-communities, efficient cities and liveable cities). All such developments underscore once again the complex context of contemporary urban water management across the cultural continuum of developed to developing countries. Without doubt the new conceptual emphasis and the implementation of sustainability tools is both an exciting and a daunting prospect for all cities (Newman, 1999:20).

1.3.3 Urban water systems and sustainable cities

The previous discussion throughout Section 2 has pointed out how cities and sustainability are contributing to a profound rethinking of traditional approaches to urban water management. This is a central point in the exposition of Butler and Maksimovic where in addition to underlining the pivotal provisions of the Brundlandt Report, of Agenda 21, and of other relevant documents, the authors emphasise the importance of public participation, individual responsibility, and of sustainable services which are environmentally friendly, socially acceptable and financially viable (Butler and Maksimovic, 1999:216).

We have already some emerging guidelines on the overall sustainable development of cities with reference also to urban water systems. The publication of the European Commission on *European Sustainable Cities* (1996) refers to five guiding principles, namely: 1) the principle of urban management (with a range of tools to address environmental, social and economic concerns); the principle of policy integration (combining the subsidiarity principle with the wider concept of shared responsibility); the principle of ecosystems thinking (incorporating the notion of a complex system that is characterised by flows as continuous processes of change and development); and, the principle of cooperation and partnership (or the sharing of experiences, professional education, cross-disciplinary partnerships and networks, community consultation and participation, etc.

Such lofty and noble goals for sustainable, liveable and viable cities require operationalisation, specification, and methodological clarity and comparability. With regard to urban water management we can begin by tracing sustainability to two background notions, i.e., that there are limits to urban growth, and the utilisation of the concept of carrying capacity. There are continuous examples in the literature as to what changes in or bad management of water supplies entails. More recently, the discussion as to whether cities are proceeding along a desired path of sustainable development and management has taken three directions: a) assessing an "ecological footprint" (in terms of surface needed to support a given city's level of consumption); b) calculating a broader measure of wealth (by focusing on an assessment and comparison of produced assets); and c) evaluating critical problem areas (notably the controlling of infectious and parasitic diseases, reducing chemical and physical hazards, achieving a quality urban environment, minimising transfer of environmental costs, and moving toward sustainable consumption (Leitmann, 1999, pp. 150 ff).

We need, then, to look at some selected remarks and cases that focus attention to urban water management with sensitivity to sustainable principles of development. In the next Section we plan to address some general remarks as to the implementing of sustainable water management. The vocabulary is quite seductive but it lends itself to misinterpretation, especially if it becomes part of narrow local efforts without sensitivity to the larger context of global sustainability. In trying to meet the urban water challenges of the 21st century, globalisation and localisation provide the opportunity for a combination of effective institutions and policy initiatives. We should aim at the development of a matrix that relates urban water functions (water supply, wastewater, stormwater, eco-aesthetic elements) to a variety of policy initiatives (such as stewardship goals, infrastructure, regulatory authority, organisational mobilisation, and supporting social capital). The configurations of urban water system are, then, tied to the key expressions of sustainability, namely environmental viability, economic efficiency, and social equity, The focus of parallel efforts in the literature is evident in many official and professional guidelines that help ensure water resources systems that are designed and managed to fully contribute to the objectives of society, now and in the future, while maintaining their ecological, environmental, and hydrological integrity.

1.3.4 Implementing Sustainable Urban Water Management

1.3.4.1 New configurations of urban water management systems

Extensive theoretical discussions cannot by themselves articulate the importance of complexity, sustainability, and integration in the planning of management of urban water systems. We need to see in pragmatic settings the bridging of theory and practice via new models, new critical variables, relevant indicators, comparable, diachronic information, and attempted interdisciplinary approaches. In reviewing past work one is struck by the observation that water management in urban areas is governed by an old, traditional technical paradigm developed at the beginning of the 20th century in Europe and North America. New approaches and a shifting paradigm accentuates the change from water disposal and treatment to conservation and recycling of resources (Niemczynowicz, 1946:204).

At the same time the traditional approach to providing water supplies (as cities grow) has been to draw from increasingly distant sources and from tapping the aquifers. Other options are forwarded, such as advanced technologies for demand management through a series of incentives and disincentives; use of water conservation technologies, reduction of unaccounted for water; reuse of treated wastewater; public education and involvement; comprehensive metering and installation of water saving technologies; water auditing; and changes in pricing procedures and cost recovery.

Broader management practices combine and expand water, sanitation, and drainage into more complex operations that add improved urban water supply management efforts such as: protection and management of watersheds (as in the case of New York City's ambitious environmental protection plan of an upstate watershed that encompasses 1,970 square miles and three reservoir systems); the increasing practice of recovering costs by treating water as a commodity (with new market mechanisms, reusing of water, or willing buyer-seller purchases);

renewed water conservation efforts via consumer education; and, the establishment of appropriate standards sensitive to particular localities and circumstances.

The management of urban waters brings back previous remarks on comprehensive planning. The nature of urban environmental problems, especially water supply, wastewater and drainage are influenced, to start with, by the interaction of numerous public, private, non-profit groups, each having competing and conflicting interests and patterns of behaviour. Thus, the preoccupation with the coordination of roles and concerns of key shareholders or parties-at-interest (Leitmann, 1999:86). Even more, the necessity of greater public participation in urban water management. The complexity of such a comprehensive urban water planning and management raises also questions of jurisdiction and different levels of responsibility, especially in metropolitan areas and expanded spatial boundaries, as well as cross-sectoral issues of complex cross-media environmental perturbations. A common complaint in urban water management is that, especially in developing countries, power and decisions have been vested in the hands of a public bureaucracy that is not particularly motivated to operate in a cost-effective manner. In a special report on the coming sanitary crisis (http://www.wateraid.org.uk/research/slumstxt.html) it has been noted that a World Bank review of 120 projects in the developing world found that the water authorities were performing well in only four countries.

It is in the context of such broad concerns that one must be able to understand the larger perspective of socio-cultural aspects of present and future urban water management schemes. Technological and social responses, as well as legal mechanisms for meeting the challenges of a dynamic urban environment should combine both "structural" and "non-structural" mechanisms as parts of a comprehensive policy connecting water to growth. Either through increased supply or diminished demand we should be affecting the extent and rate of change in the surrounding environment. Given the trends and developments about water management, an integrated system of water resources involves the following major points:

- Successful development and management of water resources require much larger institutional and organisational arrangements, quite succinct from the presently prevailing highly segmented approaches.
- Norms and cultural values concerning water use must be coordinated within a larger social planning domain.
- Each proposed water resource system, independent of its level of analysis, must develop unique patterns much more responsive to the specific cultural conditions found in a given region.

- The larger the scope of the water system and the greater the scale of analysis, the more complicated the organisational arrangements and, therefore, the more the need of coordinating powers and comprehensive planning.

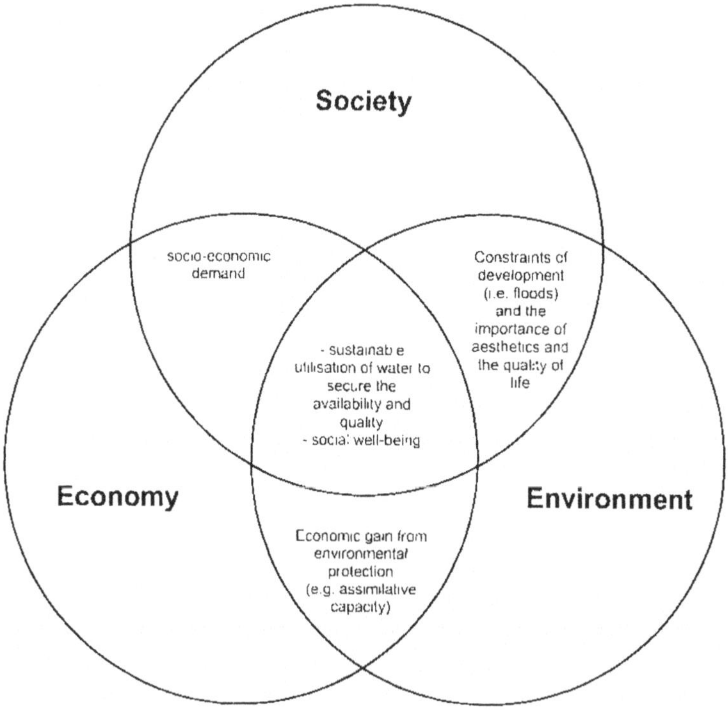

Figure 1.4 Understanding sustainable water development through the linkage of society, economy and environment (adapted from UNESCO 1995)

The most succinct observation that one can make at this point is that water resources in the past have been used to accommodate rather than shape the future. Recent developments point out that planning must be carried out with sensitivity to all affected environments, by assessing impacts and considering a wider spectrum of alternatives, and by forecasting future water resources environments in a more comprehensive manner and within a normative context such as presented in Figure 1.4.

To deal with these complexities and competing and conflicting demands, a multi-objective planning framework has been developed in order to make explicit the underlying affected values an their distribution among different groups. Such models have been particularly under development since the early 1960s, reinforced in the U.S. by the extensive environmental legislation of the 1970s and beyond. Overall, two basic approaches have been followed: the first attempted to drive single-valued answers as to what is better or worse. The second approach, largely modelled after the U.S. Water Resources Council in its 1973 guidelines, was based on a multi-objective accounting framework that traces multiple consequences of proposed actions and relates them to affected people or societal groups.

What all the above tend to indicate is that more than discussing traditional areas of concern in water resources, one must also consider conceptual, methodological, and organisational responses in managing complexity and uncertainty in interdependent natural resources systems. Again, the abundant literature suggests (as "lessons learned") similar list across many countries in terms of comprehensive approaches; coordination mechanisms; stakeholder involvement; local responsibility and accountability; sustainability and environmental ethics; social equity decision support systems and risk management; long-range, macro-engineering emphasis; and, trade-off considerations.

It is obvious, then, that water management must also incorporate new ecological values and broader criteria of sustainability. Both new socio-ecological values and the search for sustainable development require more information perhaps even knowledge and wisdom which in many regards are forcing us to consider the changes of risk and uncertainty involved in fast changing environments. What all of these remarks imply is that necessary choices must be made in the coming days, and decision would push both professionals and politicians to move beyond traditional methodologies and considerations of tradeoffs into a much more meaningful examination of underlying strategies, social goals, visions and considerations. As was succinctly stated in the ASCE/UNESCO publication the modern way of approaching planning and management problems in water resources is through systems analysis and synthesis, i.e., the identification, analysis and evaluation of the interaction of all the components of (urban) water systems in space and time. For such systems to be sustainable, they must interact smoothly with other subsystems of society, adapting, at the same time, to changes and uncertainties in supplies and demands. And the report concludes in this part by noting that today's managers of water resources systems have to consider a large number of often conflicting demands and operate under numerous social, legal and physical constraints. What we have, then, is a decision process involving multiple objectives, multiple decision-makers, multiple users, and multiple constituencies and stakeholders (ASCE/UNESCO, 1998:230).

A useful approach here is that of the integrated urban water system (IUWS) introduced by, among others, ASCE in the 1960s and 1970s as part of the Urban Water Resource Research Program (N. Grigg, unpublished ms., 2000:9). Grigg traces changes in urban water systems over the last fifty years, which he notes have been more incremental rather than revolutionary. Similarly, Butler and Maksimovic (1999) expand the argument by summarising the most pressing challenges of the 21st century around five clusters of concern, namely resources and their conservation; infrastructure expansion, improvement, innovation and integrated design; the re-introduction of water in the city as an amenity, aesthetic presence, and recreational element; institutional improvements in terms of public participation, organisational mobilisation, and water legislation and standards; and last, but not least, a "new thinking" a mind-set, if you wish, of a willingness and daring to move beyond traditional approaches and to conceive and apply new solutions to complex urban water problems.

The concept of Integrated Urban Water Management is a culminating point of part developments and of the central role of the 1992 Rio summit. Yet while developed countries have been using coordination, area-wide integration, cooperative management, and functional integration of water services, in developing countries the challenge is much greater. Here limited centralised water and sewerage, fragmented services, environmental assaults, absence of funding, and rapidly expanding population, all starkly contrast to the noble vocabulary of integrated management and sustainable development.

An integrated perspective on water resources implies both development and preservation as necessary ingredients of a collective search for survival and fulfilment. This is extremely important in large urban areas such as Sao Paulo (Figure 1.5) with significant inter-basin water transfer. Thus, the end result and long-term commitment of a integrated management approach involves: the capacity for environmental scanning (monitoring trends and developments); organisational mobilisation in terms of improved and responsive management; development of Decision Support Systems (DSS) so as to promote implementable action; and, contingency planning through a richer menu of options and a wider range of alternatives.

A final comment in the area of implementing sustainable urban water management has to do with appropriate institutional arrangements that maximise efficiency and effectiveness. Current practices seem to point out that a good institutional arrangement for successful urban water management is one that ultimately facilitates social choice against a wide range of alternatives. Institutional arrangements must also reflect in some reasonable manner aspects of political efficacy, i.e., the willingness of all political actors and units to assume risk and incur costs that may look at first glance as quite unrealistic. Decisions about institutional arrangements should take into account the preference and interests of those affected by particular policy actions. Finally, the design of institutions for managing urban water systems must consider a move away from the preoccupation with sectoral planning and reductionist analysis, to a systemic orientation that incorporates multi-objective, multi-disciplinary considerations.

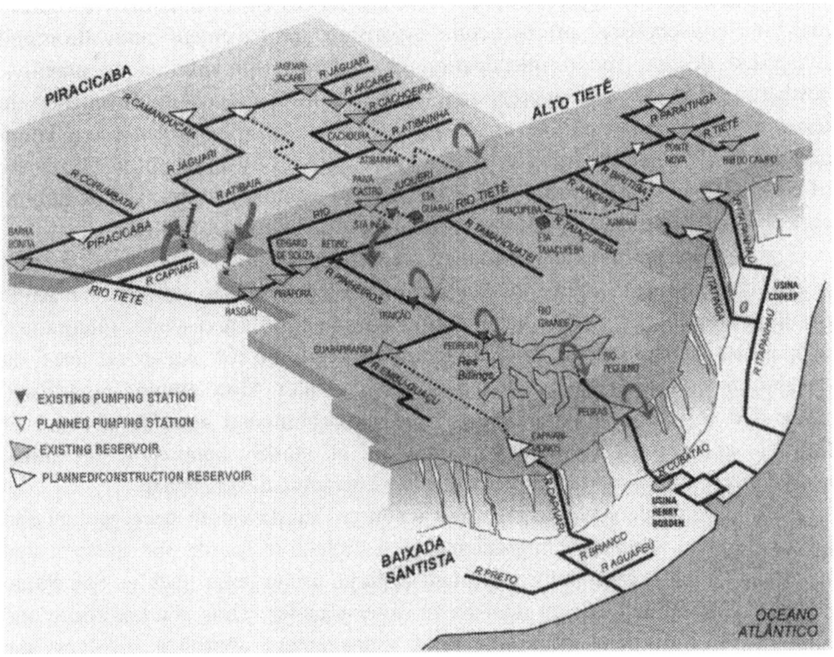

Figure 1.5 Inter-basin water transfer in the metropolitan region of Sao Paulo (Braga 2000)

1.3.5 Towards a new planning and management horizon

The reaction to the shortcomings of traditional water and other resources planning has brought forward a series of improved planning and implementation processes and approaches. They include Participatory Planning, Action Planning, Comprehensive Planning, and Strategic Planning, with such noticeable characteristics as cross-sectoral coordination and integration, interdisciplinary approaches, linkages to and from national policies, normative orientation, extensive public participation, mechanisms for conflict resolution, regular monitoring and evaluation, as well as futures-oriented approaches. This new planning and management horizon reinforces the premises of sustainability, reflects new paradigms and practices, and exemplifies the challenges and opportunities for innovative approaches in urban water management. And, in this context water resources management requires continuous coordination between such conflicting and complementary purposes as: resource planning for substantial economic output; regional planning for successful human habitat; facility planning for technical efficiency; ecological planning for biotic fitness; and, social planning for community integration.

Recent history, rapid socio-economic changes and the transitions necessitated by the turbulent decades of the 1980s and 1990s underscore the increasing emphasis on the variety of environmental challenges, the search for sustainable development, the promotion of integrated planning and management, and the attempts to apply structural and non-structural solutions to persistent water resources problems. If we were to summarise four broad areas of on-going transformations in water resources, we should emphasise: a) *conceptual breakthroughs*, including shifting paradigms in terms of ecosystems, sustainability, heterarchisation, complexity, uncertainty, turbulence, and interdependence of surrounding environments; b) *methodological advances*, especially multi-purpose/multi-objective approaches, Decision Support Systems, Risk Analysis, and the implications of rapidly expanding computational prowess; c) *organisational mobilisation*, in terms of new administrative mechanisms, institutional arrangements, renewed interest in river basins contingency planning, Alternative Dispute Resolutions, etc.; and, d) *contextual changes*, signifying the entire gamut of on-going and future quantity and quality problems, new areas of concern, shifting priorities, as well as potential socio-political intervention mechanisms and comprehensive resource policies. Essentially, we need to understand the interactions that take place within urban communities and between settlements and their hinterlands, as well as the commitment on a long-term vision rather than short-term, corrective solutions.

The improved design and operation of urban water systems and of environmental infrastructure and services would require first of all an innovative combination of important instruments for achieving water objectives, such as regulatory provisions (technology standards, permits and licenses, land use controls, etc.); economic instruments (such as tax credits and subsidies, water pricing, etc.); property rights, surface and groundwater rights, spatial planning, and environmental zoning; and Geographic Information Systems (GRS), environmental impact assessments, public information, and research and monitoring. Second, improvements in service delivery through better sector management and strategic investments and financing, as well as public/private partnerships. And, finally, building both local capacity in integrated planning and management, together with more effective coordination of urban water actions through clearly defined institutional arrangements.

There are two major problems for the implementation of the concept of sustainable urban water resources development in the urban environment of the 21st century. The first is related to finding the appropriate financing for the heavy infrastructure investments required in the developing world and the rehabilitation of existing water supply and wastewater networks in the developed world. The second is related to an appropriate institutional framework that integrates the transportation, housing and land use sectors into the water resources sector. A third associated issue is related to education, increased social awareness of population towards the caring for the urban water resources, and the balancing of professional analytical studies and expert opinion against the concerns of various affected constituencies. The megacities of the 21st century will exhibit a variety of social niches from the most conservative technological to the most conservative environmental. With the democratisation of established power, all these groups, from different backgrounds and lobbying on totally different issues, will have greater and greater participation in the decision-making process. Consequently, in order to have proposed actions implemented and monitored in a long range and sustainable basis, the future urban water resources master plans must contemplate the effective involvement of the civil organised society and water users at the early stages of the planning process. This will require a major change in the existing institutional and legal arrangements for water management in most countries around the world, regardless of their degree of social and economic development.

Leitmann (1999) concludes his work on sustaining cities by referring most centrally to urban environmental planning and management as a mechanism for bridging the gap between traditional urban development and environmental rural preoccupation. Such an effort requires once again a paradigm shift and changes in mindsets from which goals, rules and feedback structures arise. The samples of such transformation are increasing as e.g. in the case of Metropolitan Sao Paulo (with a population of approximately 15.5 million people) and the recent efforts for coordinating actions of various public agencies and public participation; in the Colorado Plateau and Metropolitan Denver (part of a Front Range megalopolis) with innovative land use management planning, economic incentives and water markets, and coordination mechanisms among the various municipalities; or, in Singapore's water catchment approach for conserving and managing drinking water through nature conservation activities, collection and treatment of urban stormwater as a source of raw water, careful metering of water consumption, and a sophisticated monitoring system for measuring and modelling water distribution, pressure and leakage.

The conclusion, then, at this point becomes apparent. For integrated urban water management, we refer more broadly to normative planning and to water policies driven by larger values, goals and objectives. Such a normative emphasis and the securing of the future of cities can be achieved through the process of *envisioning* (the sharing of goals), the need for *empowerment* (joint decision making), and *enactment* (praxis and implementation). Once again, this must be carried out in the combination of more complex approaches, of further analytical capabilities, and of alternative organisational and administrative arrangements.

1.4 SPECULATING ABOUT URBAN WATER FUTURES

By now the final remarks to the quest for integrated urban water management should be brief and to the point as they relate closely to the series of changes, transformations and complexities outlined in the previous pages. As it has been amply said in many books and reports the degree of sustainability of water resources will involve costs or require some reduction in the immediate benefits of those living today. We are being asked upon to develop institutional patience, long-range vision, balanced decision making, and new professional capabilities.

The common thread in any discussion of sustainable water development emphasises how new strategies are needed because water (and for that matter natural resources) problems are becoming highly complex and globalised. Some even argue that the traditional spatial environmental envelope has collapsed and project boundaries – and their impacts and consequences – are now much more diffuse. Thus, there is a need for bringing forward an environmental approach that requires drastic measures of ecological rehabilitation, innovative institutional mechanisms, and a balance between autonomy and cooperation. Such global approaches entail also improvement of environmental monitoring and information by expanding the factual basis of comprehensive urban water management models. In addition, they also imply a framework for negotiations that stresses the importance of comprehensive institutional formats and clarity in local and national decision making processes.

The questions of scale, boundary, and of geographical planning units are of central concern for efforts of problem definition, identification of parties-at-interest, and of implementation mechanisms. Whereas many authors favour the theoretical advantages of comprehensiveness, integration, and unified planning and management, successful practices seem to lag behind. It is important to recall that an on-going process of *complexification* creates a great diversity of interests, differences in resources accessibility, complex organisational arrangements, and competing and conflicting water policies. Finally, diachronic considerations complicate matters further by a shifting over time of goals, preferences, priorities, political consensus, leadership, sponsors and a variety of costs.

One of the major requirement for complex planning and management approaches is the central role of "participatory planning," especially in the form of public involvement. In the past, water planning was conceived as proceeding from clearly stated goals whose achievement was viewed as the optimal solution. Without entering into any lengthy discussion of this vast and important topic of the centrality of public participation in water resources planning, we may contrast in a summarising form the transformation from:

- elitist to participatory planning;
- deterministic to contingency planning;
- reactive to anticipatory planning;
- predetermined to an interactive process;
- presumed goals and objectives to ill-defined problems;
- fixed to a learning process.

It has already been stated that more sensitive methodologies for planning water resources systems must involve three time perspectives, namely the traditions of the *past*, incorporating the trends and events that have shaped water development; the conditions of the *present*, reflecting the constraints/facilitators that characterise the current state of affairs in a given system; and the visions about the *future*, as part of the forces that propel us toward potential or preferable future environments.

Whatever specific strategies and tactics are devised, the challenges of integrated water management raise larger issues of comprehensive urban planning that involve, among others:

- An articulation and explication of a shared vision of the desired urban water system.
- Baseline conditions for urban ecosystem functioning and sustainability and mechanism for environmental scanning and for monitoring change in terms of present and emerging trends and developments.
- Coordinated approaches among all concerned urban water management units and continuous participation of all stakeholders in the community as well as empowerment of many groups so that they can participate in discussions and support decision making.
- The development of the capacity to assume risks in terms of commitment and courage to undertake innovative actions.

- The encouragement of flexibility and the capacity to respond to both surprises and new emerging trends and developments.
- A combination of structural and non-structural solutions especially in terms of both physical infrastructure changes but also institutional and behavioural adaptations to changing urban water systems.
- Baseline conditions for urban ecosystem functioning and sustainable development against which meaningful change can be measured and determination can be made if goals and objectives are being achieved.

It is at this point, that many individual, local organisations, and international bodies have asked for needed research on such broad areas as financing schemes (especially cost-recovery and privatisation); improved management (in terms of streamlined and competent administration, networking, coordination, new mindsets, participatory and anticipatory emphasis, etc.); the appropriate use of techñical advances (in the form of technological innovations, renovations, rehabilitation and conservation); and, institutional streamlining (with a mobilisation of socio-cultural resources, promotion of sustainable development ethos, and citizen empowerment).

According to the World Bank World Development Report 1999/2000, in meeting the urban challenges of the 21st century, the most effective institutions and policy initiatives will exploit the opportunities that globalisation (an impetus for economic growth) and localisation (empowered communities) present (p. 140). We should note, however, that this tempting thought can be related to two common policy recommendations of how to cope with growing urban water demands, i.e., the raising of existing fees and the promotion of privatisation as a means of or enhancing the efficiency of urban water utilities. The last, attractive as it may sound, should give pause for reflection, especially when privatisation to foreign firms gives rise to many political issues. The sobering events of resistance to privatisation in Jakarta, in Cali, Columbia, in the riots of Cochabamba, Bolivia or the unrest in Johannesburg, S.A. and Bahia Blanca, Argentina, should be contrasted to relative successes in Mexico City and Buenos Aires and in cities in Europe and North America. Blind transfer of technology and institutions and lack of sensitivity to particular cultures and local conditions, as well as inappropriate timing, could have disastrous consequences. The obvious conclusion is that site-specific and cost-effective strategies need to be implemented.

The comprehensiveness of water resources planning has been the subject of controversy and debate in the literature. It has been recognised, however, that in order to be able to maximise the benefits from any urban water resources project a much larger systemic analysis of the surrounding environment is needed, a broadening of the traditionally narrow planning and management approaches, and an increased sensitivity to decision-making problems associated with multi-objective and multi-purpose actions. The need is increasingly expressed for a comprehensive framework for analysing policy and options where water scarcities exist, inefficiencies persist, and environmental damages are becoming apparent. Comprehensive approaches imply local and national strategies for urban water management that include elements for formulating public policies, regulations, incentives, public investment plans and environmental protection (as well as the inter-linkages among them). Technological and social responses to such approaches, as well as implementation mechanisms revolve around strong incentives for efficient water use, including better calculation and description of economic costs and benefits; significant structural changes such as innovative organisational arrangement, creation of new water agencies, etc.; "regulatory counter-incentives," especially in terms of stricter enforcement and pricing policies; improved data, information and monitoring systems; and, last but not least changes in "water intensive" lifestyles (certainly in developed countries) and adaptive cultural practices.

To paraphrase and expand the generic priority areas in the strategy of change proposed by Butler and Maksimovic (1999) as the five Is (or now seven Is) we should emphasise: Integration (the development of coordinated and integrated urban water models); Interaction (or the search for the additive, cumulative synergistic effects of complex urban systems); Interfacing (especially with the public and the environment); Instrumentation (in terms of real time control, sensors and non-invasive techniques); Intelligence (and the expansion of data, information, and knowledge through GIS, and hydroinformatics); Interpretation (the complementarity of data and judgement, and a wise combination of structured reasoning and disciplined imagination); and, finally Implementation (the capacity for true and concerted action and the transformation of policies into reasonable practice that combines corrective and proactive strategies and tactics).

There are, obviously, inherent dilemmas in addressing the complex urban water systems, especially of the rapidly expanding megacities of our world. These dilemmas are the result of multiple objectives, of expanded geographical scope and of conflicting and competing demands and choices. Yet, there is a discernible convergence into three main themes:

- The need for *new paradigms* that would incorporate increasing preoccupation with sustainability, social indicators of development and ecosystem maintenance; the shift from multi-disciplinary integration; a perceived transformation from crisis (reactive) to risk (proactive) management; and imaginative approaches of heterarchical considerations rather than exclusive preoccupation with hierarchical, linear systems.

- The understanding of *new contexts*, reflecting complexity and rapidity of change (or what some futurists have been summarily describing as "raplexity"); new transfrontier regimes, globalisation and interdependence; global change and climatic anomalies requiring new conceptual models and theoretical propositions; and different educational milieu, approaches and shifting social values.
- The emergence of *new methodologies*, addressing cumulative, synergistic, diachronic impacts and consequences of natural and human factors; emerging Decision Support Systems (such as GIS and expert systems) that combine data, information and judgment; benefiting from rapidly increasing computational prowess and new complex multi-factorial models; improved risk assessment and vulnerability analysis; integrated, holistic, comprehensive, heterarchical management schemes, network analysis and organisational mobilisation; and anticipatory scenarios, contingency planning, and expanded policy options.

As we conclude, it has been continuously pointed out that there is a need for establishing a broader perspective in urban water management. Such a perspective would enable us to account for the variety of environments affected by urban water by considering more disciplines; by incorporating an extended horizon in our thinking; and by concerning ourselves with the far reaching consequences of human actions in the surrounding environment. The last points out also that there must be a fundamental shift from a prevailing crisis management approach (short-range preoccupation and technological fixes) to a more anticipatory risk management that allows us to concentrate on contingency planning and reasonably foreseeable futures. We should consider the development of a new "social calculus" which would enable us not only to meet the present exigencies, but also to consider the meaning of our actions in the context of long-range plans and preferable futures.

The vocabulary of policy formulation, management and implementation, entails fundamental changes in outlook, visionary and goal-oriented commitment, as well as acceptance of the central premise that social, technical, economic and environmental problems are intertwined and most be resolved together. In essence, a commitment to water stewardship and the incorporation of sustainable development as a moral principle of what *should* be in water resources planning and management. And as it was succinctly stated in a popular pamphlet, a "Blue Revolution" in water supply is needed as much today as the "Green Revolution" in food production was needed earlier to feed expanding populations. Simply stated, the transition through the 21st century will require an institutional order of cooperation, of comprehensive management principles, and of sharing of experiences gained in the practice of ecosystemic principles in water resources projects. The main problem will be how to achieve integrated planning and management within institutional frameworks that have evolved under different socio-political conditions and for needs which are incongruous with the present and will most probably differ from projected futures.

1.6 REFERENCES

ASCE/UNESCO Task Committee (1998) Sustainability Criteria for Water Resource Systems. American Society of Civil Engineers, Reston, Virginia.

Balbo, M. (1993) Urban planning and the fragmented city of developing countries. *Third World Planning Review* **15**:23–35.

Bartelmus, P. (1994) Environment, Growth and Development: *The Concept and Strategies of Sustainability*. Routledge, London.

Beatley, T. (2000) Green Urbanism: *Learning from European Cities*. Island Press, Washington, D.C.

Braga, B.P.F. (2000) The Management of Urban Water Conflicts in the Metropolitan Region of Sao Paulo, *Water International*, **25** (2), 1–6

Brennan, E. M. (Summer 1999) Population, urbanisation, environment, and security: a summary of the issues. Environmental Change and Security Project Report 5, 4–14.

Brockerhoff, P. Martin (2000) An Urbanizing World *Population Bulletin,* **55** (3).

Butler, D. and C. Maksimovic (1999) Urban water management - challenges for the third millennium. *Progress in Environmental Science* **1** (3), 213-235.

Carew-Reid, J. et al. (1994) *Strategies for National Sustainable Development*, Earthscan Publications, Ltd., London.

Corbett, M. and J. Corbett (2000) *Designing Sustainable Communities*, Island Press, Washington, D.C.

Costanza, R. and Patten, B.C. (1995) Defining and predicting *sustainability Ecological Economics* **15** (3), 193–196.

Diewald, W.J. (1990) Trends and issues in urban water resources and the identification of needed research. *In Urban Water Infrastructure* (ed. K. E. Schilling and E. Porter), pp. 275–279, Kluwer Academic Publishers, Dordrecht.

Engelman, R. and LeRoy, P. (1993) Sustaining Water: Population and the Future of Renewable Water Supplies, *Population Action International*, Washington, D.C.

Falkenmark, M. (1998) Sustainable development as seen from a water perspective. In *Perspectives of Sustainable Development* (1), 71–84, Stockholm Studies in Natural Resources Management, Stockholm, Sweden.

Grigg, N.S. (1986) *Urban Water Infrastructure: Planning, Management, and Operations*, John Wiley & Sons, New York.

Hamm, B. and P.K. Muttagi eds. (1998) *Sustainable Development and the Future of Cities, Centre for European Studies,* Intermediate Technology Publications, Ltd., London, UK.

Hinrichsen, D. and Robey, B. (2000) Population and the Environment: The Global Challenge. *Population Reports, Series M,* No. 15. Johns Hopkins University School of Public Health, Baltimore, Population Information Program, Fall.

Inoguchi, T. et al. (eds.) (1999) *Cities and the Environment: New Approaches for Eco-Societies.* U.N. University Press, Tokyo.

IUCN - The World Conservation Union (1991) *Caring for the Earth: A Strategy for Sustainable Living,* Gland, Switzerland.

Leitmann, J. (1999) Sustaining Cities*: Environmental Planning and Management in Urban Design.* McGraw-Hill, New York.

Lele, S.M. (1991) Sustainable development: a critical review. *World Development* **19**(6), 607–621.

Lo, F. and Y. Yeung (eds) (1998) *Globalization and the World of Large Cities.* United Nations University Press, Tokyo.

Low, N., Gleeson, B., Elander, I., and Lidskog, R. (2000) *Consuming Cities: The Urban Environment in the Global Economy after the Rio Declaration.* Routledge, London and New York.

Mazmanian, D.A. and M.E. Kraft, eds. (1999) *Toward Sustainable Communities: Transition and Transformation in Environmental Policy.* The MIT Press, Cambridge, Massachusetts.

Munasinghe, M. and W. Shearer (1995) *Defining and Measuring Sustainability: The Biogeophysical Foundations* The World Bank, Washington, D.C.

Newman, P. and Kenworthy, J. (1999) *Sustainability and Cities: Overcoming Automobile Dependence* Island Press, Washington D.C.

Niemczynowicz, Janusz (1996) Megacities from a water perspective *Water International* 21, 198–205.

Pezzoli, K. (1997) Sustainable development: a transdisciplinary overview of the literature, *Journal of Environmental Planning and Management* **40** (5), 549–574.

Pezzoli, K. (1998) *Human Settlements and Planning for Ecological Sustainability: The Case of Mexico City,* MIT Press, Boston.

Roseland, M. (1998) *Toward Sustainable Communities: Resources for Citizens and Their Governments* (1998) New Society Publishers,

Rogers, R. (1997) *Cities For a Small Planet* (ed. P. Gumuchdjian), Westview Press.

Serageldin, I. (1996) Sustainability as opportunity and the problem of social capital. *The Brown Journal of World Affairs* **3** (2), 187–203.

Simonovic, S.P. (1996) Decision supported systems for sustainable management of water resources: 1. General principles. *Water International* **21**, 223–232.

Stren, R. et al. eds. (1992) Sustainable Cities*: Urbanisation and the Environment in International Perspective,* Westview Press, Boulder, Colorado.

Trzyna, T.C., ed. (1995) A Sustainable World*: Defining and Measuring Sustainable Development.* California Institute of Public Policy, Sacramento.

Tyson, J.M., Guarino, C.F, Best, J.J. and Tanaka, H. (1993) Management and institutional Aspects. *Water Science and Technology* **27** (12), 159–172.

Uitto, J.I. and Biswas, A.K. eds. (2000) *Water for Urban Areas: Challenges and Perspectives.* United Nations University Press, Tokyo.

UNESCO (1995) *Integrated Water Resources Management in Urban and Surrounding Areas.* UNESCO, Paris.

United Nations Centre for Human Settlements (1996) *An Urbanising World: Global Report on Human Settlements,* 1996. Oxford University Press,

United Nations Development Programme et al. (2000) *World resources 2000–2001: People and Ecosystems.* The Fraying Web of Life, World Resources Institute, Washington, D.C.

van den Bergh, J.C.J.M. and van der Straiten, J. eds. (1994) *Toward Sustainable Development.* Island Press, Washington, D.C.

Vlachos, E. (1997) Importance of indicators for policy formulation: sustainability and water resources policy. Paper presented at the *Water 21 Workshop "Sustainability Indicators and Criteria for Water Policy Formulation*", WRc Medmenham, UK.

Wilson, P.A. (1997) Building social capital: a learning agenda for the twenty-first century. *Urban Studies* **34** (5–6), 745–760.

The World Bank (1994) *Making Development Sustainable.* World Bank, Washington, D.C.

World Commission on Environment and Development (1987) *Our Common Future.* Oxford University Press.

2

Urban water as a part of integrated catchment management

Jiri Marsalek, Quintin Rochfort and Dragan Savić

2.1 INTRODUCTION

Since the Earth summit in Rio de Janeiro in 1992, it has become obvious that there can be no global sustainable development without sustainable settlement. With respect to land occupancy, urbanisation is characterised by concentration of large populations in relatively small areas. General statistics indicate that of the current (2001) world population of about 6 billion people, more than 54% live in urban areas, and in many countries this proportion reaches 90% or more. This concentration of population is even worse in developing countries that account for a high number of "mega" cities with many millions of inhabitants. Thus, at present, the issue of urban environmental sustainability is becoming

Edited by Čedo Maksimović and José Alberto Tejada-Guibert. ISBN: 1 900222 76 0

critical, because urbanisation and its associated environmental impacts are occurring at an unprecedented rate.

Urbanisation causes increased demands for energy and raw materials, and results in pollution and production of waste. The key activities of modern cities – transportation, electricity supply, water supply, waste disposal, heating, supply of services, manufacturing, etc. – are all characterised by the aforementioned problems. Thus, high concentration of people in urban areas dramatically alters material and energy fluxes in the affected areas, with concomitant changes in landscape; altered fluxes of water, sediment, chemicals, and micro-organisms; and, increased release of waste heat. These changes then impact on urban ecosystems, including urban waters, and result in their degradation. Such circumstances make provision of water services to urban populations even more challenging.

When water resources are threatened and their quality and utility deteriorate, the need for comprehensive management of water resources becomes much stronger. Without such management, further development of the society is impeded and its well-being and existence are threatened (Marsalek 1988). These concerns have long been recognized by UNESCO, addressing in the past the role of water in urban areas, effects of urbanisation on the hydrological cycle and water quality, and certain aspects of integrated water management in urban areas (e.g. Geiger *et al.* 1987, Marsalek 1995, Geiger and Hofius 1995). This initiative has been extended and focuses on water management. While urbanisation impacts on many facets of the urban environment, the discussion in this book is more or less limited to water resources and their management. The material discussed in this chapter starts with the effects of urbanisation on water resources, followed by the discussion of integrated catchment management and planning procedures for integrated management, and finishes with conclusions and references.

2.2 EFFECTS OF URBANISATION ON WATER RESOURCES

Combined effects of urbanisation, industrialisation and population growth alter natural landscape and hydrological response of the affected areas, and cause environmental and ecological impacts. During this process, many elements of the natural environment are replaced by man-made ones, with concomitant changes in the hydrological cycle. The principal changes caused by urbanisation include climatic changes, runoff/streamflow changes, water quality changes, and the resulting changes in the environment and ecosystems, including reduced biodiversity. Typically, these changes are caused by transformation of undeveloped land into urban land (including transportation corridors), increased energy releases, and increased demands on water supply, wastewater management, and drainage services. The major processes affecting air, water and soil quality in urban areas are shown in Fig. 2.1.

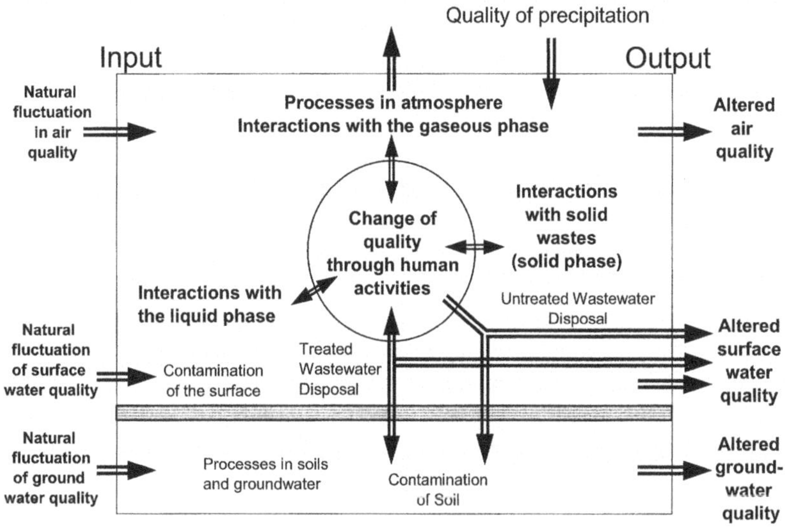

Figure 2.1 Major air, water and soil quality interactions in urban areas (after Geiger 2000)

2.2.1 Climatic effects

The effects of large urban areas on local microclimate have long been recognized and occur as a result of changes in the energy regime, air pollution, air circulation patterns (caused by buildings and/or transformations of the land surface), and by release of radiatively active (greenhouse) gases. These factors contribute to changes in the radiation balance and the amounts of precipitation and evaporation, and consequently to changes in the hydrological cycle. Many of these effects have been documented in urban areas and indicate that total annual precipitation in large industrial cities is generally 5–10% higher than in the surrounding areas, and for individual storms, this increase in precipitation can be as high as 30% (Geiger *et al.* 1987). Greater rainfalls are usually observed on the downwind side of urban areas. Similarly, urban heat islands produce air temperatures over urban areas exceeding those in the surrounding localities by as much as 4–7 °C. These thermal phenomena then explain higher evaporation rates (5–20%) in urban areas (Geiger *et al.* 1987).

More recently, attention focused on greenhouse effects and increasing concentrations of radiative gases (particularly CO_2) in the atmosphere, with concomitant changes of climatic conditions. As concentrations of greenhouse gases in the atmosphere continually increase, further climatic changes can be expected. Simulations with global circulation models, at relatively large scales, indicate that in most regions of the world, temperature increases of 3–5 °C can be expected in the next 50 years, but precipitation may either increase or decrease (Van Blarcum *et al.* 1995). Furthermore, greater variability in climate and more pronounced extremes are also predicted. Releases of radiative gases are caused primarily by the industry and transportation, and their control should be an international priority that clearly extends beyond the scope of urban water management.

With respect to sustainability of water supplies, some researchers suggest that the magnitude of effects of changing climate on water supplies may be much less important than changes in population, technologies, economics or environmental regulations (Lins and Stakhiv 1998). Furthermore, most well-designed water supply systems have a built in buffering capacity that ensures a stable supply of water despite seasonal and annual variations in climate.

2.2.2 Urban wastewater effluent impacts on receiving waters

Urban impacts on receiving waters are caused mostly by discharges of municipal wastewater effluents, which include stormwater runoff, combined sewer overflows (CSOs), and wastewater treatment plant (WTP) and industrial effluents. The urban effluent impacts reported in the literature include physical habitat and water quality changes, sediment and toxic pollutant impacts, impacts on biological communities, groundwater impacts, and public health impacts caused by discharges of faecal pollution, particularly during wet weather (Ellis and Hvitved-Jacobsen 1996).

The assessment of urban wastewater effluent impacts has to reflect the appropriate temporal and spatial scales, among which the temporal scales should correspond to the acute or cumulative nature of impacts, and the spatial scales should reflect the spatial extent of impacts in the receiving waters (Lijklema *et al.* 1993). Acute impacts are exerted quickly, within minutes of the onset of effluent discharges, and typically are caused by flow, biodegradable matter, acutely toxic concentrations of ammonia and heavy metals (Lijklema *et al.* 1993) or total residual chlorine (Orr *et al.* 1992), and, faecal bacteria (impacts on public health). In characterisation of acute impacts, pollutant concentrations and fluxes, and the frequency and duration of occurrence of pollutant concentrations are of interest.

Transport dynamics in the receiving waters, including pollutant mixing, dispersion and decay, is important and affects flows, fluxes and ambient concentrations of pollutants in the receiving waters (Marsalek 1998).

Cumulative impacts result from gradual changes in the receiving waters and become apparent only after such cumulative changes exceed some critical threshold value. Typical examples of cumulative impacts include releases of nutrients and toxicants from accumulated sediments (Harremoës 1988), or geomorphological changes in urban streams. For cumulative impacts, fine time scale dynamics is not important and the main interest is in loads or accumulation changes integrated over extended time periods.

Spatial and temporal scales of urban wastewater effluent impacts on receiving waters are displayed Fig. 2.2 (Lijklema *et al.* 1989) and various types of impacts are discussed below.

2.2.2.1 Physical impacts

Urbanisation increases surface runoff volumes and peak flows, and such increases may lead to flooding, sediment erosion and deposition, habitat washout (Borchardt and Statzner 1990), geomorphological changes (Schueler 1987), and reduced recharge of groundwater aquifers. Both acute (e.g., flooding, stream channel incision) and cumulative (morphological changes, the lowering of groundwater tables) impacts may be encountered. Ecological impacts include those on food web, critical species and ecosystem development; fishing is the most affected primary beneficial water use (Lijklema *et al.* 1993).

In warm weather, urban surfaces collect heat, which contributes to heating of stormwater runoff and increased temperatures in receiving waters by up to 10 °C (Schueler 1987; Van Buren *et al.* 2000). In highly developed catchments, these processes may lead to algal succession from cold-water species (mainly diatoms) to warm-water filamentous green and blue-green species, impacts on cold-water invertebrates, and cold-water fishery succession by a warm-water fishery (Galli 1991). Ecological impacts of thermal enhancement include those related to energy dynamics, food web, genetic diversity, and dispersal and migration; the most impacted beneficial water use is fishing (Lijklema *et al.* 1993).

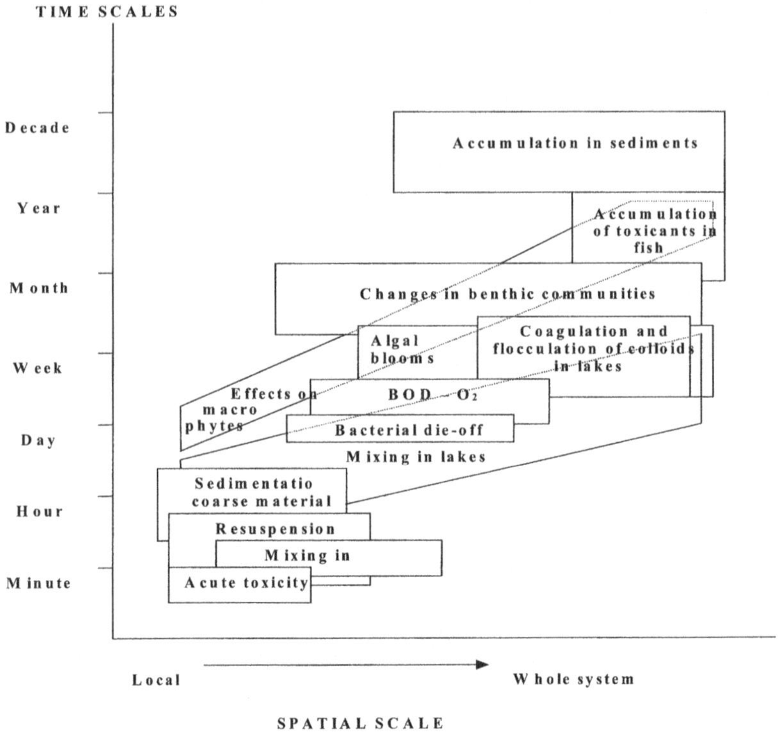

Figure 2.2 Spatial and temporal scales of impacts of urban wastewater effluents on receiving waters (after Lijklema *et al.* 1989)

Urban stormwater ponds, or small lakes, become chemically stratified during winter months (Marsalek *et al.* 2000), mostly by chlorides originating from road salting (Environment and Health Canada 2000). The resulting environmental effects include high levels of dissolved solids and densimetric stratification, which inhibits vertical mixing and transport of oxygenated water to the bottom layers, and may also enhance the release of metals bound in the sediments (Marsalek *et al.* 2000). The ecological impacts include those on the food web, genetic diversity and ecosystem development; the affected beneficial water uses include water supply, fishing, and irrigation (Lijklema *et al.* 1993).

2.2.2.2 *Chemical impacts*

Urban wastewater effluents may cause acute impacts resulting from toxic concentrations of chemicals, and chronic impacts caused by cumulative water quality stress and by pollutant accumulation in aquatic sediment affecting the organisms living in or on the bed (Horner *et al.* 1994). Ecological impacts of chemicals include those on the food web, biodiversity, critical species, and ecosystem development. Beneficial water use impairments include impaired fishing and degradation of the receiving water ecosystem (Lijklema *et al.* 1993). The discussion of chemical impacts starts with individual types of chemicals and follows by the discussion of classes of urban wastewater effluents.

Dissolved oxygen. Discharges of urban wastewater effluents may lead to biomass accumulation in receiving waters and reductions in dissolved oxygen (DO). In this respect, CSOs with high loads of oxygen demanding substances (including ammonia) are particularly of concern (Harremoës 1988); well-treated WTP effluents and stormwater discharges are less significant. The associated environmental impacts occur either as short-term impacts caused by dissolved biochemical oxygen demand (BOD)/chemical oxygen demand (COD) and ammonia, or intermediate-term impacts, which are caused by the sediment oxygen demand (Hvitved-Jacobsen 1982). These impacts affect both the ecology of receiving waters and beneficial water uses; ecological impacts include those on biodiversity and critical species; the affected water uses include water supply, bathing, fishing and industrial water supply (Lijklema *et al.* 1993).

Nutrients. Nutrient enrichment and eutrophication of receiving waters is caused by nitrogen and phosphorus found in various forms in WTP effluents, CSOs, stormwater and some industrial discharges. Nutrient loadings may cause eutrophication, which is characterised by an overall increase of planktonic biomass, and changes in the composition of algal communities from one-celled diatoms to filamentous green forms, followed by blue–green forms. Nutrient additions increase primary production, and the enhanced plant growth may accentuate diel swings in DO (during daytime, oxygen is produced by photosynthesis, but consumed during night time by respiration) and increased sediment oxygen demand, because of dead plant material decomposition at the sediment–water interface (Chambers at al. 1997). Eutrophication effects manifest themselves over longer time periods, at least a season or even longer (Harremoës 1988). Ecological impacts include those on energy dynamics, food web, critical species, and ecosystem development; the impaired beneficial water uses include water supply, bathing, recreation, fishing, industrial water supply, and irrigation (Lijklema *et al.* 1993).

Toxic Substances. Toxic impacts of urban wastewater effluents may be caused by toxic concentrations of ammonia, total residual chlorine, chlorides, cyanide, sulphides, phenols, surfactants, heavy metals, hydrocarbons (particularly polycyclic aromatic hydrocarbons, PAHs), and some other trace organic contaminants, including pesticides (Chambers *et al.* 1997; Hall and Anderson 1988; Dutka *et al.* 1994a, b). Toxicity responses are classified as acute (lethality), genotoxic (damage to genetic material) and chronic (reproduction or development). Toxicity of urban effluents can be measured by bioassays that determine the effluent concentrations that are lethal or cause physiological or behavioural effects (Chambers *et al.* 1997).

Endocrine Disrupting Substances (EDSs) and Other New Chemicals of Concern. Drugs and hormones excreted in human waste can pass through sewage treatment systems without being degraded. Only recently, concerns have been expressed over the fact that these substances may be the primary source of a significant (measurable) level of endocrine disrupting chemicals found in the receiving waters (Ternes *et al.* 1999). Even at low levels, EDSs can result in estrogenic effects in receiving waters by binding to estrogen receptors in the male fish, which results in the production of vitellogenin (an egg protein) as well as intersex (ova-testes). These factors can significantly jeopardise fish reproductive abilities. Many of EDSs are also found in industrial effluents and can exert both toxic effects and estrogenic effects, even at low levels (Servos 1999). Newly identified chemicals of concern include pharmaceuticals and personal care products (such as antibiotics, blood lipid regulators, analgesics, anti-inflammatory drugs, beta-blockers, fragrances, skin care products, disinfectants and antiseptics) (Daughton and Ternes 1999) and some other substances found in complex chemical mixtures in wastewater effluents.

Stormwater and CSOs. In the assessment of toxicity of runoff from various land use, highway runoff (particularly in winter) and industrial runoff discharges were generally found to be the most toxic (Marsalek *et al.* 1999a). Higher toxicity of winter runoff follows from extended accumulations of toxicants in the snowpack (Viklander 1997, Oberts *et al.* 2000), winter road maintenance involving applications of road salts and anti-skid materials, and higher mobility of toxic metals in chloride-laden runoff (Marsalek *et al.* 1999a). During runoff events, the highest toxicity appeared to coincide with the first flush describing high concentrations of pollutants during the early stages of runoff (Marsalek *et al.* 1999b). In residential areas, runoff samples were mostly non-toxic.

CSOs. CSOs contain conventional substances found in sewage and also toxic substances (including metals and trace organic chemicals) found in stormwater (Rutherford *et al.* 1994; Golder 1995a b). High flows during storm events result in high levels of total suspended solids (TSS), BOD, nutrients (N and P), various toxic chemicals, faecal bacteria and other pathogens. During the first flush, combined sewers are purged, and the scouring of sludge (accumulated sediment) from sewer inverts results in high concentrations of TSS, chemicals, and bacteria. Benthic communities in receiving waters may be particularly affected by this first flush, because of the associated large hydraulic disturbance and influx of sediment (Borchardt and Stazner 1990). Fish spawning areas and shellfish beds may also be affected by PAHs, sediment and bacteria, resulting in a reduced fisheries use in the impacted waters. Long-term effects can include permanent species change and loss of biodiversity (Chambers *et al.* 1997).

WTP effluents. WTP effluent composition can vary considerably, depending on the source of wastewater. In areas without industrial inputs (or with industrial inputs well pre-treated), municipal sewage can be non-toxic. The two most frequent causes of WTP effluent toxicity were identified as unionised ammonia and total residual chlorine (Orr *et al.* 1992). Other chemicals of concern include endocrine disrupting substances, pharmaceuticals and personal care products, and treatment by-products, which are not necessarily less toxic than the original substances. Disinfection of wastewater effluents is a common practice employed to reduce the contamination of the receiving waters by pathogens. Disinfection methods include chlorination (sometimes followed by de-chlorination), bromination, ultra-violet (UV) irradiation and ozonation (O_3). All these methods may induce chemical changes in WTP effluents, although UV irradiation appears to have the least impact on effluent chemical structure (Blatchley *et al.* 1997). Because of these changes, common disinfection methods may increase the effluent toxicity. Residual chlorine (where dechlorination is incomplete or not performed at all) can be extremely toxic.

Industrial effluents. The chemical composition of industrial discharges is more predictable than that of other types of municipal effluent. Each type of industry produces a characteristic type of effluent, whose composition will, however, somewhat fluctuate on a daily, weekly or seasonal basis. Loads and concentrations of contaminants in industrial effluents depend on the level of treatment applied, the efficiency of process water management (including water re-use, or losses), and maintenance procedures. The range of chemicals found in industrial effluents is rather broad, and includes heavy metals, hydrocarbons (including polycyclic aromatic hydrocarbons, PAHs), suspended and dissolved solids, toxic substances (including cyanides, chlorides), as well as endocrine disrupting substances (Chambers *et al.* 1997).

2.2.2.3 Impacts on public health

Urban wastewater effluents impact on public health in both developing countries and developed countries with modern infrastructures. In developing countries, there is significant pressure on authorities and urban infrastructure to provide clean water, sanitation and solid waste collection to ever increasing populations. At the same time, these countries find it difficult to obtain the capital to improve and expand the existing systems to service the surrounding peri-urban environment. The result is often an overburdened infrastructure, which operates improperly or becomes degraded, and causes significant human health impacts (Birley and Lock 1999).

Concerns in developing countries. The human health impacts associated with the expansion of cities in developing countries are many, including vector borne diseases, bacterial infections, drinking water contamination and water reuse concerns. Malaria and dengue fever (spread by mosquitoes) are common diseases in developing countries, as well as schistosomiasis, filariasis, chagas disease, cholera, meningitis, ear, eye and skin infections, and hepatitis. These diseases are often caused by poor sanitation (open sewers) and the presence of large breeding areas for disease vectors (reservoirs, poor drainage, flooded fields). Pathogenic infections usually result from contaminated food and drinking water supplies, and include tapeworms, roundworms, hookworms, *Shigella*, *Giardia*, *Salmonella* and *Cryptosporidium*.

As urban populations expand, more food must be generated. In order to increase production, farmers generally irrigate, and apply pesticides and fertilisers to fields to improve yields and reduce losses to insects and disease. Water used in irrigation may be provided from streams or reservoirs (usually dammed for storage purposes), and these water bodies may be breeding grounds for disease vectors, or may become contaminated with untreated or partially treated wastewater, toxic leachates (from landfill operations), or industrial effluents. Also, fertilisers may be derived largely from animal, human and industrial solid wastes, and inadequate pre-treatment may result in crop contamination. Field workers and farmers may also be exposed to high levels of these contaminants.

Wastewater treatment in developing countries is generally simplistic and consists largely of waste stabilisation ponds. In some cases, wetlands, reed beds, hyacinth ponds or similar facilities are used to improve the quality of wastewater effluents, however, the final effluent is rarely disinfected. This poorly treated wastewater often contaminates the receiving waters, and may be reused (without prior treatment) by others living in downstream areas (and particularly those without access to potable water or sanitation), for drinking, washing or irrigation. In many wastewater reuse operations, wastewater is diverted to fishponds or shellfish beds to provide nutrient rich waters and improve production in fisheries. Without proper treatment, fish and crops may become contaminated with toxic chemicals and pathogens, and passed on to humans for consumption (Birley and Lock 1999).

Concerns in developed countries. In developed countries, the concerns about the impacts of urban wastewaters on public health are mostly related to restricted recreational use of waters (beach closures), contamination of shellfish and closure of harvesting areas, and some rare impacts on drinking water supplies representing isolated incidents caused by serious equipment breakdowns, or human errors. The microbiological pollution impacts on human health and biomass are associated primarily with untreated CSOs and stormwater discharges, and to a lesser extent with treated WTP effluents.

Both stormwater and CSO discharges carry high loads of faecal bacteria, which are typically described by such indicators as *Escherichia coli* or faecal coliform. After discharge into receiving waters, bacterial concentrations start to diminish through effluent mixing and bacterial dieoff. The actual health risk associated with exposure to contaminated waters is not well known (lack of epidemiological data) and depends on the nature of recreational activities - the highest being for swimmers and the smallest for those engaged in wading. Water quality guidelines specifying permissible indicator bacteria concentrations are used to protect the public using recreational waters (e.g., in Ontario, 100 *E. coli*/100 mL). When such guidelines are exceeded, swimming beaches are closed. Many challenges are encountered in applying these guidelines.

Recognising that typical concentrations of *E. coli* in CSOs (up to 10^7 EC/L) and stormwater (up to 10^6 EC/L) exceed the recreational water quality guidelines by several orders of magnitude, exceedance of these guidelines in wet weather is not surprising. Many beaches in urban areas are frequently closed during and immediately after rainfall events, because of faecal bacteria contamination caused by stormwater and CSOs (Dutka and Marsalek 1993; Marsalek *et al.* 1994). Efforts to control stormwater and CSO discharges include source controls, storage and reuse (Vaes and Berlamont 1999), real-time control systems (Stirrup 1996; Petruck *et al.* 1998), and disinfection. Ecological impacts of microbiological pollution include those on energy dynamics, food web, and ecosystem development. The impacted water uses include water supply, bathing, and fishing (Lijklema *et al.* 1993).

2.2.2.4 Combined impacts

Municipal wastewater effluents impact on receiving water ecosystems through a combination of many factors, including habitat destruction, thermal enhancement, and pollutant discharge. Consequently, it is sometimes impossible to separate the effects of individual factors and a better understanding of the combined effects can be obtained by assessing the biological community performance, as shown in Fig. 2.3, in conjunction with the assessment of physical, chemical and biological factors, (Horner *et al.* 1994). Typical biological effects, observed in connection with urban wastewater effluents, include adverse impacts on food web, biodiversity, critical species, genetic diversity, dispersal and migration, and ecosystem development (Lijklema *et al.* 1993).

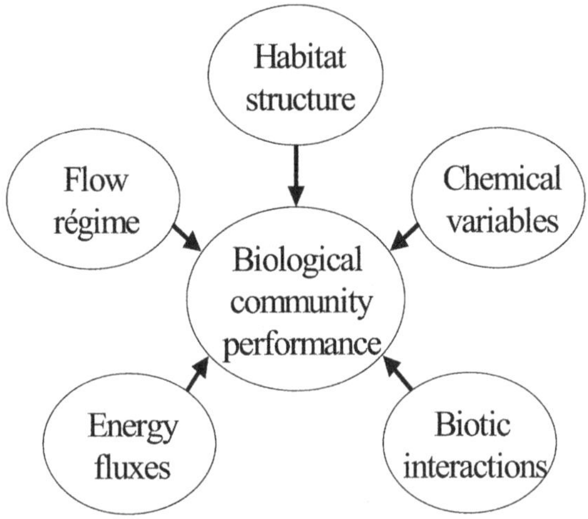

Figure 2.3 Combined impacts on biological community performance

2.3 INTEGRATED CATCHMENT MANAGEMENT

Water management planning should reflect the objectives of urban water management, characteristics of urban water resources, multiple functions (beneficial uses), interdependencies, local disconformities, scale economies and multiple use benefits, and natural monopoly and pricing (Marsalek 1988). Multiple uses of urban waters are self-evident and their impact on water management is twofold: it implies the reuse of water for various purposes, and individual management measures may provide benefits for multiple uses. The multiple uses follow certain priorities, with the highest priority assigned to water supply. High mobility of water and physical connectivity of water resources, both natural and man-made, contribute to interdependencies of actions of independent users or developers. Interdependencies of water resources and their utilisation become particularly strong under intensive use and provide an impetus for integrated water management (Fig. 2.4). Managed interdependencies lead to spreading the benefits arising from such actions (Marsalek 1988).

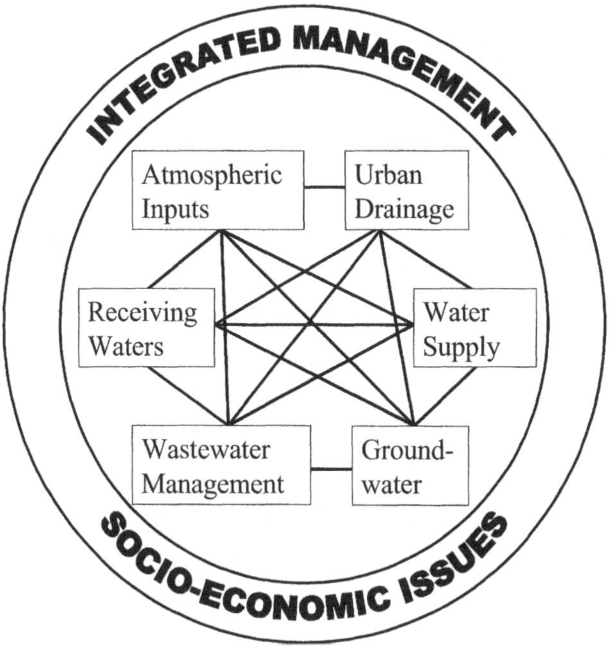

Figure 2.4 Integrated catchment management

Water resources development generally requires large investments and significant savings can be achieved, both in the development of water resources and provision of services, by employing large scale facilities and management systems. Such economies may not be feasible under the existing fragmented jurisdictions, and the movement towards large-scale regional facilities is sometimes opposed, because it is perceived as leading to the loss of autonomy and environmental impacts of larger facilities may be more difficult to manage. It is recognised at the same time that the ability to assign a monetary value to an urban water management project does not determine its importance. In this analysis, a distinction needs to be made between price and value, where price is a current, transient factor and not an accurate measure of ultimate value. Conventional project cost considerations should be extended for environmental costs, which consider positive environmental effects as well as environmental damages caused by project implementation (Geiger and Hofius 1995).

General values of water in densely urbanised areas are well recognised, particularly with respect to water supplies for domestic use, industry and agriculture; and other water uses, such as fishing, navigation and recreation. Less evident and less well recognised are environmental and ecological goals of integrated water management, which support the protection and enhancement of ecological integrity. The appropriate management strategy must consider the history and the current characteristics of the ecosystem to be protected (Geiger and Hofius 1995). The ecosystem characteristics that are critical for ecosystem maintenance have been defined by Herricks and Schaeffer (1987) and can be used for identifying critical ecosystem integrity attributes, which include elemental dynamics, energy dynamics (physical), food web (trophic dynamics), bio-diversity, critical species, genetic diversity, dispersal and migration, natural disturbance, and ecosystem development:

- habitat for desired diversity and reproduction of organisms,
- phenotypic and genotypic diversity among organisms,
- robust food chain supporting the desired biota,
- an adequate nutrient pool for desired organisms,
- adequate nutrient cycling to perpetuate the ecosystem,
- adequate energy flux for maintaining the trophic structure,
- feedback mechanisms for damping undesirable oscillations, and
- the capacity to temper toxic effects, including the capacity to decompose, transfer, chelate or bind anthropogenic inputs to a degree that they are no longer toxic in the system.

The main objective of water management is to provide a wide range of goods and services, including water itself and the related services and benefits. Such services include protection against harmful effects of water, be they physical (e.g., flooding), or chemico-biological (contamination of water resources by chemicals or micro-organisms). The related objectives can then be defined as the need to select the combination of water services to be produced from specific water resources and the requirement to provide those services efficiently within the overall goals and policy constraints developed at the formative level of management. To meet these two objectives, it is required to appraise demands and supplies, assess benefits and costs of solution alternatives, select the best alternatives meeting the demands most efficiently within the given policy constraints, implement the chosen alternatives, and conduct continuing sustainable operations as required.

With respect to integrated management, there is a need to promote a broader, multidisciplinary approach to integrated catchment management, where the environment to be managed is a complex entity of interacting ecosystems that operate within a landscape. Integrated water management accounts for system complexity and inter-connectivity of its elements. It is holistic in its approach, which is characterised by involving local and regional authorities, employers, environmentalists and decision makers, politicians, as well as all the people affected. It is cross-sectorial by its nature, and transcends societal structures, divisions and institutions. The greatest challenge encountered in implementation of integrated water management is the creation of cross-sectorial co-operation and integrated multi-disciplinary actions (Geiger and Hofius 1995). This type of water management has also been referred to in the U.S. and Canadian literature as an ecosystem based approach to water management, in which the interests of society, economy and environment receive equal consideration (Hartig and Vallentyne 1989).

Thus, the ultimate aim of integrated water management is to "*achieve the sustainable, co-ordinated management of water resources within a region, with the objectives of controlling and conserving water, minimising adverse effects, and achieving specified and agreed water management and social objectives*". In this context, sustainable water systems are described as those which "*are designed and managed to fully contribute to the objectives of the society, while maintaining their ecological, environmental and hydrological integrity*," and meeting the demands to the system without its degradation, now and in the future (ASCE and UNESCO 1998).

In urban waters, the goal of sustainability is probably best supported by the total management of the water cycle in the affected areas (Lawrence *et al.* 1999). Urban water management categories representing components of a total water cycle can be defined as (a) re-use of treated wastewater, as a basis for disposing of potential pollutants, or as a substitute for sub-potable water supply; (b) integrated stormwater, groundwater, water supply and wastewater based management, as the basis for economic and reliable water supply, environmental flow management (deferment of new capital outlays, return of water to streams and aquifers), urban water scape/landscape provision, substitute for sub-potable sources of water (wastewater and stormwater reuse, protection of downstream waters from pollution); and, (c) water conservation based approaches (by demand management) including more efficient use of water (water saving devices, irrigation practices), substitute landscape forms (reduced water demand), and substitute industrial processes (reduced demand, recycling). This sustainability has to be achieved in the context of the climate change caused by the greenhouse effect (Marsalek, 2000a).

Typical services provided to urban populations include water supply, wastewater treatment and disposal, flood protection and drainage, and provision of water as a general amenity. These services are discussed first, followed by discussion of planning techniques for integrated water management, with reference to these specific services.

2.3.1 Water supply

The gulf between society's needs and the water provision capability of nature is ever-widening. The requirement for water is growing and with the rapid expansion of urban areas, many communities are or will be entering a state of continuous water stress. Water supply is a part of the overall water services cycle which is bridging the gap through an artificial 'water cycle' (Fig. 2.5).

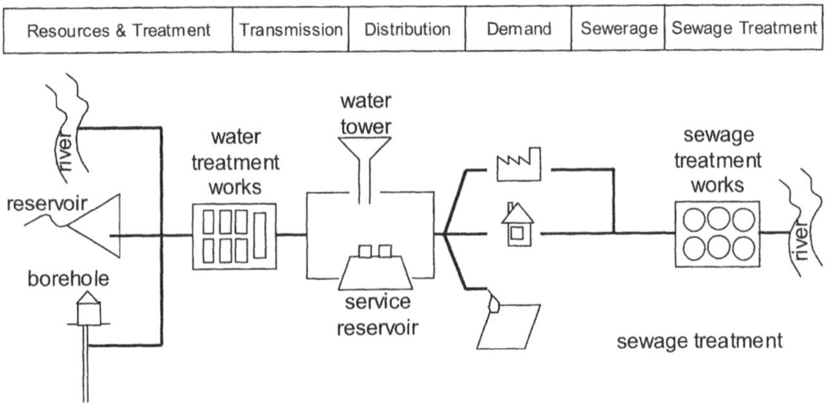

Figure 2.5. Water services cycle (after Latham 1990)

Basic human needs for water were articulated at the Mar de Plata conference, by declaring that:

> "*... all peoples, whatever their stage of development and their social and economic conditions, have the right to have access to drinking water in quantities and of a quality equal to their basic needs*" (United Nations 1977).

This right was reaffirmed during the 1992 Earth Summit in Rio de Janeiro and in the Comprehensive Assessment of the Freshwater Resources of the World, prepared for the UN General Assembly (United Nations 1997). However, the reality is such that in many countries (representing a population of about 1 billion people) domestic per capita water withdrawals range from 4.5 to 48.2 litres/person/day and fall below the basic water requirement (BWR) of 50 litres/person/day recommended by Gleick (1998). This rate was derived by considering 25 l/p/d of clean water for drinking and sanitation, and adding minimum quantities of water for bathing and cooking. The proposed BWR is not too different from the target of 20–40 l/p/d set by the U.S. Agency for International Development, the World Bank, and the World Health

Organisation, and it is also comparable to the recommendations of the United Nations International Drinking Water Supply and Sanitation Decade and Agenda 21 of the Earth Summit (Gleick 1998).

The failure to meet the basic water needs is rather costly with respect to both social and economic aspects. Direct annual medical expenses and lost work time were estimated at $125 billion (late 1970 dollars) by Pearce and Warford (1993). Estimates of costs of meeting water and sanitation needs in developing countries range from about $26 to $50 billion per year (Gleick 1998), which is significantly less than the somewhat dated estimate of direct losses.

Inadequate water supply and sanitation severely impact on human health, through water-related diseases. The World Health Organisation estimated that there may be on the order of 250 million of water-related diseases annually and 5 to 10 million deaths (Nash 1993). Water related disease cases fall into four classes; waterborne, water-washed, water-based, and water related insect vectors. Some progress is being made in fighting these diseases, as is the case of dracunculiasis (Guinea Worm), which is on the verge of eradication. With respect to other diseases, such as cholera, the progress in disease prevention is much more limited (Gleick 1998).

Water supply is a dominating concern of urban water management. In many countries, this concern is relatively successfully addressed by providing relatively inexpensive and reliable service, using such approaches as water conservation by demand management and distribution loss management, optimal utilisation of the available water supplies, and water reuse. These measures can be applied to each of the four stages of the water services cycle, from the abstraction of water (from aquifers, reservoirs, boreholes, rivers or even the sea), through treatment and transmission, to delivery of water to domestic and industrial customers.

The old approach to water supply management by developing new water projects and transfer schemes, in order to keep ahead of the ever-increasing requirements due to the growing population and improved living standards, is a practice which cannot be sustained in the future. Optimal utilisation of the available water sources includes measures to control abstractions, artificially recharge groundwater aquifers or encourage the conjunctive use of multiple water sources, which can be sustained. The joint use of two or more sources according to a carefully planned rule, can lead to a cheaper supply than that gained by their independent use (Walsh 1971). Demand management together with distribution loss management through pressure regulation, metering and asset management is seen as a promising way of reducing future water requirements within distribution systems.

The quality of water supplies is governed by guidelines that are specified for various water uses (e.g., drinking water, raw potable water, irrigation water, livestock water and industrial waters). The most stringent are the guidelines for drinking water, which require that such water is free of health hazards, as specified by permissible concentrations of various chemicals and microorganisms. Drinking water guidelines usually include microbiological parameters (e.g., algal toxins, coliforms, coliphages, parasites, pathogens, protozoa and viruses), physical parameters (colour, odour, taste, pH, temperature and turbidity), radiological parameters (e.g., isotopes of caesium, iodine, radium, strontium and tritium), inorganic parameters (trace metals, nitrate/nitrite, chloride, sulphate, sulphide, total dissolved solids), and organic parameters (including pesticides, chlorinated phenols, trihalomethanes, polycyclic aromatic hydrocarbons, and others). The drinking water guidelines are typically developed at the national level. Extensive documentation for guideline development and surveillance and control of community water supplies has been published by the World Health Organisation (WHO 1993, 1996, 1997, 1998).

In a large percentage of urban developments, a single potable water supply is used to provide water for drinking, washing, irrigation and toilet flushing, even though water does not need to be of potable standards for many of these activities (e.g., flushing toilets), and it may not be sustainable to continue to use such high quality water for all these activities. Provision of a non-potable water supply (e.g., untreated or partly treated surface water or stormwater, water recycling/treatment and reuse) is an option investigated in some areas (Pratt 1999). The provision of a secondary water supply (for toilet flushing and clothes washing) may help offset the cost of treating and pumping more potable water, and could increase the sustainability of the potable water supply in the future (van der Hoek *et al.* 1999). This type of dual water supply system may not be economically or logistically feasible in areas with abundant water resources and/or in older developments, where the high cost of retrofitting can be prohibitive and the associated risks too high (Mikkelsen *et al.* 1999). In other cases, where the quality of water in the distribution system may be at risk at times (e.g., because of infiltration into leaking systems), another type of "dual water supply" is applied by using bottled water for drinking.

Significant progress has been made in reducing domestic water usage by technological devices, including water saving toilet flushers, shower heads, washing machines and dishwashers; and, reuse of water in washing machines and dishwashers. Further reductions/reuse are feasible with water reuse for different processes (moving from the highest quality towards the lowest), and use of additional (sub-potable) water supplies (Terpstra 1999). In the latter case, systems are tested which use rainwater or grey water for sub-potable water supply, usually after some pre-treatment, e.g., filtration and disinfection. Finally, black waters are also reused, mostly in agriculture and landscape irrigation.

Industrial operations offer many opportunities for water reuse and recycling. Small to moderate scale industrial operations, which are located within municipal boundaries, may account for only 20% of the total potable water demand, although this percentage varies depending on local conditions and products manufactured. Significant improvements in process efficiency and water recycling have been made by most industries to provide sustainable solutions with respect to water usage. Rising costs of, and demands on, the potable water supply have also driven recycling efforts (WBCSD and UNEP, 1998).

Many reuse opportunities exist with respect to cooling water, which accounts for a very high percentage of the overall water use in industry. This water, which is normally discharged as waste, can easily be treated (if required) and reused in other aspects of the industrial process. Other opportunities for reuse depend on the type of industry. For example, a large amount of water in textile manufacturing process can also be reused, resulting in cost savings, up to half of the cost of using potable water. With appropriate planning and process analyses, minimal treatment may be required for reuse (Van Riper and Geselbracht 1999). Where treatment is required, simple and low-cost treatment processes including ultrafiltration, activated carbon filtration or reverse osmosis systems are preferred. The treatment design depends on the quality of the effluent being treated and the requirements on the recycled water (Roeleveld and Maaskant 1999).

Agriculture exerts high demands on water resources, particularly for such operations as irrigation and dairy farming (large amounts of water are used for cleaning). This subject is beyond the scope of this report, but nevertheless it should be noted that agriculture water demands compete with urban demands. Conservation strategies applied in agriculture provide more water for other sectors. The means of increasing the sustainability of freshwater supplies in agriculture include the use of rainwater for washing and animal drinking, and treatment of wash water for reuse in irrigation and manure flushing (Willers *et al.* 1999).

Hydroponic (soil-less) growing systems have been traditionally designed with water replacement after just a few cycles, rather than treating and reusing it. New methods call for the use of simple filtration (sand filters or biofilters), followed by disinfection (generally by ultraviolet light irradiation), to promote reuse of water over many cycles, thus reducing the effluents and increasing the sustainability of this type of agricultural practice (Van Os 1999).

Finally, the best practice in integrated water supply requires integration of water resources and development of interconnecting transport networks, water conservation measures based on public education and development of water saving techniques in all sectors, and utilisation of alternative water resources. Such resources include wastewater reuse (e.g., in Israel, up to 70% of wastewater collected in sewers is reused for agricultural purposes), collection of surface runoff/rainwater for groundwater recharge and other purposes, and desalination of brackish waters (IWA Water Reuse Committee 2000).

2.3.2 Wastewater treatment and disposal

Wastewater treatment and disposal are widely practised in developed countries, with a high percentage of the total population (more than 90%) being connected to central or regional wastewater treatment facilities. Many chemical, physical and biological processes are used to treat wastewaters (Metcalf and Eddy Inc. 1991) to an acceptable quality for discharge to rivers, lakes, estuaries and seas. Although most of the wastewater treatment technology has been used for many years and is well proven, research continues to further improve the performance and efficiency of these treatment processes. The situation is different in developing countries, where with respect to wastewater treatment, the most important issue seems to be the choice of appropriate technologies.

Environmental sanitation has been addressed by WHO with respect to environmental factors (disposal of human excreta, sewage, household waste and other waste with infectious agents, water drainage, domestic water supply, and housing) and sanitation practices (personal hygiene, household cleanliness and community cleanliness). It is recognized that good sanitation is critical for reducing the exposure to infectious agents and the risk of contracting diseases. In the Global Water Supply and Sanitation Assessment 2000, it is stated that while the percentage of population served by 'improved' water supply rose from 4.1 billion in 1990 to 4.9 billion in 2000, and during the same period, the access to excreta disposal facilities increased from 2.9 billion people to 3.6 billion people served, still about two-fifths of world population (2.4 billion people) lacked access to 'improved' sanitation in 2000 (WHO 2001). WHO defined the 'improved' sanitation as any of the following technologies: connection to public sewers or septic systems, pour-flush latrine, simple pit latrine, and ventilation improved pit latrine. 'Not improved' technologies include service or bucket latrine (where excreta are manually removed), public latrine, and open latrine. With respect to selecting and adopting appropriate sanitation technologies, WHO emphasizes strongly that these processes depend on social and cultural practices and beliefs, which must be duly considered when planning the implementation of such technologies (WHO 2001).

The development of a water pollution control strategy to increase the sustainability of urban receiving waters through the use of appropriate technologies can be broken down into smaller components or projects. This allows the overall strategy to be realised over a limited period of time, and to support this strategy by incremental expenditures with the achievement of the desired objectives (Furguson and Horsefield 1999). The use of low-cost methods of treatment or pre-treatment, particularly in developing countries, can aid the development of a sustainable wastewater system capable of protecting the receiving waters. For example, the "low-technology" upflow anaerobic sludge bed (UASB) bioreactors can be used to pre-treat high-strength industrial wastes (El-Gohary and Nasr 1999), as well as be used in de-centralised treatment of residential or agricultural wastes (Zeeman and Lettinga 1999). These systems can work very well, provided that the proper design and construction methods are applied, and overloading does not occur (Kalker *et al.* 1999).

Upgrading wastewater treatment plants to provide secondary or tertiary treatment does not guarantee that effluents will attain an acceptable quality. While upgrading facilities from primary to secondary and from secondary to tertiary treatment generally results in a reduction in concentrations of total suspended solids (TSS), BOD, COD and ammonia in the effluent, the actual concentrations depend upon the composition of the wastewater being treated (Chambers *et al.* 1997). As technology advances, many new treatment methods such as membrane filtration have been applied to wastewater treatment (van der Graaf *et al.* 1999). However, these high-performance systems are expensive to install and maintain, and, consequently, their use may be warranted only in the protection of highly sensitive receiving waters of extreme importance.

To enhance the sustainability of wastewater management, separation of wastewaters at the source was proposed. This separation improves the efficiency of the end-of-pipe treatment processes, and facilitates wastewater recycling, and retention and extraction of nutrients from solids, with subsequent reuse in agriculture (Hedberg 1999). Dilution of solids with other wastewaters makes recovery more difficult, increases the amount of wastewater that has to be conveyed and treated, and thereby increases the risk of pollution of the receiving waters. When wastewaters are separated at the source, blackwater is treated using aerobic bioreactors, while greywater can be treated in such simple systems as wetlands or sand filters, and reused (Skjelhaugen 1999). Treating wastewater at the source (often termed "source control") may allow a more efficient reuse of resources, while increasing the sustainability of the system (Otterpohl *et al.* 1999).

Research is progressing on developing new methods of efficient and sustainable sanitation, aiming for a high hygienic standard and full reuse of resources. One of such methods employs no-mix toilets (also called separating or sorting toilets), in which yellow water (urine) is conveyed by a separate pipe to a storage tank and eventually used as a fertiliser. A minimum storage of six months is recommended to degrade any medical residues in urine. The brownwater (faeces) is collected either separately or with greywater and discharged into a one or two chamber composting tank. The products are removed from the composting tank after a year and used for soil improvement, or subject to further composting (Otterpohl 2001).

In another decentralised scheme, vacuum toilets, vacuum sewers and a biogas plant were implemented in a new urban development in Northern Germany. In this design, blackwater is mixed with shredded biowaste and used to produce heat and power for this housing estate. Surface runoff is infiltrated through pervious swales and greywater is treated in constructed wetlands. The material and energy use intensities are less than half of those for the conventional central treatment system serving a medium density urban area. For the central treatment system, most of the material use intensity results from the construction of the sewer system. The predicted effluent quality values compared well to those of a modern, well-performing treatment plant with advanced nutrient removal (Otterpohl 2001).

Besides the technological improvements in wastewater treatment technology, advances in the disposal or reuse of treated effluents and sludge are also of great interest. There are methods for pre-treating the wastewater for reuse or disposal, many of which are similar to domestic sewage-treatment processes (Rae 1998). The reuse of cooling water, which was often disposed with wastewaters, represents one of the simplest methods of wastewater reuse. Other examples of reuse include applications in agriculture and irrigation, enhancement of wetlands habitat, groundwater recharge, aquaculture and fisheries, and sub-potable water supply. During the past three decades, there has been a notable increase in the use of wastewater for crop irrigation, especially in arid and seasonally arid areas of both industrialised and developing countries. Suitability of the wastewater for the selected application and the required level of pre-treatment must be carefully considered, particularly with respect to food crops, in order to reduce the chances of poisoning or the spread of infectious agents.

Other approaches to selecting appropriate wastewater treatment and disposal methods use life-cycle evaluation to determine the most suitable treatment methods. This approach allows each treatment method to be ranked according to its cost, environmental sustainability and potential impacts (Ashley *et al.* 1999; Tao and Hills 1999).

Wastewater treatment generally produces large quantities of sludge, which has to be processed, treated and properly disposed. Sludge processing methods include preliminary operations (e.g., sludge grinding, degritting, blending and storage), thickening, stabilisation, conditioning, disinfection, dewatering, heat drying and thermal reduction. The most common methods of ultimate disposal are land application, distribution and marketing, chemical fixation, landfill and lagooning (Metcalf and Eddy, Inc. 1991).

2.3.3 Flood protection and drainage

Protection of the society, material property and business activities against flooding and inundation is one of the basic goals of water management, and as such, flood protection and drainage have been practised in urban settlements since ancient times. However, only recently, flood management has been addressed in a more holistic way, generally referred to as the ecosystem approach. In this approach, flood management planning is connected with land use planning and addresses not only flood defence, but also full protection of water resources and their ecosystems. Floods impact on the society, economy and environment. The first two impacts are usually characterised by loss of human life, direct costs (property damage, disruption of economic activities) and intangible impacts (impairment of human health, loss of memorabilia and non-marketed goods, disruption of life, etc.). In recent years, direct cost of damages caused by weather related disasters, including floods, have been rising dramatically (Fig. 2.6). Only recently, flood impacts on the environment and ecosystems have been also recognized. The best-

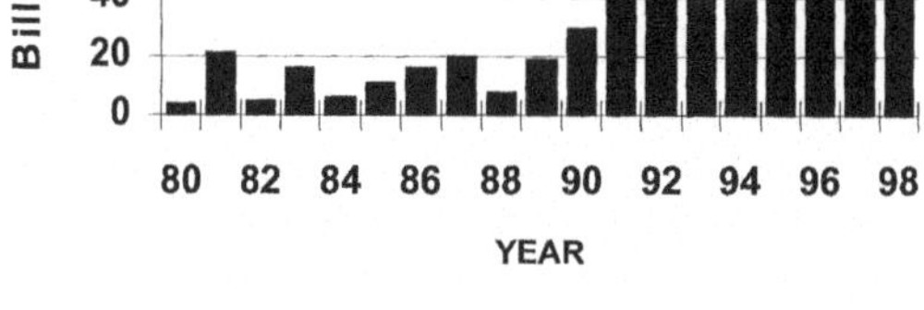

Figure 2.6 Direct costs of damages caused by weather related disasters from 1980 to 1998 (*The Toronto Star* 1998).

known among such impacts are those on morphological processes, which affect the formation of natural habitat for floodplain flora and fauna serving as essential links in many food chains (Marsalek 2000b). Floods also cause changes in population dynamics of fish, animals and insects, but often these changes are just temporary. It can be concluded that floods strongly affect the riverine and floodplain natural environments, and the resulting changes are likely to affect an area well beyond the river corridor (Arnold 1976).

Environmental impacts are also caused by structural flood protection measures, particularly where large reservoirs or embankments are constructed. Such impacts include ecological destruction of inundated land, displacement of population from the area to be flooded by the reservoir, interference with sediment transport, and increased discharges in channelised river beds. Such problems should be minimised by planning flood protection in conjunction with runoff source controls, and considerations of water quality, water resources conservation, and aquatic ecosystem protection (Marsalek 2000b). Ecosystem protection and sustainability require co-operation of all stakeholders and public involvement.

Flood management measures are generally classified into three categories, living with floods, non-structural measures, and structural measures. The measures in the first group are based on learning about flood impacts and their characteristics, and using this information to minimise losses caused by floods, while maximising their benefits. This generally requires avoiding floodplain occupancy and restricting floodplains use for wildlife habitat, land conservancy, and limited agricultural activities (e.g., cattle grazing). Non-structural measures comprise regulations and policies, flood defence and flood insurance. These measures reduce flood impacts, by increasing the threshold flows, for which no damages occur. Regulations control the land use and occupancy of floodplains and inundation-prone areas. Flood defence is based on flood forecasting, warning and evacuation. Flood insurance programs range from public disaster relief to private insurance programs. Non-structural flood management measures greatly benefited from recent advances in hydroinformatics (Marsalek 2000b).

Structural flood management measures are designed to reduce flood volumes and peak flows. Typical structures include storage facilities, river channel embankments, and earthen platforms and polders built in floodplains. The selection of flood management measures should support sustainable development of the entire river watershed, while maximising the economic efficiency of land use. This may be achieved by integrated land use and water resources management, integrated hydrological cycle management, and economic analysis of all impacts of the available options (Marsalek 2000b).

Urban drainage serves to prevent or reduce local flooding and water ponding, and to control pollution of receiving waters, by applying stormwater management and CSO control, storage and treatment. Urban areas are characterised by large impervious areas (e.g. roads, roof surfaces), which generate high rates and volumes of surface runoff and contribute to reduced recharge of groundwater. Furthermore, urban surface runoff is polluted and may transport large quantities of solids. Thus, surface drainage serves not only to provide for convenience of urban dwellers, but also to maintain the recharge of groundwater and protect receiving waters. Great advances have been achieved in modern drainage design by implementation of stormwater management, which prevents or mitigates potential impacts of urbanisation on surface runoff and the receiving waters, and by using storage and treatment to control CSOs.

Stormwater management measures are classified in four categories referred to as policies and source controls, site best management practices (BMPs), community BMPs, and watershed-level measures. Policies and source controls are non-structural measures which typically include public education; planning management (preventing problems by progressive zoning ordinances and buffers for streams and wetlands); material use, exposure and disposal controls; spill prevention and cleanup; prevention/elimination of illegal dumping and illicit connections; and, street drainage and stormwater facilities maintenance (Camp Dresser & McKee *et al.*, 1993).

Site BMPs include lot-level source controls, biofiltration structures (swales and filter strips), local (minor) infiltration facilities, stormwater filters, and water quality inlets. Lot-level source controls include enhanced rooftop detention, flow restrictions at catch basins to enhance local storage/detention, reduced lot grading to slow down runoff flow and enhance infiltration, redirecting roof leader discharges to ponding areas or soakaway pits, and sump pumping of foundation drains (MOEE, 1994). Biofiltration of stormwater by grass filters and swales reduces runoff volume by infiltration and enhances runoff quality by such processes as settling, filtration, adsorption and bio-uptake. These measures are feasible in low-density developments with small contributing areas with diffuse runoff, suitable soils (good sorption), and lower groundwater tables (Schueler 1987 and Schueler *et al.* 1992). Infiltration facilities serve to reduce the volume and rate of runoff, reduce pollutant transport and recharge groundwater. They are designed as wells (pits), trenches, basins, and perforated pipes and drainage structures. The main difficulties with their applications include potential contamination of groundwater and uncertain longevity of these structures. Water quality inlets provide stormwater treatment by sedimentation and skimming of floatables (and oil), and are particularly well suited for parking lots, and commercial or industrial land. Oil/grit separators function similarly as water quality inlets, but can be also installed in-line further downstream from inlets. Well-maintained stormwater sand filters are effective in removal of pollutants. Biofilters (i.e., filters employing a coarse medium with biofilm on granular surfaces) show good promise for removal of dissolved heavy metals (Lau *et al.* 2000).

Community-level BMPs include larger infiltration facilities, stormwater ponds, constructed wetlands, extended detention basins, and multiple systems. Infiltration trenches are generally designed for contributing areas of less than 2 ha, and infiltration basins serve areas of up to 5 ha. Stormwater management ponds provide flow control (reduction of flow peaks), sedimentation (removing sand, and some silt and clay), and removal of dissolved pollutants by aquatic plants (Van Buren *et al.* 1997). Pond components include an inlet (spreading the influent), sediment forebay (easily accessible for maintenance), outlet (preferably a perforated riser), outfall (protected by riprap) and emergency overflow. Regular maintenance of ponds is required, including the removal of accumulated sediments. Constructed wetlands (Fig. 2.7) provide stormwater detention and treatment by such processes as filtration, infiltration, and biosorption, and remove both particulate and dissolved pollutants (Rochfort *et al.*, 1997). Extended detention (dry) basins provide stormwater settling in areas where it is difficult to maintain wet facilities. In multiple systems, two or more BMPs may be stacked vertically or in a series, to increase the system performance or reliability, or to reduce maintenance.

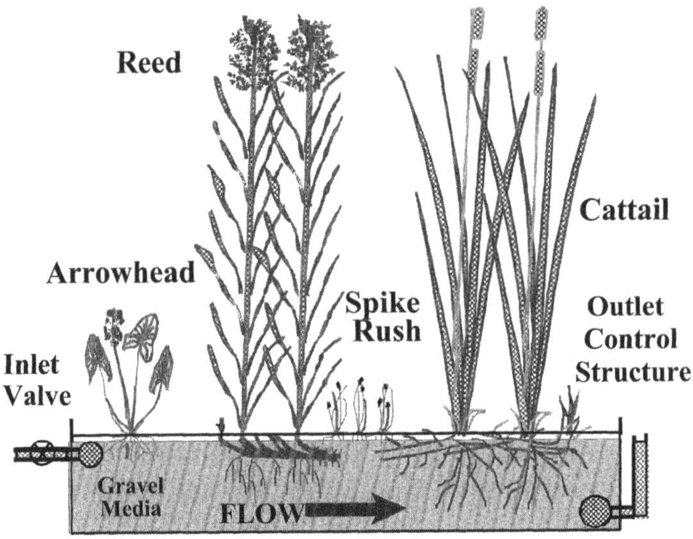

Figure 2.7. Schematic of a constructed wetland cell (after Rochfort *et al.* 1997)

Watershed-wide planning employs watershed-level BMPs to mitigate cumulative impacts, protect specific features and resources, support land-use decisions, improve source-control BMPs, and assist in BMP siting (i.e., local vs. regional facilities). Site resources to be protected in watershed-wide stormwater management include wetlands (provide habitat, water storage and treatment), floodplains (provide flood conveyance, habitat, and recreation opportunities), riparian (forested) buffers (moderate stream temperatures and dissolved oxygen variation, protect stream banks and wildlife habitat), meadows (function as buffers), and soils (impact on water quality) (DDNR&EC and EMCBC 1997).

In protection of receiving streams, a clear connection between "the degree of development" (quantified by the total impervious area, TIA) and stream quality (described by the biological integrity, which is usually defined by diversity and population richness of benthic communities) is recognised. There are three classes of biological stream integrity, high (TIA < 5–10%), intermediate (5–10% < TIA < 35–45%), and low (TIA > 35–45%). For each of these classes, the overall strategy in stormwater management is somewhat different. High biological stream integrity streams exhibit abundant, diverse aquatic population. The main goal is to preserve and protect this resource by minimisation of the development, application of a conservation approach, and preservation and enhancement of natural features. Streams with intermediate biological integrity are characterised by reduced abundance and diversity of aquatic life, but generally their water quality complies with water quality guidelines. The main strategy here is to enhance the habitat and ensure non-degradation of water quality by a high level of stormwater management, with implementation of various BMPs, stream restoration (using natural design), and applying flexible design allowing future upgrading. Finally, streams with low biological integrity are characterised by low abundance and diversity of aquatic life, and non-compliance with water quality guidelines. The overall strategy is to prevent further degradation of water quality by BMPs, but without enhancing wildlife habitat to prevent contaminant uptake by wildlife. Where feasible, stream restoration should be considered.

The stormwater management strategy is included in watershed plans, which are developed in a hierarchical manner, employ the ecosystem approach and provide a basis for the development of more detailed plans. In general, plan goals must be attainable, endorsed by the public and economically responsible.

A high degree of pollution may come from the peri-urban areas where drainage systems, sanitation and water supply systems are not fully developed, or where the expansion exceeds the existing design capacity (Pegram *et al.* 1999). The advantage of integrated catchment management is that drainage can be interrelated with such functions as groundwater management, water supply, waste disposal, use of water as a general amenity and others.

2.3.4 Water as a General Amenity

A relatively new dimension in urban water management is the protection of receiving waters as a recreational, aesthetic, environmental and ecological amenity. Each of these water uses imposes specific requirements on water quality and on measures that prevent impairment of such beneficial uses.

Urban waters are used for various types of water-based recreation, including boating, water skiing, fishing, swimming, and wading. Public health risks associated with those activities vary. Much attention has been paid to recreational water quality, particularly where such waters are used for swimming. Towards this end, specific guidelines were defined for various water quality parameters, with perhaps the most difficult ones to attain being those for indicator bacteria and pathogens. These guidelines usually specify the permissible indicator organism limits, such as e.g., faecal coliform or *Escherichia coli*, in samples collected from recreational waters. These limits are often exceeded, particularly in wet weather and during the period of after-effects, and temporary closures of recreational activities are implemented. Improvements should be achieved by more stringent control of source faecal bacteria, particularly CSOs (Marsalek *et al.* 1994).

Aesthetic features of urban waters are generally related to preserving or restoring the natural form and appearance of urban waters, often by trying to return to the original layout of urban streams with their natural features and riparian vegetation and corridors. This also includes restoration of aquatic habitats and ecosystems, which may extend well beyond the reach of urban areas (e.g. in the case of migratory birds). Such an objective is closely related to the goals of sustainable development. Considerable research on public perception of water quality has been undertaken (House *et al.* 1993) and identified indicators of 'good' and 'bad' water quality. Typical indicators included physical indicators (e.g., water clarity), the presence or absence of biota (e.g., fish), and presence or absence of user groups, such anglers or swimmers (House *et al.*, 1993). The public has a good understanding of indicators of 'bad' water quality, such as scum on the surface, water of unusual colour or smell, foam on water, and protruding sediment deposits or garbage. Identification of clean water is much less clear. Improved aesthetics and visual amenity of urban waters has a number of economic benefits, including increased satisfaction of existing users of urban waters, gains resulting from additional users, additional amenity gains to property along the watercourse (Green and Tunstall 1992), and increased tourism and associated benefits.

Urban receiving waters are also valued for their environmental and ecological functions, providing aquatic and wildlife habitat, and educational resource for urban dwellers. Consequently, various criteria of biological integrity of receiving waters were introduced and support of such integrity becomes one of the goals of urban water management. Karr and Dudley (1981) defined biological integrity as "*the ability of an aquatic ecosystem to support and maintain a balanced, integrated, adaptive community of organisms having a species composition, diversity and functional organisation comparable to that of the natural habitats within a region.*" Thus, the delineation of ecoregions and reference sites, and their assessment by bioassessment methods are essential tools of urban water management. Bioassessment of impairment of biological integrity complements the earlier described procedures dealing with human health, recreation, and aesthetics.

Recognising difficulties with full assessments of biological communities, rapid bioassessment protocols have been introduced by the U.S. EPA (1989) and focus on benthic macroinvertebrates and fish. In these protocols, an integrated assessment is conducted by comparing habitat (physical structure and flow regime) and biological measures with empirically defined reference conditions. Reference conditions are established through systematic monitoring of actual sites (ecoregion reference sites) that represent the natural range of variation in "least disturbed" water chemistry, habitat and biological condition (Horner *et al.* 1994). In this process, one has to overcome difficulties with characterising ambient water quality.

The assessment of macroinvertebrate communities can be supplemented by benthic sediment toxicity testing. Four such tests were described by Reynoldson et al (1999), including *Hyalella azteca* 28-day survival and growth test, *Chironomus riparius* 10-day survival and growth test, *Hexagenia* spp. 21-day survival and growth test and *Tubifex tubifex* 28-day survival and reproduction test. Applications of these tests by Rochfort et al (2000) in the vicinity of storm sewer and CSO outfalls produced little evidence of confirmed toxicity, thus indicating that many chemicals found in urban sediments are not bioavailable. At CSO sites, nutrient enrichment was found. The Chironomidae, Tubificidae and Naididae, which are known to dominate in areas with elevated organic pollution, represented the most abundant families at almost all urban sites.

2.3.5 Integration of water management requirements

Integrated management is approached in different ways by different societies. While the traditional societies emphasise endogenous knowledge and time-tested, successful approaches with inherent integration of resources and demands (Geiger and Hofius 1995), in developed countries, the emphasis is placed on technology, including planning techniques, and computer software and hardware. Recognising the large amount of information required to consider full spatial, temporary and functional integration in planning urban water management, this task can be expedited by the use of computers and hydroinformatics, which merges the benefits of information technology with environmental modelling. While the literature on water management planning is very extensive and includes examples of many elegant methods and tools, post-audit is rarely practised or reported. Yet is appears inevitable that future progress will depend on learning more about the results of water management schemes and practising adaptive water management, in which water project design is based on the best available data and knowledge, but as more information becomes available on the newly adopted system, further changes are implemented to meet the original objectives, or to accommodate changes in objectives (Phillips *et al.* 1999).

2.4 PLANNING AND IMPLEMENTATION OF INTEGRATED URBAN WATER MANAGEMENT

Integrated management of urban waters requires comprehensive planning techniques that will address all management objectives and limitations of the resources available. Such techniques are addressed in this section, starting with description of the planning process, principles of hierarchical planning, the role of urban water planning within the catchment planning, and special considerations in planning new developments and upgrading the old ones.

2.4.1 Planning process

Integrated water management planning follows the system analysis procedure, starting with problem definition, followed by development of solutions, selection of the best alternative, and post-audit (Fig. 2.8). A brief overview of such a planning process follows.

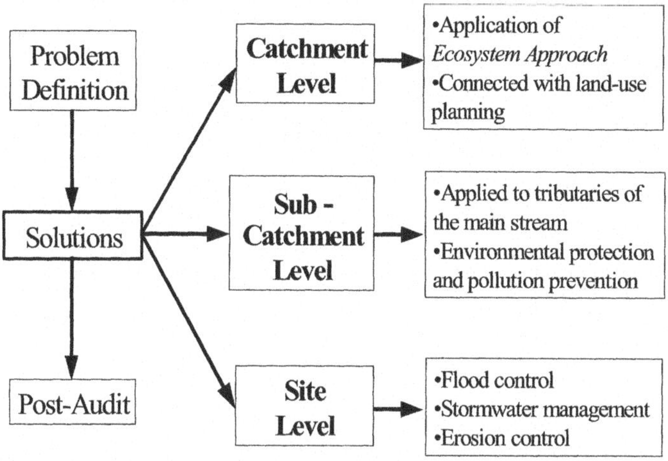

Figure 2.8 Planning process for integrated water management

2.4.1.1 Problem definition

Geiger and Hofius (1995) listed the following tasks to be undertaken during the problem definition stage:

- Investigation of all water resources and specification of their capacities
- Detailed identification of water pollution levels in surface waters and groundwater
- Specification of the effects of degraded water quality on the society and aquatic life
- Inventory of aquatic life in all components of the surface water system
- Evaluation of the water budget of the area, including dry, wet, and average conditions
- Inventory of all discharge points of stormwater, CSOs, industrial and municipal wastewaters
- Inventory of existing treatment facilities, including the type of treatment and treatment efficiencies and capacities
- Inventory of commercial and industrial wastewater discharges into municipal sewers

- Inventories of ecological potential in preserved areas of the catchment
- Inventory of land use in the area
- Specification of dust fallout and rainwater contamination
- Flood plain maps for various return periods
- Inventory of soil contamination and mapping of the contaminated areas
- Definition of the groundwater movement.

This information needs to be assembled at various spatial and temporal scales, and archived for further use and processing. In modern practice, data storage and processing is achieved by means of Geographical Information Systems (GIS), which provide their users with information about objects and features in their geographical context.

2.4.1.2 Hierarchical planning of problem solutions

Various elements of problem solution were discussed in more detail in Section 2.3 with respect to specific services provided in urban water management. These solutions need to be developed at various hierarchical levels, usually progressing from the catchment level, to sub-catchments, city-wide master plans, site plans, and plans and design and operation of individual structures.

A catchment is defined in general as an area that contributes surface runoff to a single point along a stream. Other synonymous or similar expressions in the literature include the terms “watershed” and “basin”. With respect to water management, the catchment appears to be a logical spatial unit to be employed. However, the administrative units do not always correspond to catchment boundaries and that creates problems with fragmented jurisdictions. These problems can be avoided where special boards, or river authorities are created for individual river catchments.

The development of water management plans is conducted at various spatial scales, ideally starting at the catchment plan scale, and developing further detail at the subcatchment, or site scales. Guidance documents for preparing such water management plans have been produced by many authorities and should be consulted when working on new plans. Sustainable water management begins with the application of an ecosystem approach on a catchment basis, and moves towards finer spatial scales: subcatchment, and eventually site development in municipal planning documents. Thus, catchment management is approached from the ecological perspective, in which the water ecosystem comprises all the water, the processes and factors and natural cycles that affect it, and the organisms in water. By its nature, water management is closely connected to land use planning, which typically falls under the municipal authority. For example, the concept of catchment management was recognised by the Province of Ontario (Canada) in

1946, when the first conservation authority was established for a particular catchment and currently there are 38 such authorities.

As currently practised, the catchment plan presents the overall catchment management strategy, identifies the form and function of natural systems, defines the relationship of the catchment plan to subcatchment plans, lists the catchment issues (including public health and safety, aquatic life, resource management, floodplain management, and urban, agricultural and other uses), and presents plan recommendations, including the specification of special areas designated for protection, rehabilitation, and/or enhancement.

Typical catchment plans provide policies and directions for a wide range of issues including the following:

- Management of water quantity and quality (both surface water and ground water)
- Ecological integrity and carrying capacity
- The protection of valley systems and green space planning
- Fisheries management
- Rehabilitation/enhancement programs
- A framework for implementation of watershed policies and programs
- Regional opportunities and constraints
- Documented servicing needs/availability of water/sewerage
- Delineate subcatchment planning areas
- Present targets, goals and objectives for individual subcatchments.

The plan goals must be attainable, endorsed by the public and economically responsible. Management goals must be practical, explicit and verifiable, result-focused and based on accountability. The overall goal should be to reduce impacts on the environment and promote sustainable use of the catchment (Braden and van Ireland 1999).

Subcatchment planning provides a step between the catchment planning and site development. The essential principles are the same as for catchment planning, but carried out for tributaries of the main stream. It has many advantages, including environmental protection/pollution prevention, better planning, guidance for infrastructure decisions and spending, streamlined approvals, major overall cost savings (especially remediation costs), local public involvement, agency credibility, enhanced economic viability of specific areas, and economic benefits of a clean environment. These plans are developed either with or without catchment plans in place.

Site management plans are developed in accord with catchment and subcatchment plans. A variety of site-specific plans are prepared as part of draft proposal for plans of subdivisions and other development applications, which require the use of site control. Well-known examples are stormwater management plans, flood control plans, sediment and erosion control plans, and plans for the servicing of roads, water and sewers. When such plans are formulated in accord with the policies listed here, they are more effective, provide practical applications for the objectives of subcatchment plans, and contribute to environmental benefits. Where this planning process is implemented, the review agencies may consider it unnecessary to review individual site plans, because they have been prepared in accord with the higher-level plans. Approvals for the construction and operation of facilities indicated in these plans, however, may still be required under specific legislation.

Where no catchment (subcatchment) plans exist, it is difficult to assess overall cumulative impacts of land use on water and related resources. For small scale development proposals processed as severances, site planning and spot-zoning amendments, water resources management issues will be limited to the identification of specific stormwater, erosion and sedimentation control design and construction measures. Local bylaws should address routine site management requirements including topsoil protection, urban forests, and sensitive terrestrial habitats.

Indicators can be used to assess the sustainability of a development project. These weighted factors can help to isolate the critical areas that should receive the highest priority when considering development plans (e.g., protection of a species at risk, a sensitive groundwater aquifer or a wetland) and allocating costs (Icke *et al.* 1999; Lundin *et al.* 1999).

Water management plans developed at all levels usually contain alternative solutions and their evaluations. From such evaluations, recommended action, in the form of specific steps to be taken, can be formulated. The key to integrated planning is the positive interaction among various planning levels. In a recent UNESCO report, three planning levels, A–C, were proposed with the following objectives (Geiger and Hofius 1995):

Level A: basic water quality concepts are developed, including ecological functions of the water system, possible water uses, definition of water quality criteria and guidelines, and specifications of user groups and characteristics of the specific water system.

Level B: represents the actual planning of individual measures, which meet the minimum requirements and are adaptable to meet higher requirements. These measures include treatment plants, drainage networks including stormwater management and CSO treatment facilities, and management measures applied in receiving waters. At this level, the costs and financing are also addressed.

Level C: presents of the control of the effects of proposed measures, locates monitoring stations, defines operational programs and maintenance, and determines new trends and tendencies in water system parameters (particularly water quality). Level C will generally produce improvements of the measures planned at Level B, and the findings of levels B and C may lead to adoption of water quality goals not anticipated at Level A. Experience further shows that prescriptive management (i.e., achieving the planned goal in one step) is not feasible, and interactive step-wise planning, allowing future adjustments and improvements, has a better chance of success. This whole procedure should be developed specifically for a targeted water system (Geiger and Hofius 1995).

At all levels of planning, public participation is essential. In this process, various groups are targeted, including friends (supporters), affected parties, local elected officials, government agencies, and the general public. The methods used include printed materials, field trips, public meetings, media, and public opinion polling.

2.4.1.3 Integrated urban water planning within the catchment master planning

Two main issues surround the integration of urban water planning within the catchment master planning. These issues include:

- urban areas represent sub-elements of the whole catchment, and
- urban water planning typically encompasses provision of water services, including water supply and transfers, wastewater management, and the management of receiving waters.

Urban areas, with their specific water management problems and solutions, represent only sub-elements of river catchment and their management has to fit into the overall catchment management planning. This is ensured through catchment management plans, which are developed by river boards or authorities. These plans provide boundary conditions for the development of city-wide plans. Exceptionally, there can be cases where the city area encompasses several small river catchments (e.g., the City of Toronto, with four catchments, draining into Lake Ontario). In this particular case, the Toronto Region Conservation Authority administers the whole area, including all catchments.

The provision of urban water services affects not only urban areas, but also other areas, which may be located a great distance away, sometimes even in a different catchment. This is particularly the case of municipal catchments serving to provide drinking water, which is then conveyed over a great distance to the urban area (Marsalek 1988). Furthermore, the conditions in streams outside of the urban area may impose additional constraints on urban water management, particularly with respect to the protection of downstream waters and their aquatic ecosystems.

Urban water supply may require water withdrawals from both surface and groundwater sources. In that sense, the urban water management impacts on the hydrological cycle of other areas. It is generally beneficial to reduce such impacts by water demand management and water reuse. One of the more effective tools in water demand management is full pricing, and public awareness/education. The total water cycle management, discussed earlier (Section 2.3), offers a good rationale for water use management and conservation.

Urban populations generate large quantities of various types of wastewater, including domestic sewage, industrial sewage, urban stormwater (surface runoff), and CSOs in areas served by combined sewers. Such wastewaters are subject to treatment in most countries, though the type of treatment and extent of services (in terms of the population served) greatly varies. Wastewater treatment strategies develop in a certain progression, usually starting with primary treatment of municipal sewage (i.e. domestic and industrial sewage), followed by secondary treatment, high control of combined sewer overflows, stormwater management/treatment, and tertiary treatment. The goal of this strategy is to minimise impacts of the urban area on the receiving waters and protect their beneficial uses.

2.4.1.4 Approaches in planning new developments and upgrading old ones

In urban waters, the goal of sustainability is well supported by the total management of the water cycle in the affected areas (Lawrence *et al.* 1999), as discussed earlier in Section 2.3. In this connection, two types of developments are encountered in urban water management, new or greenfield developments and reconstruction of the existing areas, or brownfield developments. Perhaps the most difficult task is to justify upgrading and retrofit of the existing developments, which have functional but obsolete infrastructures.

Environmental protection in greenfield developments is easier to implement, because space (land) and funding required for such protection can be secured during the planning stage. The land required for management measures is reserved early during the planning stage and the implementation of measures is financed from lot levies, which are assessed against individual properties. The typical management measures were described earlier (Section 2.3) with reference to total water cycle management (TWCM) and further details can be found in numerous development manuals.

Brownfield developments are much more challenging. In retrofit situations (as opposed to a complete redevelopment), lack of funding (older areas, with lower taxation base) and lack of space for management measures make this job rather difficult. Environmental controls have to be retrofitted to these areas, often at relatively high costs imposed by numerous site constraints. A complete redevelopment is less challenging from the technical point of view, because of opportunities to make large changes in water services and the overall layout of the area. However, costs are very high and difficult to justify.

2.4.1.5 Considerations for integrated water management planning in developing countries

The acceptance of the principles of integrated or holistic approach to urban water management has risen globally during the past 25 years or so. Such principles have been adopted in projects operated by the International Monetary Fund, World Bank, and Food and Agriculture Organisation, and often are used by national aid programs as well. Successful examples of such projects can be listed in all major areas of development (agriculture, industry, infrastructure programs), with one particular distinction of integrated water management—its benefits materialise over a longer time period (Johanns and Hagsma 1995). Thus, integrated water management is a long-term policy for developing water resources use in a sustainable and socially balanced fashion. The main prerequisites for introducing integrated water management include (Johanns and Hagsma 1995):

- Commitment to integrated water management at the government level (political acceptance)
- Introduction of regulatory measures in support of integrated water management (management instruments)
- High local awareness of the benefits of integrated water management (economic and social acceptance), usually achieved through:
    - local community information programs
    - local control functions implemented through local community
    - direct feedback between agencies and community
- Balancing community needs with integrated water management needs and goals (social acceptance), through:
    - counselling of locals for advice and on matters of property and claims

• highest community involvement in planning and actual execution of policies
• development through adoption from local resources.

2.4.2 Implementation strategy

Successful implementation of integrated water management requires meeting the prerequisites (such as e.g., those listed in Section 2.4.1.4) and a well-prepared strategy. Nine steps of such a strategy were listed by Marsalek (1994) and Geiger and Hofius (1995):

- Establish a common cause of all interest groups
- Establish a stakeholders group
- Develop a plan of action, by selecting the best alternative
- Develop a funding mechanism
- Engage in political lobbying to gain political support
- Involve politicians in promoting the plan of action
- Revise institutional arrangements to support the plan implementation
- Implement the plan
- Verify the plan success (or shortcomings) by post-audit.

Further comments on these steps follow.

Successful water management requires that all interest groups sharing a particular water resource, feel that they are part of the ecosystem in the studied area and that their interests are well served by the proposed course of action. This common cause is also sometimes referred to as "enlightened self-interest". (Christie *et al.* 1986). Since all citizens in a river catchment face common challenges, they must be committed, in their own interest, to find solutions. The definition of this common cause is often provided by citizen groups, political parties or government agencies.

In every river catchment, there are several agencies, organisations, associations and programs, which are involved in various aspects of water management. Depending on their responsibilities and jurisdictions, these entities are often depicted in schematic diagrams developed in both vertical and lateral directions. The challenge of the integrated plan is to rearrange these parties into a circle, in which all parties carry the same weight and communicate directly among themselves (Hartig and Vallentyne 1989). This is generally achieved under newly created bodies, referred to as stakeholders groups or committees. It is essential that such committees involve concerned citizens and voluntary associations, because their involvement will facilitate the financing and implementation of the plan.

The next step is to develop a plan of action, which is usually done in several stages, as described earlier. The first stage usually represents the problem definition and plan goals and objectives, and the subsequent stages deal with the actual solutions and management measures. It is essential that this plan is based on ecosystem science describing the catchment, its water resources, environmental/ecosystem conditions, and socio-economic aspects in full comprehensiveness, reflecting the interconnectivity of various actions. Where scientific data are missing, their collection may be part of this plan, particularly in the problem definition stage. New technologies involving knowledge support systems and GIS systems are highly suitable for presentation of ecosystem data.

The plan must be specific in terms of objectives, correction of existing problems, the listing of remedial options, identification of preferred options, and definition of the final state of the catchment with all the desirable water uses and functions restored. The plan should also include the means of verification that the objectives have been attained and can be sustained. An important part of the plan are the estimates of costs of various remedial options, reflecting both the initial (capital) costs and operating and maintenance costs.

Remedial activities can be costly and their funding may be beyond the means of stakeholders. Consequently, it is required to develop a funding mechanism that would support the financing of the remedial plan over a long-term period. Differences in political systems, institutions and general perceptions do not allow the establishment of general principles in this field. However, a discussion of possible options is of interest. In general, the user or polluter pays principle should be applied as much as possible. Some financing through general taxation, though unpopular, is probably unavoidable. Other funding mechanisms include taxes on consumptive resource use, developing full cost pricing by internalising pollution costs and full cost pricing for new developments, and developing options for charges and tax incentives that shift environmental costs from the taxpayer to the consumer. While many of these options may seem attractive, political realities may render them unfeasible in various political systems and situations (Marsalek 1994).

Plan implementation, involving large funds and responsibilities, may call for revisions of the existing institutional arrangements to support the plan implementation. Examples of such institutions can be found in the literature, but are rarely transferable to other jurisdictions, because they generally reflect local socio-economic conditions (Lord 1984). Plan implementation is an important stage, which involves extensive activities often dispersed throughout the catchment. These activities are conducted by various contractors retained by stakeholders and often it is necessary to hire professional managers for these activities. Important issues include co-ordination of remedial activities so that they proceed in a timely manner and cost overruns are avoided. Following the implementation, a post-audit is conducted. The essential steps comprise the comparison of predicted and actual results and evaluation of the planning methodology.

2.4.3 Legislative and administrative support

Regulations concerning water were among the early regulations introduced by various societies, and water found its place in religious as well as legal frameworks, but without much consistency. As a result of historical developments, different situations can be found in various countries with respect to the ownership of water, including free ownership, public ownership, restrictions on upstream uses, no restriction on upstream uses, and riparian rights. New methods of integrated water management require appropriate administrative and legislative support, which may or may not be available in individual cases. In general, there is no single model policy that would adequately fit conditions in various countries. Instead, political systems in individual countries encourage development of various models that support implementation of integrated water management. The most significant difference is perhaps the division of legislative powers between the central and regional levels of government.

In some countries, water management represents one of the central government responsibilities and this affects the introduction and implementation of water management plans. This system encourages a top-down approach, in which the main initiative comes from the central authority. In other systems, water management is a territorial, or regional government responsibility with much greater emphasis on participation of local citizens in the formulation and implementation of management plans. It should be emphasised, however, that integrated water management can be accomplished in either system, if there is enough political will and commitment to the principles of integrated water management. It should be also recognised that integrated water management is unlikely to provide a strong impetus for changing the division of power among various government levels and it may be more important to look for ways of implementation of management plans within the existing system rather than trying to introduce its changes.

Examination of specific legislation in various countries indicates that integrated water management can be successfully implemented with or without the legislative creation of special bodies put in charge of individual river catchments. The former case is well documented by examples of such authorities as Water Boards in the Netherlands, or River Authorities in the United Kingdom. The second case, is documented by the experience from the Canadian Great Lakes Basin, where special planning/implementation teams (stakeholder committees) were created within the existing legislative framework by co-operation of various agencies and involvement of the public. It is important for developing countries to examine both systems and develop their own policies, particularly where policies governing water management have not yet been fully developed. Regardless of the model adopted, it is important to prevent fragmentation of management efforts, where separate government agencies operate more or less independently (sometimes even in competition) and operators of systems are not involved in the planning process.

2.5 CONCLUSIONS

Rapidly progressing urbanisation impacts dramatically on water resources in the affected areas. As these water resources become depleted or their poor quality impairs beneficial uses, the need for integrated management of urban waters becomes strong. Such management is generally accomplished within the framework of sustainable development, by providing the maximum support for the society's needs without degradation of water resources and their ecosystems now and in the future. The planning and implementation of urban water management require a full spectrum of tools and procedures, ranging from communications and empowerment to sophisticated computer tools and techniques. It is more and more recognized that successful management of urban waters cannot be accomplished with the conventional prescriptive water management. Adaptive management, proceeding with the best scenarios, while collecting additional data and monitoring the system performance and making further adjustments as required, has a better chance of success. Many aspects of urban water management (land use, industrial development, environmental laws) reach outside of the authority of urban water managers, and consequently, the protection and restoration of urban waters will require co-operation of the entire society.

2.6 REFERENCES

American Society of Civil Engineers (ASCE), Water Resources Planning and Management Division & UNESCO International Hydrological Programme IV Project M-4.3 Task Committee on Sustainability Criteria, (1998). *Sustainability criteria for water resource systems*. ASCE, Reston, Virginia, USA.

Arnold, M. (1976). Floods as man-made disasters. *The Ecologist* **6**, 169–172.

Ashley, R.M., Souter, N., Butler, D., Davies, J., Dunkerley, J., & Hendry, S. (1999). Assessment of the Sustainability of Alternatives for the Disposal of Domestic Sanitary Waste. *Wat. Sci. Tech.* **39** (5), 251–258.

Birley, M.H., & Lock, K. (1999). *The Health Impacts of Peri-Urban Natural Resource Development.* A report prepared for the Department of International Development, U.K.. Liverpool School of Tropical Medicine, Liverpool, U.K..

Blatchley, E.R. (III), Hunt, B.A., Duggirala, R., Thompson, J.E., Zhao, J., Halaby, T., Cowger, R.L., Straub, C.M., & Alleman, J.E. (1997). Effects of Disinfectants on Wastewater Effluent Toxicity. *Wat. Res.*, **31** (7), 1581–1588.

Borchardt, D., & Stazner, B. (1990). Ecological Impact of Urban Stormwater Runoff Studied in Experimental Flumes: Population Loss by Drift and Availability of Refugial Space. *Aquat. Sci.*, **52**, 299–314.

Braden, J.B., & van Ierland, E.C. (1999). Balancing: The Economic Approach to Sustainable Water Management. *Wat. Sci. Tech.* **39** (5), 17–23.

Camp Dresser McKee, Larry Walker & Associates, Uribe & Associates, and Resource Planning Associates. (1993). California Stormwater Best Management Practices Handbooks. Stormwater Quality Task Force, California.

Chambers, P.A., Allard, M., Walker, S.L., Marsalek, J., Lawrence, J., Servos, M., Busnarda, J., Munger, K.S., Adare, K., Jefferson, C., Kent, R.A., & Wong, M.P. (1997). Impacts of Municipal Wastewater Effluents on Canadian Waters: a Review. *Wat. Qual. Res. J. Canada*, **32**, 659–713.

Christie, W.J., Becker, M., Cowden, J.WE., Vallentyne, J.R. (1986). Managing the Great Lakes Basin as a Home. *J. Great Lakes Res.*, **12**, 2–17.

Daughton, C.G. & Ternes, T.A. (1999). Pharmaceutical and Personal Care Products in the Environment: Agents of Subtle Change? *Environmental Health Perspectives*, **107**, Suppl. 6, 907.

Delaware Department of Natural Resources and Environmental Control and The Environmental Management Centre of the Brandywine Conservancy. (1997). Conservation Design for Stormwater Management. September, 1997.

Dutka, B.J., & Marsalek, J. (1993). Urban Impacts on River Shoreline Microbiological Pollution. *J. Great Lakes Res.*, **19**, 665–674.

Dutka, B.J., Marsalek, J., Jurkovic, A., Kwan, K.K., & McInnis, R. (1994a). Ecotoxicological Study of Stormwater Ponds under Winter Conditions. *Zietschrift fur angewandte Zoologie,* **80** (1), 25–43.

Dutka, B.J., Marsalek, J., Jurkovic, A., Kwan, K.K., & McInnis, R. (1994b). A seasonal Ecotoxicological Study of Stormwater Ponds. *Zietschrift fur angewandte Zoologie,* **80** (3), 361–383.

El-Gohary F. & Nasr, F.A. (1999). Cost-Effective Pre-Treatment of Wastewater. *Wat. Sci. Tech.* **39** (5), 97–103.

Ellis, J.B., & Hvitved-Jacobsen, T. (1996). Urban drainage impacts on receiving waters. *J. Hydraulic Res.*, **34** (6), 771–783.

Environment Canada & Health Canada, (2000). Canadian Environmental Protection Act 1999. *Priority Substances List Assessment Report - Road Salts*. Draft Report for Public Comments. Government of Canada Publication.

Furguson, B., & Horsefield D. (1999). Improving Urban Waterways in Emerging Countries: An Action Plan for Madras. *Journal of the American Water Resources Association*, **35**, 923–937.

Galli, F.J. (1991). *Thermal Impacts Associated with Urbanisation and BMPs in Maryland.* Metropolitan Washington Council of Governments, Washington, D.C.

Geiger, W.F., & Hofius, K. (1995). Integrated water management in urban and surrounding areas - findings of the International Workshops in Essen 1992 and Gelsenkirchen 1994 by the German-Dutch IHP Committee to UNESCO, Project M3-3a. In: J. Niemczynowicz and K. Krahner (eds.), *Preprints of the Int. Symp. "Integrated Water Management in Urban Areas"*, Sept. 26–30, Lund, Sweden, pp. 127–151.

Geiger, W.F., Marsalek, J., Rawls, W.J., & Zuidema, F.C. (ed.) (1987). *Manual on Drainage in Urbanized Areas. Volume 1: Planning and Design of Drainage Systems.* A contribution to the International Hydrological Programme. United Nations, Educational, Scientific and Cultural Organization, Paris, France.

Geiger, W.F. (ed.) (2000). *Integrated Water Resources Management in Urban and Surrounding Areas. Conservation strategies, technological developments, ecological values, socio-economic conditions, institutional arrangements.* Draft document of a contribution to the International Hydrological Programme. United Nations, Educational, Scientific and Cultural Organization, Paris, France.

Gleick, P.H. (1998). *The World's Water 1998–99. The Biennial Report on Freshwater Resources.* Island Press, ISBN 1-55963-592-4, Washington, D.C.

Golder Associates Ltd. (1995a). *Joint Industry-Municipal North Saskatchewan River Study.* Prepared for AT Plastics Inc., Capital Region Sewage Commission, Celanese Canada Inc., City of Edmonton - Goldbar WWTP, Dow Chemical Canada Inc., Dupont Canada Inc., Geon Canada Inc., Imperial Oil Products Division, Petro-Canada Products, Shell Canada Products Ltd., Edmonton, Alta., Sherritt Inc., Fort Saskatchewan, Alta., and Redwater, Alta.

Golder Associates Ltd. (1995b). *Joint Industry-Municipal North Saskatchewan River Study. Appendices.* Prepared for AT Plastics Inc., Capital Region Sewage Commission, Celanese Canada Inc., City of Edmonton - Goldbar WWTP, Dow Chemical Canada Inc., Dupont Canada Inc., Geon Canada Inc., Imperial Oil Products Division, Petro-Canada Products, Shell Canada Products Ltd., Edmonton, Alta., Sherritt Inc., Fort Saskatchewan, Alta., and Redwater, Alta.

Green, C.H., & Tunstall, S.M. (1992). The amenity and environmental value of river corridors in Great Britain, In: P.J. Boon, P. Calow and G.E. Petts (eds.), *River conservation and management*, John Wiley and Sons Ltd., pp. 425–441.

Harremoës, P. (1988). Stochastic Models for Estimation of Extreme Pollution from Urban Runoff. *Water Res.*, **22**, 1017–1026.

Hall, K.J., & Anderson, B.C. (1988). The Toxicity and Chemical Composition of Urban Stormwater Runoff. *Can. J. Civ. Eng.*, **15**, 98–106.

Hartig, J.H. & Vallentyne, J.R. (1989). Use of an Ecosystem Approach to Restore Degraded Areas of the Great Lakes. *Ambio* XVIII, 423–428.

Hedberg, T. (1999). Attitudes to Traditional and Alternative Sustainable Sanitary Systems. *Wat. Sci. Tech.* **39** (5), 9–16.

Herricks, E.E. & Schaeffer, D.J. (1987). Selection of Test Systems for Ecological Analysis. *Water Sci. Tech.* **39** (5), 47–54.

Horner, R.R., Skupien, J.J., Livingston, E.H., & Shaver, H.E. (1994). *Fundamentals of Urban Runoff Management: Technical and Institutional Issues.* Terrene Institute, Washington, D.C.

House, M.A., Ellis, J.B., Herricks, E.E., Hvitved-Jacobsen, T., Seager, J., Lijklema, L., Aalderink, H., & Clifforde, I.T. (1993). Urban drainage–impacts on receiving water quality. *Wat. Sci. and Tech.*, **27** (12), 117–158.

Hvitved-Jacobsen, T. (1982). The Impact of Combined Sewer Overflows on the Dissolved Oxygen Concentration of a River. *Water Research,* **16**, 1099–1105.

Icke, J., van den Boomen, R.M., & Aalderink, R.H. (1999). A Cost-Sustainability Analysis of Urban Water Management. *Wat. Sci. Tech.* **39** (5), 211–218.

IWA Water Reuse Committee Newsletter (2000). IWA, London, UK, June.

Johanns, R.D. & Haagsma, I.G. (1995). Integrated Water Resources Management in the Information Age. In: J. Niemczynowicz and K. Krahner (eds.), *Pre-Prints of the Int. Symp. "Integrated Water Management in Urban Areas"*, Sept. 26–30, Lund, Sweden, pp. 169–179.

Kalker, T.J.J., Maas, J.A.W. & Zwaag, R.R. (1999). Transfer and Acceptance of UASB Technology for Domestic Wastewater: Two Case Studies. *Wat. Sci. Tech.* 39, 219-225.

Karr, J. & Dudley, D. (1981). Ecological perspective on water quality goals. *Environ. Management* **5**, 55–68.

Latham, B. (1990). *Water Distribution.* Institution of Water and Environmental Management, London, U.K.

Lau, Y.L., Marsalek, J. & Rochfort, Q. (2000). Use of a Biofilter for Treatment of Heavy Metals in Highway Runoff. *Water Qual. Res. J. Canada* **35**, 563–580.

Lawrence, A.I., Ellis, J.B., Marsalek, J., Urbonas, B., & Phillips, B.C. (1999). Total urban water cycle based management. In: Joliffe, I.B. and Ball, J.E. (Eds.), *Proc. of the 8th Int. Conf. on Urban Storm Drainage, Sydney, Australia*, Aug. 30 – Sept. 3, 1999, pp. 1142-1149.

Lijklema, L., Roijackers, R.M.M., & Cuper, J.G.M. (1989) Biological assessment of effects of CSOs and stormwater discharges. *In* J.B. Ellis (ed.), *Urban discharges and receiving water quality impacts*, Pergamon Press, Oxford, p. 37–46.

Lijklema, L., Tyson, J.M., & Lesouf, A. (1993). Interactions Between Sewers, Treatment Plants and Receiving Waters in Urban Areas: a Summary of the INTERURBA '92 Workshop Conclusions. *Water Science and Technology*, **27** (12), 1–29.

Lins, H.F., & Stakhiv, E.Z. (1998). Managing the Nation's Water in an Changing Climate. *Journal of the American Water Resources Association*, **34**, 1255–1264.

Lord, W.B., (1984). Institutions and Technology: Key to Better Water management. *Water Resour. Bull.* **20**, 651–656.

Lundin, M., Molander, S., & Morrison, G.M. (1999). A Set of Indicators for the Assessment of Temporal Variations in the Sustainability of Sanitary Systems. *Wat. Sci. Tech.* **39** (5), 235–242.

Marsalek, J. (1988). Integrated water management in urban areas. *In* Proc. Int. Symp. on Hydrological processes and water management in urban areas, Invited Papers Volume, Duisburg, Germany, Apr. 24–29, 1988.

Marsalek, J. (1994). Vision of a New Integrated Approach. A contribution to the UNESCO Project M3.3a. on Integrated Water Management in Urban and Surrounding Areas. Nat. Water Res. Institute, Burlington, Ontario.

Marsalek, J. (1995). The ecosystem approach to water management in urban areas. New World Water, Sterling Publications Ltd., London, U.K., 27–30.

Marsalek, J. (1998). Challenges in urban drainage. *In* Marsalek, J., Maksimovic, C., Zeman, E., & Price, R.(Eds.), *Hydroinformatics tools for planning, design, operation and rehabilitation of sewer systems.* Kluwer Academic Publishers: Dordrecht, The Netherlands, p. 1-24.

Marsalek, J. (2000a). Urban drainage: past achievements and future challenges. *Journal of Sewerage, Monthly*, **23**: 78–82 (in Japanese). Also available as NWRI Contribution No.00-001, Burlington, Ontario, Canada.

Marsalek, J. (2000b). Overview of flood issues in contemporary water management. In: Marsalek, J., Watt, W.E., Zeman, E. & Sieker, F. (eds.), *Flood issues in contemporary water management, NATO Science Series Vol. 71*, Kluwer Academic Publishers, Dordrecht/Boston/London, ISBN 0-7923-6451-1, pp. 1–16.

Marsalek, J., Dutka, B.J. & Tsanis, I.K. (1994) Urban impacts on microbiological pollution of the St.Clair River in Sarnia, Ontario. *Wat. Sci. Tech.*, **30** (1), 177–184.

Marsalek, J., Rochfort, Q., Brownlee, B., Mayer T., & Servos, M. (1999a). An Exploratory Study of Urban Runoff Toxicity. *Wat. Sci. Tech.*, **39** (12), 33–39.

Marsalek, J., Rochfort, Q., Mayer, T., Servos, M., Dutka, B. & Brownlee B. (1999b). Toxicity testing for controlling urban wet-weather pollution: advantages and limitations. *Urban Water* **1**, 91–103.

Marsalek, P.M., Watt, W.E., Marsalek J., & Anderson, B.C. (2000). Winter flow dynamics of an on-stream stormwater management pond. *Water Qual. Res. J. Canada*, **35**, 505–523.

Metcalf & Eddy (Inc.), (1991). *Wastewater Engineering: Treatment, Disposal and Reuse, (Third Edition)*. (Tchobanoglous G., & Burton, F.L. Eds.). McGraw Hill Publishing Company, Toronto.

Mikkelsen, P.S., Adeler, O.F., Albrechtsen, H.-J., & Henze, M. (1999). Collected Rainfall as a Water Source in Danish Households – What is the Potential and What are the Costs? *Wat. Sci. Tech.* **39** (5), 49-56.

Ministry of Environment and Energy. (1994). Stormwater Management Practices Planning and Design Manual. Ontario Ministry of the Environment and Energy, Toronto.

Nash, L. (1993). Water Quality and Health. In: P.H. Gleick, (ed.), *Water in Crisis: A Guide to World's Freshwater Resources*. Oxford University Press, New York, 25–39.

Oberts, G.L., J. Marsalek and M. Viklander. (2000). Review of water quality impacts of winter operation of urban drainage. *Water Quality Research Journal of Canada*, **35** (4): 781–808.

Orr, P., Craig, G.R., & Nutt, S.G. (1992). *Evaluation of Acute and Chronic Toxicity of Ontario Sewage Treatment Plant Effluents*. Prepared for MISA Municipal Section, Ontario Ministry of the Environment, Toronto.

Otterpohl, R. (2001). Options for alternative types of sewerage and treatment systems directed to improvement of the overall performance. In: *Interactions Between Sewers, Treatment Plants and Receiving Waters in Urban Areas - Interurba II*. 2nd Int. Conference, Feb. 19–22, 2001, Lisbon, Portugal.

Otterpohl, R., Albold, A., & Oldenburg, M. (1999). Source Control in Urban Sanitation and Waste Management. *Wat. Sci. Tech.* **39** (5), 153–160.

Pearce, D.W. & Warford, J.J. (1993). *World Without End: Economics, Environment and Sustainable Development*. Oxford University Press, New York, N.Y.

Pegram, G.C., Quibell, G., & Hinsch, M. (1999). The Nonpoint Source Impacts of Peri-Urban Settlements in South Africa: Implications for Their Management. *Wat. Sci. Tech.* **39** (12), 283–290.

Petruck, A., Cassar, A., & Dettmar, J. (1998). Advanced Real Time Control of a Combined Sewer System. *Wat. Sci. Tech.*, **37** (1), 319–326.

Phillips, B.C., Lawrence, A.I. & Boon, B. (1999). City of Clarence (Tasmania) stormwater management strategy. In: I.B. Jollife and J.E. Ball, *Proc. of the 8th Int. Conf. on Urban Storm Drainage, Sydney, Australia*, Aug. 30 – Sept. 3, 1999, pp. 801–808.

Pratt, C.J. (1999). Use of Permeable, Reservoir Pavement Constructions for Stormwater Treatment and Storage for Re-Use. *Wat. Sci. Tech.*, **39** (5), 145–151.

Rae, T. Wastewater Treatment (1998). *Chartered Institution of Water and Environmental Management,* Booklet 8, London, UK.

Reynoldon, T.B., Logan, C., Milani, D., Pascoe, D. & Thompson, S.P. (1999). Methods Manual: II. Sediment Toxicity Testing, Field and Laboratory Methods and Data Management. NWRI Contribution No. 99-212, National Water Research Institute, Burlington, ON.

Rochfort, Q., Anderson, B.C., Crowder, A.A., Marsalek, J. & Watt, W.E. (1997). Field-Scale Studies of Subsurface Flow Constructed Wetlands for Stormwater Quality Enhancement. *Water Qual. Res. J. Canada* **32**, 101–117.

Rochfort, Q., Grapentine, L., Marsalek, J., Brownlee, B., Reynoldson, T., Thompson, S., Milani, D. & Logan, C. (2000). Using Benthic Assessment Techniques to Determine Combined Sewer Overflows and Stormwater Impacts in the Aquatic Ecosystem. *Water Qual. Res. J. Canada* **35**, 365–397.

Roeleveld, P.J., & Maaskant, W. (1999). A Feasibility Study on Ultrafiltration of Industrial Effluents. *Wat. Sci. Tech* **39** (5), 73–80.

Rutherford, L.A., Doe, K.G., Wade, S.J. & Hennigar, P.A. (1994). Aquatic Toxicity and Environmental Impact of Chlorinated Wastewater Effluent Discharges from Four Sewage Treatment Facilities in the Atlantic Region, p. 179–195. *In* van Collie, R., Roy, Y., Bois, Y., Campbell, P.G.C., Lundahl, P., Martel L., Michaud, M., Riebel, P., & Thellen, C., *Proceedings of the 20th Annual Aquatic Toxicity Workshop, October 17-21, 1993, Quebec City.* Can. Tech. Rep. Fish. Aquat. Sci. No. 1989.

Schueler, T. (1987). Controlling Urban Runoff: A Practical Manual for Planning and Designing Urban BMPs. Publication no. 87703. Metropolitan Washington Council of Governments. 275pp.

Schueler, T., Kumble, P., & Heraty, M. (Anacostia Restoration Team) (1992). *A Current Assessment of Urban Best Management Practices, Techniques for Reducing Non-Point Source Pollution in the Coastal Zone*, Washington D.C.: Metropolitan Washington Council of Governments, USA.

Servos, M.R. (1999) Review of the Aquatic Toxicity, Estrogenic Responses and Bioaccumulation of Alkylphenols and Alkylphenol Polyethoxylates. *Water Qual. Res. J. Canada*, **34**, 123–177.

Skjelhaugen, O.J. (1999). Closed System for Local Reuse of Blackwater and Food Waste, Integrated with Agriculture. *Wat. Sci. Tech.* **39** (5), 161–168.

Stirrup, M. (1996). Implementation of Hamilton-Wentworth Region's Pollution Control Plan. *Water Qual. Res. J. Canada*, **31** (3), 453–472.

Tao, Y.X., & Hills, P. (1999). Assessment of Alternative Wastewater Treatment Approaches in Guangzhou, China. *Wat. Sci. Tech.* **39** (5), 227–234.

Ternes, T.A., Stumpf, M., Mueller, J., Haberer, K., Wilken, R-D., & Servos, M. (1999). Behaviour and Occurrence of Estrogens in Municipal Sewage Treatment Plants – I. Investigations in Germany, Canada and Brazil. *Sci. Total. Environ.* **225**, 81–90.

Terpstra, P.M.J. (1999). Sustainable Water Usage Systems: Models for the Sustainable Utilization of Domestic Water in Urban Areas. *Wat. Sci. Tech.* **39** (5), 65–72.

The Toronto Star (1998). Toronto, Nov. 28, 1998.

United Nations. (1977). Report of the United Nations Water Conference, Mar del Plata, March 14–25, 1977. United Nations Publication E.77.IIA.12, New York.

United Nations. (1997). *Comprehensive Assessment of the Freshwater Resources of the World.* World Meteorological Organisation and the Stockholm Environment Institute, Geneva, Switzerland.

U.S. Environmental Protection Agency. (1989). Rapid Bioassessment Protocols for Use in Streams and Rivers: Benthic Macroinvertebrates and Fish. EPA/444/4-89-001. Washington, D.C.

Vaes, G., & Berlamont, J. (1999). The Impact of Rainwater Reuse on CSO Emissions. *Wat. Sci. Tech* **39** (5), 57–64.

Van Blarcum, S.C., Miller, J.R. & Russell, G.L. (1995). High Latitude River Runoff in Doubled CO_2 Climate. *Climatic Change* **30**, 7–26.

Van Buren, M.A., Watt, W.E. & Marsalek, J. (1997). Removal of Selected Urban Stormwater Constituents by an On-Stream Pond. *J. Env. Plan. Manage.* **40**, 5–18.

Van Buren, M.A., Watt, W.E., Marsalek, J., & Anderson, B.C. (2000). Thermal Enhancement of Stormwater Runoff by Paved Surfaces. *Water Research*, **34** (4), 1359–1371.

van der Graaf, J.H.J.M, Kramer, J.F., Pluim, J., de Koning, J., & Weijs, M. (1999). Experiments on Membrane Filtration of Effluent at Wastewater Treatment Plants in the Netherlands. *Wat. Sci. Tech.* **39** (5), 129–136.

van der Hoek, J.P., Dijkman, B.J., Terpstra, G.J., Uitzinger, M.J., & van Dillen, M.R.B. (1999). Selection and Evaluation of a New Concept of Water Supply for "Ijburg" Amsterdam. *Wat. Sci. Tech.* **39** (5), 33–40.

Van Os, E.A. (1999). Closed Soilless Growing Systems: A Sustainable Solution for Dutch Greenhouse Horticulture. *Wat. Sci. Tech.* **39** (5), 105–112.

Van Riper, C., & Geselbracht, J. (1999). Water Reclamation and Reuse. *Wat. Environ. Res.*, **71**, 720–728.

Viklander, M. (1997). Snow quality in urban areas. Ph.D. Thesis, Luleå University of Technology, Luleå, Sweden.

Walsh, P.D. (1971). Designing Control Rules for the Conjunctive Use of Impounding Reservoirs. *J. Instn. Wat. Eng.*, **25** (7), 371.

Willers, H.C., Karamanlis, X.N., & Schulte, D.D. (1999). Potential of Closed Water Systems on Dairy Farms. *Wat. Sci. Tech.* **39** (5), 113–119.

World Business Council for Sustainable Development (WBCSD) & United Nations Environment Programme (UNEP) (1998). *Industry, Fresh Water and Sustainable Development.* A joint publication of the World Business Council for Sustainable Development and the United Nations Environment Programme.

World Health Organisation. (1993). Guidelines for Drinking-Water Quality, Vol. 1: Recommendations. ISBN 92 4 1544660 0, Geneva, Switzerland.

World Health Organisation. (1996). Guidelines for Drinking-Water Quality, Vol. 2: Health Criteria and Other Supporting Information. ISBN 92 4 154480 5, Geneva, Switzerland.

World Health Organisation. (1997). Guidelines for Drinking-Water Quality, Vol. 3: Surveillance and Control of Community Water Supplies, 2nd ed. ISBN 92 4 154503 8, Geneva, Switzerland.

World Health Organisation. (1998). Guidelines for Drinking-Water Quality, Addendum to Vol. 1: Recommendations. ISBN 92 4 154514 3, Geneva, Switzerland.

World Health Organisation. (2001). Web Site: www.who.int/water_sanitation_health/Environmental_sanit/envindex.htm

Zeeman, G. & Lettinga, G. (1999). The Role of Anaerobic Digestion of Domestic Sewage in Closing the Water and Nutrient Cycle at Community Level. *Wat. Sci. Tech.* **39** (5), 187–194.

3

Interactions with the environment

David Butler and Čedo Maksimović

3.1 INTRODUCTION

The current urban water management paradigm consists of intricately engineered systems to collect and supply potable water, collect and dispose of wastewater, and drain urban areas of stormwater taking care of receiving water bodies and urban streams and providing water features in urban areas. This chapter is concerned with ways in which these systems interact with the natural environment from which water is abstracted and to which used water is returned. This mainly concerns the aquatic environment. Both water *quantity* and *quality* aspects are covered. In this "state-of-the art" review, we look at the sources most commonly used for supplying water, provide an outline of the elements that go to make up the infrastructure for supply of clean and disposal of dirty and storm water and discuss their interaction with the urban water environment. We also look in more detail at the releases from such systems and their impacts, together with discussion of the relevant legislation and modern informatic tools.

Edited by Čedo Maksimović and José Albert Tejada-Guibert. ISBN: 1 900222 76 0

3.1.1 Urban water cycle

Urban water systems are required in developed urban areas because of the interaction between human activity and the natural water cycle. This interaction has two *main* forms: the abstraction of water from the natural cycle to provide a supply for human life and the covering of land with impervious surfaces that divert rainwater away from the local natural system of drainage.

But what is the extent of this interaction at the global level? Approximately 2.5% of the earth's total water is fresh, although only about 5% of that is directly usable by humans and other organisms (see Figure 3.1). Of this freshwater resource, approximately 9,000–12,000 km^3 per annum is practically available for human use (municipal, industrial and agricultural) (Postel, 1996). In 1995, world human consumption was an estimated 3,800 km^3(Cosgrave & Rijsberman, 2000); 2,100 km^3 was consumed and the remainder returned to water bodies, usually with significant reductions in quality. This amounts to a large fraction of total available resources, and is on the increase. Direct municipal use accounts for about 10% of consumption.

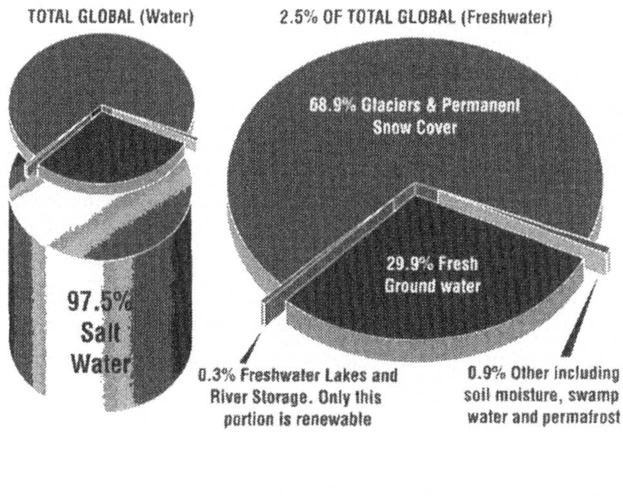

Figure 3.1 Balance of fresh water available for human use at the global level (Shiklomanov, 1993)

The rise in world population has been startling, to say the least, with a rise in numbers from a few thousand, 200,000 years ago, to today's estimate of 6 billion individuals. Figure 3.2 shows the global reach of humankind in 1995, with urban areas and cultivated land now covering 50% of the globe's surface (Harrison & Pearce, 2001). Even more dramatic has been the recent growth in cities at the global scale such that by the middle of this century, 70% of the world population is predicted to reside in urban areas. The number of mega cities (>10 million inhabitants) will increase to over 20, 80% of which are in developing countries (Niemcynowicz, 1996).

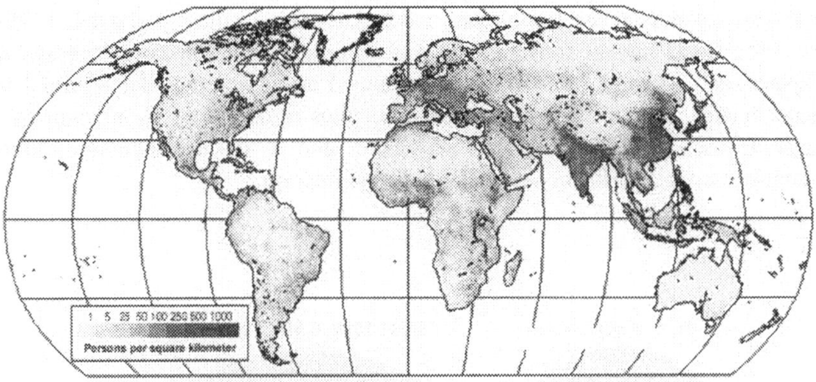

Figure 3.2 Population density using UN adjusted 1995 data[1]

Yet although the figures for human water consumption and urbanisation on a global scale are disturbing, the greatest *effects* (that is, interactions) are felt at the local and regional scale. Figure 3.3 illustrates, by example, the dramatic case of Sao Paulo, which has grown from a city of less than 1 million in 1930 to 15 million at the turn of the millennium, concomitant with a 10-fold increase in area (Porto, 1999). Almost all the fresh water in the upstream catchment of the River Tieté is required to supply the still growing city, leaving little or no baseflow in the river for a significant part of the year.

As cities grow, even greater abstractions are required in relatively localised areas, yet the available resources are finite and may even be being reduced by pollution from the very areas they serve. Urban watercourses receive not only domestic and industrial wastewater, but also polluted storm runoff and, in many

[1] Center for International Earth Science Information Network (CIESIN), Columbia University; International Food Policy Research Institute (IFPRI); and World Resources Institute (WRI). 2000. Gridded Population of the World (GPW), Version 2. Palisades, NY: CIESIN, Columbia University. Available at http://sedac.ciesin.org/plue/gpw.

cases, solid waste dumped in the absence of reliable waste collection system. Once heavily polluted at the most upstream part of the catchment, many rivers never recover.

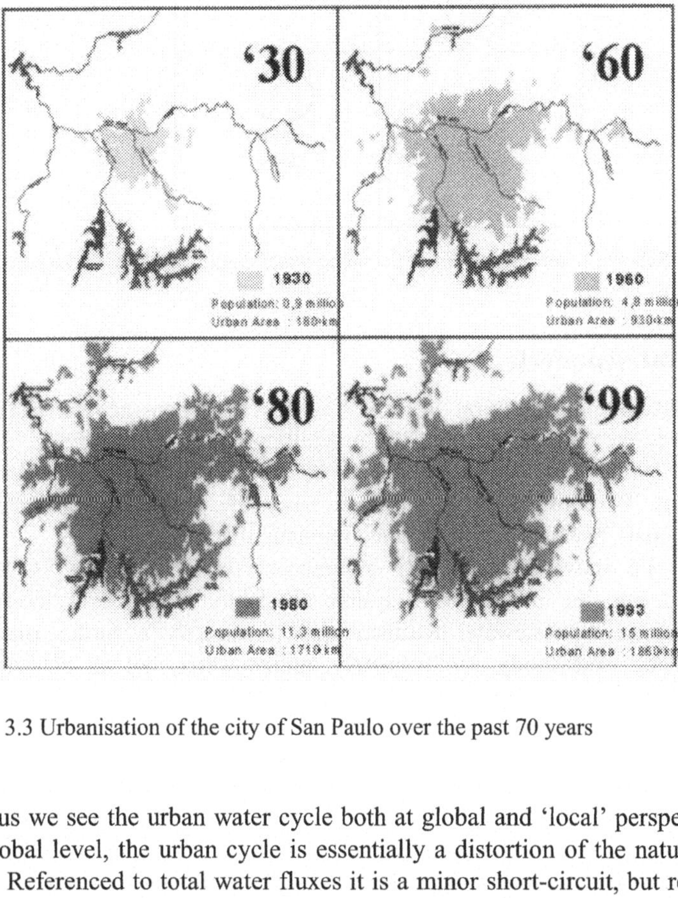

Figure 3.3 Urbanisation of the city of San Paulo over the past 70 years

Thus we see the urban water cycle both at global and 'local' perspective. At the global level, the urban cycle is essentially a distortion of the natural water cycle. Referenced to total water fluxes it is a minor short-circuit, but relative to total available fresh water resources, it is a significant and rising phenomenon. At relatively local levels, the pieces of the urban water cycle jigsaw (Figure 3.4), which form the main discussion area of this book, are progressively more interlocked and increasingly small. If we take 'natural water resources' as our measure of 'the environment', the figure also clearly indicates the main environmental interactions driven by 'water use' both to the upstream (water supply) and downstream (urban drainage).

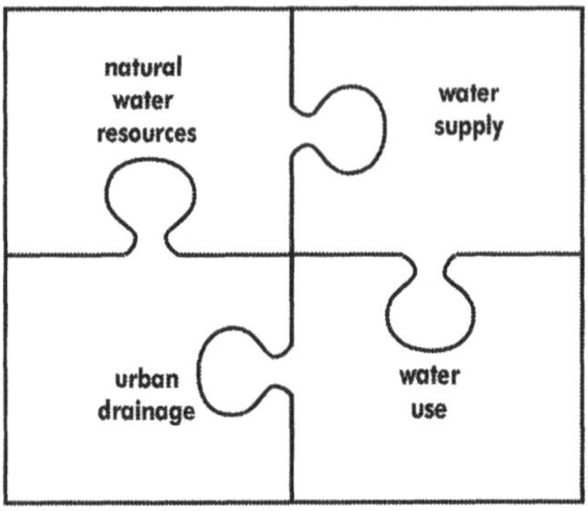

Figure 3.4 Schematic representation of the urban water cycle (Butler & Parkinson, 1997)

3.1.2 Nutrient cycle

The distortion of the natural water cycle, due to human activity, has been discussed in the previous section. But in addition, urban water use also distorts to some extent the natural geochemical cycles of carbon, nitrogen and phosphorus, in particular. On a global basis, the percentage changes are certainly small, but significant local and regional effects result.

Figure 3.5 shows how 'natural substances' (nutrients C, N, P, K) are introduced into the urban water system (basically as digested food), and transferred via the wastewater treatment plant (or directly by surface runoff) to the receiving water body. Unfortunately, unlike water, not all nutrients are renewable. Reserves of (mined) phosphorus, for example, are limited (lasting 100-1000 years, depending on the type of source and the effort involved in extracting it), and once discharged to the aqueous environment, are not readily ever reusable. Beck *et al.* (1994) consider the distorting effect of the urban water system on a range of materials (C, N, P and S-bearing, heavy metals and toxic organic chemicals, and pathogens). Their analysis points to the desirability of returning the material fluxes to the land, rather than the aqueous environment. Some call for radical solutions involving near-complete disengagement of the food and water cycles (Larsen & Gujer, 1996; Otterpohl, *et al.*, 1997).

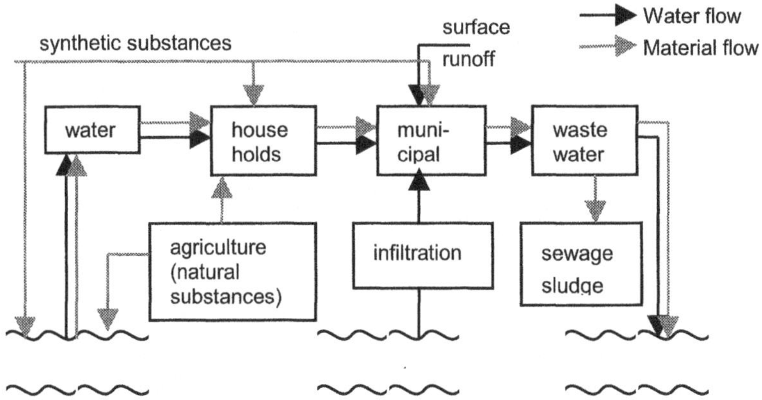

Figure 3.5 Flows of water and substances in urban water management (Larsen & Guyer, 1997)

3.1.3 Urban water systems

Urban water systems provide water to support human life; health, hygiene, safety, recreation and amenity. Water is needed by industry for cleaning, processing and sometimes for incorporation into the final product. Water systems are also commonly used to transport liquid and some solid wastes away from the city. This may be separate from or combined with the role of removing urban stormwater. Understanding and quantifying the quality and quantity aspects of the interaction between human activity and the natural water cycle, and between the urban water subsystems, is a prerequisite to the minimisation of the adverse effects caused by these interactions. In terms of water quantity, interaction takes three main forms: the abstraction of water from the natural cycle, the influence on groundwater levels in urban areas due to changes in the "vertical water balance" caused by leaking sewers and water mains, and diversion of rainwater away from the natural system of drainage due to increased coverage of the land with impervious urban surfaces (leading to increased flooding potential). In terms of water quality the interactions are seen at both points of abstraction and throughout the catchment. For example, pollution in the upstream parts of the catchment adversely affects water quality (and hence usability and cost of treatment) for human consumption further downstream. Further downstream, concentrated (point) and diffused pollution sources further degrade surface and groundwater, rendering it less suitable for various types of use.

3.2 WATER SOURCES

General aspects of water sources used for human consumption are fully covered in literature, to which the reader is referred (e.g. Twort *et al.*, 2000; Sanders and Yevjevich, 1996; Maksimović *et al.*, 1996). We have already discussed the extent of urbanisation at global and regional levels. In this section, emphasis is placed on the interaction of the source from which water is abstracted and the urban environment and urban water infrastructure systems. At river basin level, increased development leads to higher percentages of impervious areas. Increased volumes of runoff from urban areas generates additional diffuse pollution, which combined with pollution from agricultural and natural sources and point source pollution from the disposal of treated or untreated sewage, contributes to the overall degradation of drinking water sources. The main challenges are to protect and provide a high quality of water in sufficient quantity at affordable costs, whilst maintaining the various ecosystems, and to better match water demands with the availability of the resource.

3.2.1 Upland

Upland areas are generally less populated and less suitable for agricultural production. Therefore water in these parts of the catchment is less polluted by human activities and thus more favourable for direct abstraction, storage and less costly treatment. However, upland water sources water are by no means necessarily pollution free. Pollution of these water bodies mainly originates from natural processes such as biodegradation of vegetation and other biodegradable substances, washoff of atmospheric fallout (solid particles and dissolved matter) and soil erosion. In assessing the suitability of these types of water source for human consumption, in addition to water quantity considerations, the long term effects of disinfection by-products (DBPs) caused by the reaction of chlorine with non-toxic organic matter (derived mostly from decaying vegetation and fulvic and humic acids) have to be considered (National Research Council, 2000). These trihalogen-substituted single carbon organic compounds are considered to be harmful to human health and may require additional treatment (e.g. ozonation) beyond the simple processes used for many upland waters.

Additional interactions of concern are:

- Storage of upland water without providing sufficient downstream compensation flows. For example, a series of reservoirs in the Upper Tietê river basin, built to quench the thirst of Sao Paulo (Porto 1999), has left very little fresh water to sustain the habitats of downstream aquatic life. In dry weather periods, urban streams carry almost exclusively wastewater, heavily loaded by solid waste and soil eroded from the urban catchment.

- If stored water is transported through natural watercourse, before abstraction for treatment, additional urbanisation in the flood plain can be encouraged, causing the additional pollution (e.g. the Karaj dam project in Tehran, Iran)
- In the case of excessive surface water abstraction and transfers the whole environment may be changed, jeopardising habitat and causing local climate change (e.g. California).

3.2.2 Lowland

Lowland areas are generally richer in water resources, but are also more densely populated, thus both surface and ground water tend to be more polluted. Water is abstracted for potable use both from the surface watercourses and from alluvial aquifers, which interact mutually in most cases.

Consequences of anthropogenic activities in urban areas on the quantity of flow entering rivers are:

- Reduced infiltration into the underground aquifer (reduced replenishment) that may cause lowering of groundwater levels under urban areas and in the vicinity (see section 3.2.3).
- Increased volume of surface runoff that causes flooding problems both in the urban area and in downstream watercourses
- Reduced time to peak runoff (faster response, higher peak runoff) causes bed erosion, destroys aquatic habitats and generally causes harmful effects on the flora and fauna in urban watercourses.

Impacts of urbanisation on river quality include:

- High loads of dissolved pollutants and suspended sediment originating from urban runoff cause pollution of both surface streams water bodies and groundwater.

Sedimentation on riverbeds with urban washoff that differs in size and chemical content from the natural sediment not only reduces water quality, but often more significantly, clogs the interface between river and groundwater aquifer which it feeds.

3.2.3 Groundwater

Groundwater is abstracted as a major source of drinking water in many cities worldwide (Chilton, 1997). However, the magnitude of its importance varies from country to country (see Figure 3.6). This is an attractive source of water not only because it is often of a good quality (physical, chemical, bacteriological) but also because aquifers suitable for drinking water abstraction are often formed along major rivers in their flood plains, also popular locations for urban areas. Hence, abstraction can be practised in the city or its vicinity. Underground water sources are also better protected from pollution and the vagaries of the weather.

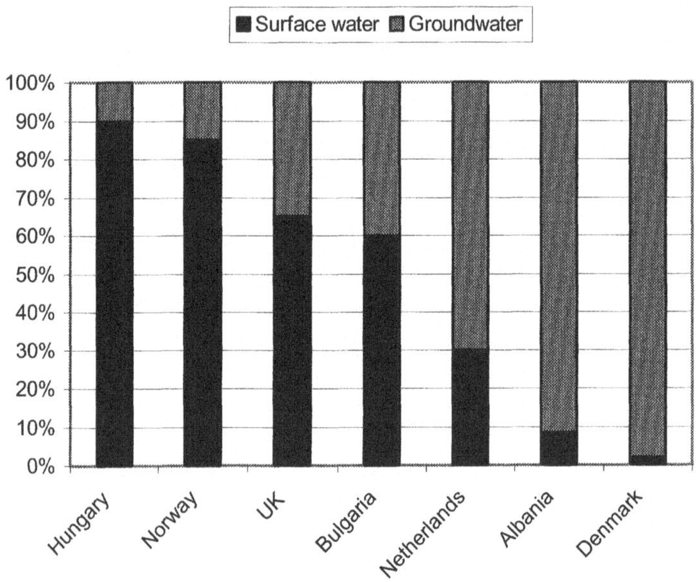

Figure 3.6 Origin of water used for drinking purposes in selected European countries (WHO, 1995)

However, perhaps the major issue surrounding groundwater as a water source is the alarming fact that many major groundwater aquifers are not being sustained (Table 3.1). Clearly, this is contingent on the balance (or lack of it) between abstraction and recharge. Yet the *availability* of water for recharge depends not only on the quantity of vertical recharge through the upper soil layers and horizontal recharge through the perimeter, but also on the properties of the "interface". The interface between aquifers and urban areas is increasingly sealed-off by impervious surfaces, built as a consequence of development. This then forms a barrier to the natural recharge process, now even more required. Additionally, the interface between the aquifer and the recharging watercourse may become clogged (Fig. 3.7), further reducing inflow.

Table 3.1 Major groundwater aquifers (After Twort *et al.*, 2000)

Aquifer / Country	**Estimated reserve 10^6 m^3**	**Estimated use 10^6 m^3 / year**	**Estimated recharge 10^6 m^3 / year**
Nubian Aquifer System / Egypt, Libya, Chad, Sudan	150 × 10^6	460	small
Great Artesian Basin /Australia	20 ×10^6	600 (1975)	1100
Libya	1.8 ×10^6	>recharge	small
Ogallalah / USA	0.36 ×10^6	3400	60
Dakota / USA	>4 ×10^6	725	>315

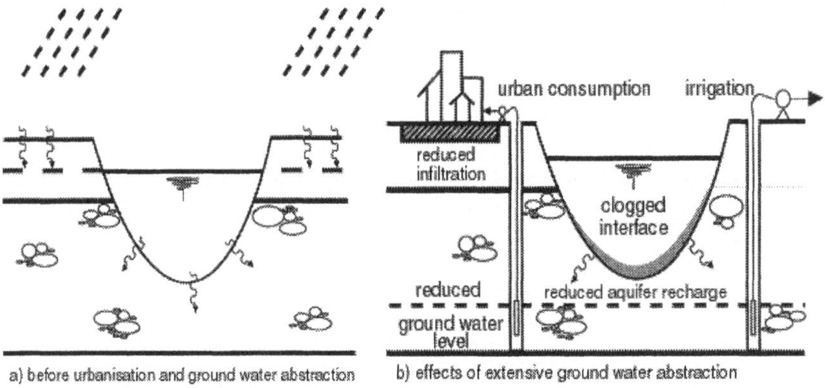

Figure 3.7 Recharge of groundwater an aquifer before and after urbanisation

Other significant problems of groundwater interacting with the urban environment include:

- Groundwater lowering due to its over- exploitation for urban, industrial, mining and other needs (e.g. Mexico City. Barrios, 2000), Bangkok, River Emscher in Germany (Geiger, 1998) or due to reduced infiltration caused by urbanisation, resulting in land subsidence.

- Rising groundwater due to decreased pumping causing problems to underground utilities (e.g. London) or due to excessive recharge by surface runoff, leakage from water mains and sewers (e.g. Tehran, Iran), causing surface flooding and posing a threat to public health.

Even well managed groundwater remains a fragile resource, vulnerable to adverse interactions with human activities causing pollution. Its vulnerability to pollution is closely related to the local hydrogeological conditions including the volume of groundwater and its recharge rate (as mentioned above), the thickness and nature of the unsaturated zone and the geology. The major groundwater pollutants arise from both rural and urban practices and include pesticides, herbicides, nutrients, heavy metals and toxic organics. The source of pollution can be local or spatially distributed. An example of a local source would be the leachate generated by poorly or minimally lined landfills or abandoned gravel pits (Fig. 3.8). At the more spatially distributed scale, aquifers in coastal areas and small islands can suffer from salt-water intrusion, which can only be controlled by careful abstraction management. In extreme cases, such as pollution of groundwater aquifers in Bangladesh by arsenic, the health of the whole population that relies on this source is endangered.

A key to maintaining the quality of groundwater sources is establishing protection zones that will reduce or prevent pollution. In the case of abstraction sites, this is through capture zones. In principle, the entire collection area, the boundaries of which do not necessarily have to coincide with surface water divides, should be considered for some sort of pollution control. Land use and agricultural practices need to be analysed and the appropriate protection zones established. Conventionally, protection zones are established in the vicinity of the point of abstraction. However, groundwater simulation models are capable of predicting the pathlines from the point of pollution to the point of abstraction or inversely, for a given point of abstraction (e.g. a well or group of wells) it is possible to identify the likely response time of a pollution incident affecting the water source depending on its location.

Plate 3.1 illustrates the concept by showing pathlines and travel times to a series of wells used for drinking water abstraction. The wells were under threat of pollution from leaking sewers and poorly built septic tanks (Pokrajac, 1999). In this case, the well capture zone was obtained by deterministic modelling. However, heterogeneity in aquifer properties, coupled with poor knowledge of the variability of these properties means that there is uncertainty in the true location of these zones. Geostatistical modelling of aquifer properties combined with Monte Carlo groundwater flow and transport simulations allow this uncertainty to be characterised (van Leeuwen, *et al.*, 2000).

Groundwater can also interact with the urban water infrastructure. Here we refer to infiltration and exfiltration from both gravity and pressure pipes. These aspects will be covered in more detail in later sections of the chapter.

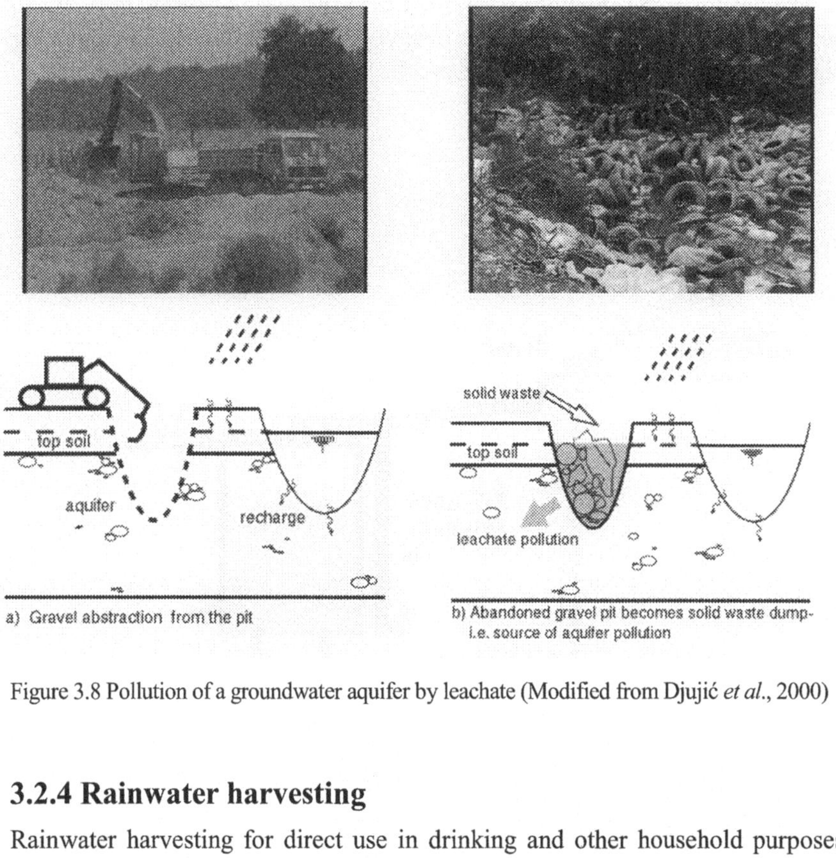

Figure 3.8 Pollution of a groundwater aquifer by leachate (Modified from Djujić *et al.*, 2000)

3.2.4 Rainwater harvesting

Rainwater harvesting for direct use in drinking and other household purposes has a long tradition in many developing countries, remote places and small islands. This technique is today seen as significant opportunity for decentralised, community based water management option to combat water shortage (Cosgrove and Rijsberman, 2000). This technology has relatively recently been introduced in developed countries as an additional measure for reducing pressure on natural resources. Herrmann and Schmida (1999), for example, report on the increase of rainwater utilisation practice in Germany, where more than 100,000 decentralised tanks have been installed in the last 10 years. Water is used mainly domestically for toilet flushing, washing machines and garden watering but also increasingly in schools, car washes and service water demanding industries. The economics of local rainwater collection are finely balanced, with the most promising scale of application being large buildings (Dixon et al., 1999).

The quality of the stormwater captured from the roof at household level can be significantly improved by simple interception of the first flush containing high concentration of the solid particles collected by the washoff (Figure 3. 9).

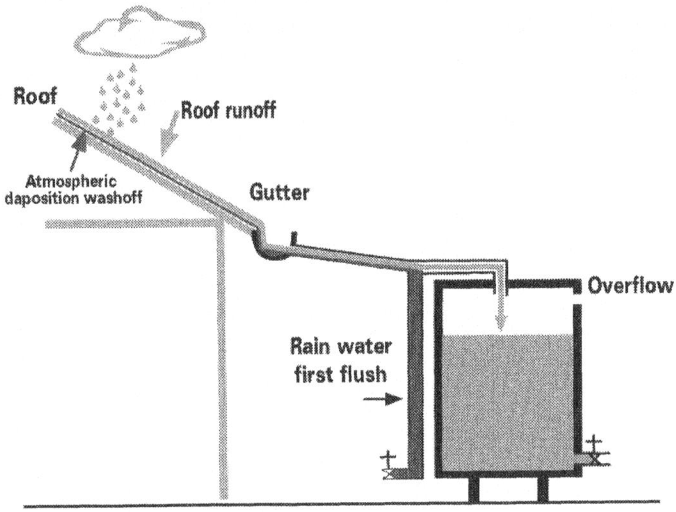

Figure 3.9 A simple system for rainwater harvesting (Maksimović & Ho, 2001).

3.3 INFRASTRUCTURE

Following the general concept of the chapter, this section deals with those aspects of urban water infrastructure where interactions with the environment are considered to be important. Although some aspects mentioned here are well-established facts, there are several innovative approaches presented throughout the text.

3.3.1 Water treatment

Potable water treatment technologies are covered in chapter 4 and in more detail in many other books (e.g. Twort *et al.*, 2000). This section outlines briefly only issues concerning the interaction of treatment plants with the environment and with components of the urban water system. We shall concentrate on two types of interaction:

- Waste generation and disposal by water treatment works and
- Technology dependent effects of the treated water on water quality processes in distribution systems.

3.3.1.1 Sludge disposal

Depending on the source of water (surface or underground), the quantity of sludge generated as a by-product of water treatment processes varies. In the case of surface water treatment, the amount and content strongly depends on land use and pollution management practices in the catchment. Depending on the specific treatment processes used, three major types of sludge are generated:

- non-chemical (unaided by chemicals)
- coagulant
- softening.

The sludge may contains suspended solids, precipitated colour, coagulant hydroxide, algae, manganese, iron, activated carbon (powdered) and impurities contained in lime and polyelectrolyte. The disposal routes of sludge and their implications are discussed further in section 3.3.4.3.

3.3.1.2 Effects of treatment on water quality in distribution networks

It is well known that each water has its own 'identity', characterised by its origin and the treatment technology used to treat it. When water enters the distribution system its quality does not remain static, but changes according to its identity and the network 'environment'. Both the pipe material and the pipe condition – extent of encrustation, biofilm thickness, sediment deposition, mutually interact with the water to alter its physical, chemical and biological quality. In order to be properly understood the quality of water managed it is essential that each case is given an individual diagnosis. There is a lack of knowledge and prediction (modelling) techniques for providing proper prevention rather that cure (dealing with consequences) of the negative effects of water quality changes such as discoloration and loose deposits. Jennings-Wrey (2000) indicates that there are strong links between treatment processes and the *rate* of water quality change, and that optimisation of treatment processes could reduce the rate of deterioration.

3.3.2 Water distribution

Providing basic water supply and maintaining an acceptable standard of drinking water is a high priority task in developing countries. In developed countries the priority is slightly different; rehabilitation of aged systems, reduction of leakage and improvement of cost efficiency of the systems is a prime task. In any case, these systems are primary consumers of investment in urban water infrastructure. Without attempting to cover all the principles of water distribution technology and management, this section emphasises several topics of current interest: the change of paradigm in design, operation and management, basic techniques in achieving the "demand managed" concept, and techniques for reducing losses and improving the efficiency of rehabilitating aged systems.

3.3.2.1 Change of paradigm

Traditionally water supply systems have been designed by assuming that the "needs" of consumers have to be met, disregarding how realistic these "needs" may be. Per capita consumption was kept unrealistically high in design, in order to be "on the safe side". Systems were often oversized, and in the absence of reliable measurement of at least two out of the three components of the water supply system (production, consumption and losses), systems were becoming ever-larger consumers of water and financial resources, placing an increasing pressure on both the environment and the economy. All this was taking place under the conditions of decreasing water availability due to increased pollution (Figure 3.10). However, in recent years, a new paradigm has emerged that replaces demand driven consumption with demand-managed consumption.

3.3.2.2. Changed role of water distribution systems

With the change of paradigm, water distribution systems have a new role to play. Instead of "unlimited" consumption, alternative methods of keeping it "manageable" are being practiced. This includes both engineering activities such as leakage reduction, pump rescheduling (to save energy), introduction of water saving devices and so on, as well as socio-economic interventions such as: raising awareness (education of customers), pricing policy change, and introduction of the innovative techniques based on resources recycling. These techniques require that the system is well documented with reliable data on such aspects as network configuration, consumption, pressure and flows. This type of information can be obtained by installing reliable sensors and by implementing data acquisition systems, GIS-supported data processing linked with hydrodynamic and quality modelling, and by providing conditions for exercising the above technical measures. (Figure 3.11)

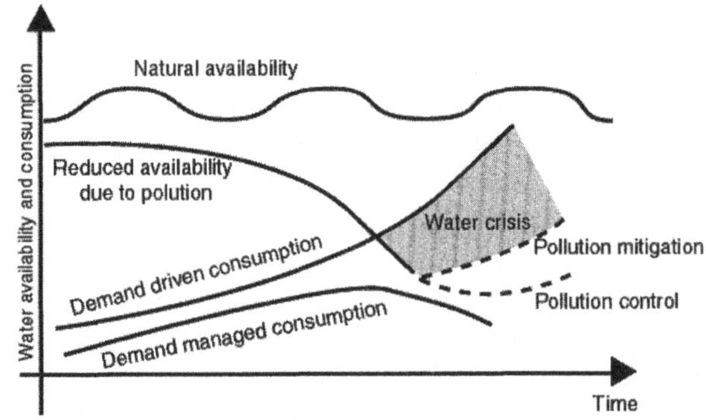

Fig. 3.10 Interactions of the demand driven and demand managed systems with natural and reduced water availability

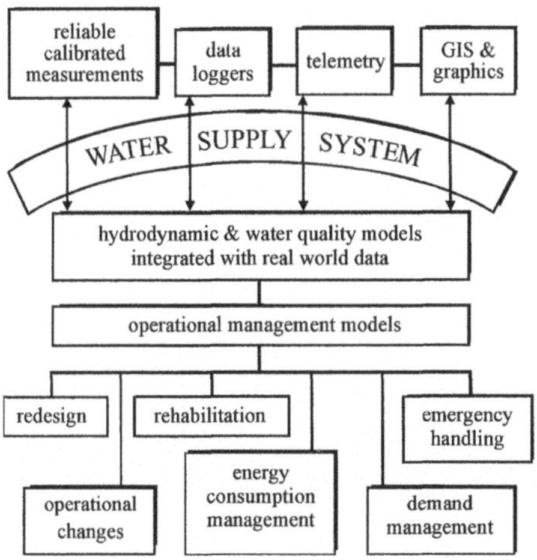

Figure 3.11 Role of a modern water supply system with internal links.

3.3.2.3 Leakage detection and management

Finding, quantifying and managing leaks is a challenge for most existing water supply systems. It is often the most important loss in the system and, even in well-managed systems, can amount to 30% of the water put into supply. Cases of leakages rates as high as 80–90% are not rare. Challenging aspects of leakage management include the fact that:

- Leakage is not constant. It increases with increased pressure in pipes
- Not all parts of supply systems are equally vulnerable to leakage. Vulnerability can be identified and preferably quantified (see, for example, Maksimović and Carmi, 1999)
- Economic evaluation of the importance of leakage can only be made if it is quantified (preferably with defined uncertainty). This requires reliable measurements to be performed in the system
- Investment in leakage repair and in leakage management pays off. Usually, it costs less to control leakage than to supply the equivalent amount of water saved from a new resource.

Although repairing leaks may still be considered to be "basic technology", leakage management increasingly requires a high level of knowledge and advanced technology in data acquisition/processing, network modelling and so on.

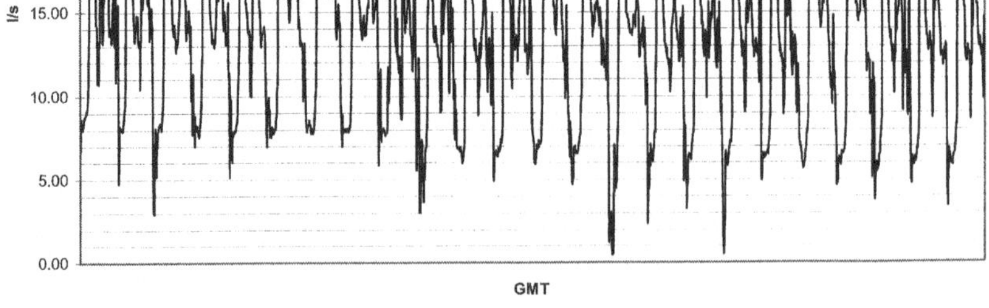

Figure 3.12 Monthly plot of an inflow to a District Meter Area

Surprisingly, almost all water supply systems lack enough reliable information to solve the basic water balance equation. At a level of a well-defined supply zone, inflow minus consumption over a defined period of time equals losses. So simple, yet difficult to materialise! This is particularly so in countries where individual consumption is not metered or the percentage of the metered properties is still relatively low (e.g. the UK). Yet it is still difficult to find a water undertaking in the world that can quantify its water balance with reliable accuracy and defined uncertainty.

Measuring inflows is the most straightforward and can be accomplished with several flow meters, at least at a level of an isolated area (for example in the UK, the DMA or District Meter Area). Leakage can then be assessed by observing minimum night flows. Interpretation of a plot such as that in Figure 3.12 is relatively straightforward, although care has to be taken to start the analysis from the data origin and to understand the data processing technique applied. In the concrete case (presented in Fig. 3.12), it looks like minimum night flow (thus leakage) is very low. However, detailed inspection of the data processing indicates that there were some zero meter readings which, when averaged over number of readings, introduced an error. Quantification of these errors can shed more light on the accuracy of the leakage assessment.

3.3.2.5 Rehabilitation of aged systems—art versus technology

Ageing of water supply systems and urban infrastructure in general is a problem for developed and developing countries alike. Old systems not only leak more, but also adversely affect water quality (and potentially human health) due to the processes described earlier in the chapter. Management of the problem, such as with pipe cleaning, can sometimes have counter effects. In fact, management and allocation of resources for rehabilitation is in a need of better quantification of vulnerability and the effects of the works. According to Saegrov *et al.* (2000), for each element in the network, the failure probability can be estimated, and the overall system vulnerability can thereafter be calculated. This is based on the assessment of its conditions of use, processes that take place within the pipe, external and internal load. The expected pattern of pipe failures is a critical part of reliability analysis. In principle, the future resistance against failure of a pipe can be estimated by deterministic and probabilistic methods, a selection of which are listed in Table 3.2.

Of course, rehabilitation technologies also play a role. Pipe network rehabilitation planning has benefited substantially from the development and application of modern heuristics, such as genetic algorithms, in the last decade (e.g. Savić *et al.*, 2000); Engelhardt *et al.*, 2000). The main advantage of these approaches is that it they allow the use of detailed hydraulic/water-quality models and global optimisation methods to support decisions in whole-life cost rehabilitation. Trenchless methods, pipe insertion and so on, combined with advanced simulation and optimisation methods, when applied wisely, can achieve significant savings in rehabilitation costs.

Table 3.2 Network failure prediction models under development in European countries (Saegrov *et al.*, 2000)

Acronym (developer)	**Objectives**	**Methods**	**Limitations**
UTILNETS (CTI, Athens *et. al*)	Forecast the structural failure of pipes in the network	Determination of mechanical load and corrosion	Fitted to cast and ductile iron
(NTNU/ SINTEF, Trondheim)	Forecast the number of failures of pipes, estimate reliability of water networks	Hydraulic/reliability model, survival data analyses, trend plots	Limited practical use
KANEW (Dresden University of Technology)	Forecast the mileage of pipe types reaching the end of their service life and long-term effects of rehabilitation strategies	Cohort survival model, interactive simulation model of rehab effects, Delphi technique for estimation of service lives and technologies	Relies on "best guess" rehabilitation needs
FAILNET (Cemagref Bordeaux)	Forecast the number of pipe failures and assess network reliability	Survival data analysis and Monte Carlo simulation, hydraulic calculation and reliability indexes	Limited practical use
AssetMap (INSA Lyon)	Forecast break rates and influence of renewal rate and risk assessment for asset stock in an urban environment	Conceptual model based on hazard functions calibrated with historical data, statistical and GIS analysis	Limited practical use
(Brno University of Technology)	Forecast the number of pipe failures and assess network reliability	Reliability model and survival data analysis	Limited practical use
(WRc, Swindon)	Determine current system failure patterns and underlying causes	Descriptive statistics, linear regression, neural network and GIS analysis	No failure prediction
(AGAC, Reggio Emilia)	Failure monitoring and analysis	Reliability analysis based on exponential distribution and questionnaire to qualified experts	Simplified failure prediction

3.3.3 Urban drainage

Urban drainage systems are designed to manage two types of 'urban' water: wastewater and stormwater. Wastewater is water that has been supplied to support life, maintain a standard of living, and satisfy the needs of industry. After use, if not drained properly, it could cause pollution and create health risks. Wastewater contains dissolved and particulate (fine and coarse) material, derived from toilet flushing and washing of various sorts, from industrial processes, and from other water uses. Stormwater is rainwater (or water resulting from any form of precipitation) that has fallen in an urban catchment area. If stormwater is not drained properly it can cause inconvenience, damage, flooding and further health risks. It contains some pollutants, originating from rain, the air, and the catchment surface (Butler & Davies, 2000).

In many urban areas, drainage is based on an engineered system of sewers: pipes and structures that collect and dispose of this water (Figure 3.13). In contrast, isolated or poor communities normally have no main drainage[2]. Wastewater is treated locally (or not at all) and stormwater is drained naturally into the ground. These sorts of arrangements have mainly existed when the extent of urbanisation has been limited; but recent thinking in relation to sustainable drainage practices encourages the use of more natural drainage arrangements wherever possible.

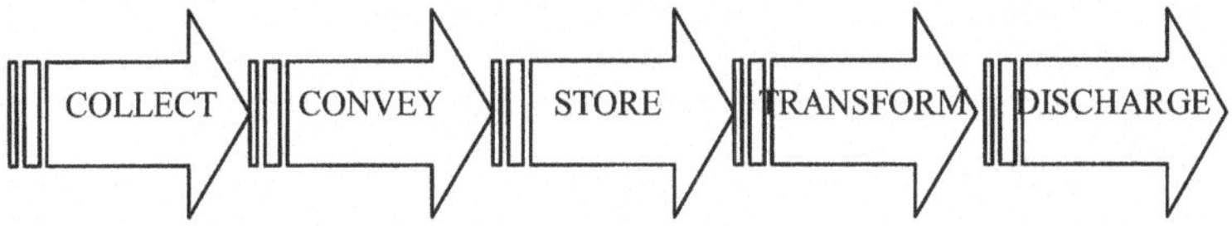

Figure 3.13 Functions of an urban drainage system

There are two main types of piped sewer system: a combined system in which wastewater and stormwater flow together in the same pipe, and a separate system in which wastewater and stormwater are kept in separate pipes. Hybrid systems are also used, for example 'partially-separate' systems, in which wastewater is mixed with some stormwater, while the majority of stormwater is

[2] On a global basis this is the case for 94% of the population (Niemczynowicz, 1997)

conveyed by a separate pipe. Hybrid systems also exist for more accidental reasons such as larger cities having combined sewers in the older central part of town with separate sewers connected in the newer areas.

In this section we assess the major interactions of the various forms of drainage system with the urban environment.

3.3.3.1 Separate foul sewerage

In theory at least, separate foul sewers should have little interaction with the environment. Their aim is to safely convey wastewater from its source to a convenient location for formal treatment. Only relatively recently has the notion of in-pipe treatment been seriously considered (Green *et al.*, 1985). But still, there are important interactions that do occur in practice: infiltration, inflow and exfiltration.

Infiltration is extraneous groundwater or water from other leaking pipes that indirectly enters the sewer system through defective drains and sewers (cracks and fissures), pipe joints and couplings, and manholes. Inflow is stormwater that directly enters separate foul sewers from illegal or misconnected yard gullies, roof downpipes or through manhole covers. Excessive amounts of infiltration and inflow are implicated in the following problems (Fiddes & Simmonds, 1985):

- reduced effective sewer capacity
- overloading of pumping stations and WWTPs
- higher frequency of CSO operation
- increased entry of sediment resulting in higher maintenance requirements.

In addition, high levels of infiltration can seriously dilute the wastewater, rendering treatment more difficult.

The extent of the infiltration problem is site-specific but is usually a result of poor design or construction and will generally increase as the system physically degrades. Inflow is almost inevitable since a drainage system does not run under pressure, and so it is hard to stop users damaging the way it operates. A survey of one separate system in the UK (Inman, 1975) found that 40 % of all houses had some arrangement whereby stormwater could enter the foul sewer. Infiltration can reach serious proportions in old systems. For example, Stanley (1972) found rates in those existing sewers subject to infiltration ranging from 15 to 50% of average dry weather flow. Remediation is an expensive option (White *et al.*, 1997).

Exfiltration is the opposite of infiltration, and consists of the wastewater leaking out of a sewer into the surrounding soil and groundwater. This can also be a problem, particularly in areas of sensitive groundwater. Factors affecting the likelihood of exfiltration are similar to those discussed for infiltration, but it is thought to be less common (Anderson *et al.*, 1996).

3.3.3.2 Separate storm sewerage (piped)

Separate storm sewers have two major interactions with the urban environment: flooding and discharge of conveyed stormwater to receiving waters.

Stormwater inlets (such as road gullies and roof downpipes) are connected to a discharge point (or outfall) at a receiving water by a network of storm sewers. The sewers themselves will stand empty for long periods of time, but will fill up during rainfall, to an extent that depends on the magnitude of the event and conditions in the catchment. During low rainfall, flows will be well below the available capacity, but during high rainfall the flow may exceed the pipe capacity inducing pressure (surcharge) flow and even surface flooding.

Storm sewers are designed to reduce the frequency of surface flooding to 'acceptable' levels. This is accomplished by choosing an appropriate, statistically relevant design storm to establish the stormwater flows to be conveyed. The choice of design storm return period determines the degree of protection from stormwater flooding provided by the system. Standard practice in the UK, for example, is to use return periods of 1 yr or 2 yr for most schemes with 5 yr being adopted where property in vulnerable areas would be subject to significant flood damage. Higher periods may be adopted for particularly sensitive sites.

In fact, specifying a 2 year return period storm, will not result in flooding with an equivalent return period. Although rainfall and runoff frequencies are approximately equal (but not identical), this is not the case for flooding frequency. Sewers are commonly laid at least 1 m below the ground surface and they can therefore accommodate a considerable surcharge before surface flooding occurs. Hence, the capacity of the system under these conditions is increased above the design capacity. Pipes designed to a 1 yr return period storm may render a 10 year return period flood protection (Butler & Davies, 2000).

Table 3.3 shows recommendations made by the relevant European Standard (BSEN 752-4:1998) for design storm return period related to the location of the area to be drained. Design flooding frequencies are also given.

Table 3.3 Recommended design frequencies (adapted from BSEN 752-4:1998)

Location	Design storm return period (yr)	Design flooding return period (yr)
Rural areas	—	10
Residential areas	2	20
City centres/industrial/ commercial areas:		
• with flooding check	2	30
• without flooding check	5	—
Underground railways/under passes	10	50

Table 3.4 Pollutant event mean concentrations and unit loads for stormwater (Ellis, 1986)

Determinand	EMC (mg/l)	Unit load (kg/imp ha.yr)*
Total suspended solids (TSS)	21–2582 (190)	347–2340 (487)
BOD_5	7–22 (11)	35–172 (59)
COD	20–365 (85)	22–703 (358)
Ammoniacal nitrogen	0.2–4.6 (1.45)	1.2–25.1 (1.76)
Total nitrogen	0.4–20.0 (3.2)	0.9–24.2 (9.0)
Total phosphorus	0.02–4.30 (0.34)	0.5–4.9 (1.8)
Total lead	0.01–3.1 (0.21)	0.09–1.91 (0.83)
Total zinc	0.01–3.68 (0.30)	0.21–2.68 (1.15)
Hydrocarbons	0.09–2.8 (0.4)	—
Faecal coliforms	400–50,000 (6430) (MPN/100ml)	0.9–3.8 (2.1) ($\times 10^9$ counts/ha)

*imp ha = impervious area measured in hectares.

The second interaction concerns the release of stormwater at the outfall, typically into receiving water. Stormwater can be heavily contaminated with a range of polluting substances, and typically contains a complex mixture of natural organic and inorganic materials with a small proportion of man-made substances derived from transport, commercial and industrial practices. Stormwater quality is influenced by rainfall and, especially, by the catchment. The major catchment sources include vehicle emissions, corrosion and abrasion, buildings and roads corrosion and erosion, bird and animal faeces, street litter deposition, fallen leaves and grass residues, and accidental spills (Butler & Davies, 2000).

Pollutants include solids, oxygen consuming materials, nutrients, hydrocarbons, heavy metals and trace organics, and bacteria. Runoff quality at any particular site will depend upon a number of factors, including:

- geographical location
- road and traffic characteristics
- building and roofing types
- weather, particularly rainfall.

Typical values and ranges of pollutant concentrations and loads are given in Table 3.4.

3.3.3.3 Combined sewerage

Despite rarely being constructed in the present day, many countries still have large numbers of combined sewers. Such sewers carry both wastewater and stormwater in the same pipe, indicating that it is not feasible to carry the full combined flow at all times to treatment. Therefore, at high flow-rates (wet weather) it is necessary for some of the flow to be discharged to a receiving water at a combined sewer overflow (CSO). Thus combined sewers interact with the environment in similar way to separate sewers, that is by: flooding, infiltration, exfiltration and release to the environment via CSOs.

The first three of these are similar to the interactions experienced by separate sewers. The only exception is that flooding from combined sewers into housing areas is likely to be more hazardous than stormwater flooding of open land, so the *type* of flooding likely to occur will influence selection of a suitable design return period.

A CSO is designed to receive an inflow, which during rainfall consists of stormwater mixed with wastewater, and divide it into two outflows. One outflow is to the wastewater treatment plant (the continuation flow, or flow retained) and one is to the watercourse (the spill flow). The normal means of achieving this is a weir. In doing so, the ideal would be that all pollutants in the flow continued to the WWTP (i.e. were retained within the sewer system), but unfortunately this is not achieved in practice. Various designs of CSO demonstrate some success in retaining larger solids, but fine suspended and dissolved material tends to be split between continuation flow and spill flow in the same proportion as the split in the flows (Saul *et al.*, 1998). Best practice guidelines are available (Balmforth *et al.,* 1994).

Table 3.5 Pollutant concentrations and unit loads for wastewater (Ainger *et al.*, 1997)

Determinand type	**Determinand**	**Unit load (g/hd.d)**	**Concentration (mg/l) Mean (range)**
Physical	Suspended solids		
	• Volatile	48	240
	• Fixed	12	60
	• Total	60	300 (180–450)
	Gross (sanitary) solids		
	Sanitary refuse	0.15	
	Toilet paper	7	
	Temperature		18 (15–20) °C: summer 10 °C: winter
Chemical	BOD_5		
	• Soluble	20	100
	• Particulate	40	200
	• Total	60	300 (200–400)
	COD		
	• Soluble	35	175
	• Particulate	75	375
	• Total	110	550 (350–750)
	TOC	40	200 (100–300)
	Nitrogen		
	• Organic N	4	20
	• Ammonia	8	40
	• Nitrites		0
	• Nitrates		<1
	• Total	12	60 (30–85)
	Phosphorus	1	
	• Organic	2	5
	• Inorganic	3	10
	• Total		15
Microbiological	Total coliforms		10^7–10^8 MPN/100ml
	Faecal coliforms		10^6–10^7 MPN/100ml
	Viruses		10^2–10^3 infectious units/100ml

The quality of CSO discharges is related to the concentration of pollutants in both the dry weather flow and in the stormwater. Table 3.5 lists a range of typical determinand concentrations and loads.

An estimate can be made of the expected concentration of CSO discharges using simple DWF conversion factors as given in Table 3.6. In some systems, higher than expected pollutant concentrations are observed in early storm flows; a 'first flush'. These are derived from sediments on the catchment surface, gully pots or in the sewers themselves, or from gully pot liquors and in-sewer wastewater (particularly near-bed solids). In some catchments (mainly larger, flatter ones) the effect is pronounced, and in others it is not observed at all.

The impact of CSO discharges on receiving waters will be considered later in the chapter. These are likely to be most serious when CSOs are poorly designed or operating ineffectively. Also, sewers that back up as a result of sediment deposition problems may cause CSOs to operate prematurely (before the inflow has reached the CSO setting), or, in extreme case, to spill even in dry weather conditions. This may cause serious pollution of receiving waters.

Table 3.6 Factors to convert DWF concentrations to average storm concentrations (Threlfall *et al.*, 1991)

Determinand	**Multiplying factor**[1]
BOD	0.5
COD	1.0
Ammonia	0.3
Suspended solids	2.0
Total dissolved solids	0.4

[1] for systems in which average sewer gradient is no steeper than 1 in 50

3.3.3.4 Storm drainage (source control)

The basic philosophy of stormwater drainage has traditionally been to solve flooding problems by transferring large quantities of water downstream as quickly as possible using piped sewers. Whilst this approach was successful in eliminating local flooding problems, the increased volumes and peak flows has caused downstream flooding problems together with pollution and erosion of natural receiving water bodies.

More recently, the philosophy has shifted from one of stormwater disposal to stormwater management, incorporating the principles of sustainable development (Butler & Parkinson, 1997). In this approach, stormwater is regarded as a resource to be managed on a catchment-scale basis including incorporation of the concepts of 'source control'. In source control (or SUDS) schemes, stormwater is not immediately discharged but is stored, treated, reused or discharged locally, close to its point of generation (Figure 3.14). The polluting effect of storm runoff has also been more fully appreciated and many methods have been re-examined or developed for their ability to improve water quality. Table 3.7 classifies the wide range of techniques available.

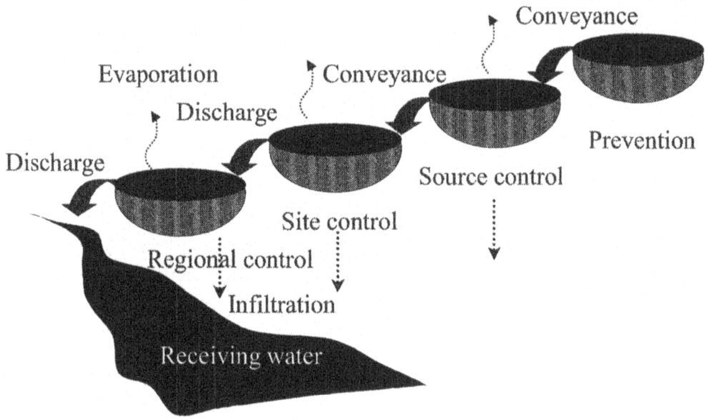

Figure 3.14 Storm water management train in an urban catchment (CIRIA, 2000)

SUDS interact with the environment in quite different ways to conventional systems, mainly beneficially, by:

- reducing peak runoff rates leading to lower frequency of flooding and/or CSO operation
- recharging soil moisture and groundwater and hence watercourse baseflow augmentation
- decreasing downstream channel erosion through flow reduction and velocity control
- reducing pollutant load to receiving waters
- preserving and enhancing natural vegetation and wildlife habitats in urban areas.

However, the methods do have some negative impacts on the environment, including:

- increased danger of local flooding due to premature system failure caused by blockage
- increased probability of basement flooding or damage to foundations due to rising groundwater levels
- contamination of groundwater with stormwater pollutants.

Table 3.7 Classification of stormwater management options (Butler & Davies, 2000)

Option	Examples	Advantages	Disadvantages
Local disposal	Infiltration devices e.g. soakaways, infiltration trenches	Runoff reduction of minor storms Groundwater recharge Pollution reduction	Capital cost Clogging Groundwater pollution
	Vegetated surfaces e.g. swales	Runoff delay Aesthetics Pollutant reduction Capital cost	Maintenance cost Groundwater pollution
	Porous pavements	Runoff reduction of minor storms Groundwater recharge Pollution reduction	Capital & maintenance costs Clogging Groundwater pollution
Inlet control	Rooftop ponding	Runoff delay Cooling effect on building Possible fire protection	Structural loading Roof leakage Outlet blockage
	Downpipe storage e.g. water butts	Runoff delay Reuse opportunities Small size	Small capacity Access difficulties
	Paved area ponding e.g. gully throttles	Runoff delay Pollutant reduction Possible retrofitting	Restricts other uses when raining Damage to surface
On-site storage	Surface ponds e.g. water meadows, detention ponds	Large capacity Runoff reduction of major storms Aesthetics Multi-purpose use Pollution reduction	Capital and maintenance cost Large footprint Pollution & eutrophication Pest breeding potential Aesthetics Safety hazard
	Underground tanks	Runoff reduction of storms Pollution reduction No visual intrusion Capital cost	Maintenance cost Access difficulties
	Oversized sewers	Runoff reduction of storms Pollution reduction No visual intrusion Capital cost	Maintenance cost Access difficulties

3.3.4 Wastewater treatment

3.3.4.1 End-of-pipe treatment

Most urban drainage systems are connected to end-of-pipe wastewater treatment facilities. The main interactions with the environment result from the two main process streams: liquid effluent and sludge and these are considered separately. The degree of treatment provided is, in principle, related to the sensitivity, type or use of the receiving water. Processes are broadly categorised as preliminary/ primary, secondary, tertiary and advanced (see table 3.8).

Table 3.8 Treatment options appropriate to receiving water types (Veenstra & Alaerts, 1996)

Preliminary/ Primary	**Secondary**	**Tertiary**	**Advanced**
Screens Grit removal Primary sedimentation Comminution Oil/fat removal Flow equalisation pH neutralisation	Activated sludge Extended aeration Aerated lagoon Trickling filter Rotating biological contactors Anaerobic treatment/UASB Anaerobic filter Stabilisation ponds Constructed wetlands Aquaculture	Nitrification Denitrification Chemical precipitation Disinfection (Direct) filtration Chemical oxidation Biological P removal Constructed wetlands Aquaculture	Chemical treatment Reverse osmosis Electrodialysis Carbon adsorption Selective ion exchange Hyperfiltration

Preliminary/primary processes consist of physical operations involving mechanical screening, grit removal and sedimentation, with the object of removing fats, oil and grease, settleable, suspended and floating solids. At the same time, at least 30% of the BOD_5 and 15% of the TN and TP are removed. Faecal coliform numbers are reduced by 1–2 log units.

Secondary processes convert biodegradable organic matter to carbon dioxide, nitrates and water by microbiological means. These aerobic processes require oxygen that is usually supplied by intensive mechanical aeration. After secondary aerobic treatment, BOD_5 and TSS removal is increased to 90–95%, TN to 40% and TP to 90%. For wastewaters with elevated temperatures, anaerobic processes can also be applied.

Primary and secondary operations produce sludges that together amount to less than 0.5% of the total wastewater flow. Heavy metals and other micro-pollutants tend to accumulate in the sludges by adsorbing onto suspended particles. Disposal of sludge is becoming more problematic than disposal of treated effluent (see below).

Extensive wastewater treatment by stabilisation ponds, constructed wetlands or aquaculture can provide adequate secondary and tertiary treatment. However, as the biological processes are not accelerated by mechanical equipment, large land areas are required.

Tertiary treatment is designed to remove nutrients from secondary treated effluents up to levels of > 80% for TN and > 90% for TP. Additional suspended solids (> 95%) and BOD removal (> 95%) is achieved by these processes. Also included in tertiary treatment is disinfection by various means (e.g. chlorination, UV adsorption, ozone).

Advanced processes include ion exchange, adsorption, hyperfiltration, and reverse osmosis. These are much less common than earlier categories of treatment, only specified where a particularly high quality effluent is required.

The impact of treatment plant discharges into the environment is discussed in section 3.5.

3.3.4.2 On-site systems

In unsewered urban and rural areas, septic tank systems are the most popular treatment system for domestic wastewater. In the US, for example, some 24% of households are connected to tank systems (US Bureau of the Census, 1990).

A basic system consists of a tank and a drainage field. The tank itself is a watertight, underground vessel that provides reasonably quiescent conditions for the settlement, storage and partial decomposition of sludge (settled waste solids). Wastewater is drained to the tank from the dwelling served and the effluent is discharged to the drainage field through the tank outlet. Sludge gradually accumulates in the bottom of the tank and has to be pumped out periodically. The organic matter in the tank may undergo anaerobic digestion depending on the size of the tank, frequency of sludge removal, and temperature.

The drainage field is equally as important as the tank for treatment and typically consists of a system of sub-surface irrigation pipes or a soakaway to drain the effluent into the surrounding soil. Appropriate design and installation of both the tank and drainage field is essential for the correct operation of the system.

The main advantages and disadvantages of septic tank systems for on-site treatment of wastewater are summarised in Table 3.9.

Table 3.9 Advantages and Disadvantages of Septic Tank Systems

Advantages	Disadvantages
Low capital cost	High space requirement
Low energy requirement	Tank effluent is unsuitable for open discharge
Low maintenance	Potential for local nuisance and health hazard
Potential for groundwater recharge	Potential for groundwater pollution
Low sludge production	Requires some owner participation

As the table shows, septic tanks systems can have both positive and negative interactions with the environment. These are similar to those for source control stormwater drainage.

The main positive environmental interaction is the low production of sludge compared to many end-of-pipe methods. The actual volume produced depends on several factors, but particularly community wealth and ambient temperature. In industrialised countries with temperate climates this is up to 600 $l.hd^{-1}.y^{-1}$, (litres per head per year), whereas values of 40 $l.hd^{-1}.y^{-1}$ are more common in developing countries (Butler & Smith, in press). Sludge may be disposed of by transporting to a wastewater treatment works or alternatively may be recycled directly to land, provided adequate controls are provided. These are discussed further below.

Negative interactions are mainly associated with backing-up or surface flooding of wastewater, and with pollution of surface or ground waters. Neither should occur in properly functioning systems. Problems that do occur are usually related to poor location or lack of maintenance (Butler & Payne, 1995). Inadequate design or installation is also important exacerbating factors.

Septic tank systems can cause local nuisance and health hazards. For example in the USA, during the period 1971–78, wastewater discharges from septic tanks or cesspools were responsible for 41% of the outbreaks of disease caused by contaminated untreated groundwater (Craun, 1981).

There are many reports of environmental problems due to contamination of ground and surface water with enteric pathogens from septic tank installations. Indeed, the potential risk of disease transmission from potable or recreational use of water contaminated by septic tank systems is a major concern in many different regions of the world (e.g. Ho & Tam, 1998; Scandura & Sobsey; 1997). The contamination of private water supplies is often the main cause of human exposure to enteric pathogens originating from septic tanks (Stilinovic *et al.,* 1995). A number of other environmental effects have also been reported (Butler & Smith, in press).

3.3.4.3 Sludge treatment and disposal

A range of sludge treatment options is available as summarised in Table 3.10. The top two processes are purely for water removal (and hence volume reduction) purposes. Of those options that transform the sludge, by far most popular is anaerobic digestion.

Residual sludge can be disposed of by landfilling, incineration and thermal drying, or be reused on agricultural land. Sludge disposal to sea is a further option, although this has been banned across Europe since 31st December 1998 (UWWT). This is despite the fact of very little evidence of detrimental effect to the marine environment (Gray, 1999). Other disposal routes include sacrificial land use, forestry, land reclamation, use as a building material and as a foodstuff (Smith, 1994).

Table 3.10 Sludge treatment & disposal options

Treatment	Disposal
Thickening Dewatering	Agriculture
Anaerobic digestion	Incineration
Aerobic digestion	Thermal drying
Lime	Landfill
Composting	Land application
	Sea disposal

Concentrating on the agricultural land application route, untreated sludge can be recycled directly onto land, but should be managed by the provision of barriers to disease transmission and minimisation of odour nuisance. Suitable measures include incorporation of the sludge into the soil by cultivation immediately after application or sub-surface injection, coupled with restrictions on the sowing, harvesting and types of crops that can be grown (DOE, 1996). Despite there being no reported cases of disease outbreak where these practices have been followed (Royal Commission on Environmental Pollution, 1996), some countries do not allow direct land application of untreated sludge, including the sludge from septic tanks. The UK, for example, is committed to phasing out the recycling of untreated sludge to agricultural land on precautionary grounds (Royal Commission on Environmental Pollution, 1996).

The relevant EU directive (CEC, 1986) is a flexible legislative vehicle, allowing for local conditions of farming, soil and climate to be taken into account. Individual European Member States have found it necessary to set additional rules to firm up on areas where the Directive does not provide guidance. This is typically related to pathogen destruction and heavy metal limits. Substantially different metal limits have been set internationally, reflecting different philosophies (Davis, 1996):

- No accumulation of metal in the soil beyond background concentrations (very precautionary)
- Minimisation of metal accumulation in the soil (precautionary)
- Risk assessment.

The first philosophy will usually result in such restrictive limits as to preclude agricultural utilisation of even the cleanest of sludges. This has the effect of displacing sludge to other less sustainable disposal routes and more vulnerable environmental compartments.

3.4 RELEASES

3.4.1 Types

Pollutant releases to the aqueous environment can be conveniently classified into point sources and non-point (or diffuse) sources. Point sources occur at specific locations where polluted water is routed directly to the receiving water, for example by discharge pipes. Non-point pollution arises from a broad group of human activities for which the pollutants have no obvious discharge location (Ongley, 1996). In practice, the detailed distinction between point and non-point sources can be unclear, although releases from the urban wastewater system are normally classified as point sources (Figure 3.15). Such releases may be continuous, largely unaffected by the prevailing hydrological conditions, or intermittent, occurring only in response to wet weather periods (see below), as below:

- Continuous releases of treated effluent from normally functioning wastewater treatment plants (WWTPs).
- Intermittent shock loads (of suspended solids and/or ammonia) from WWTPs disturbed by wet weather transient loads
- Intermittent discharges of a mixture of surface runoff, domestic, commercial and industrial wastewater with groundwater and sewer deposits from combined sewer overflows.
- Intermittent discharges of runoff from urban surfaces from separate storm sewer outfalls

The most common non-point release from urban areas is the intermittent discharge of runoff directly from roofs, roads and other paved areas.

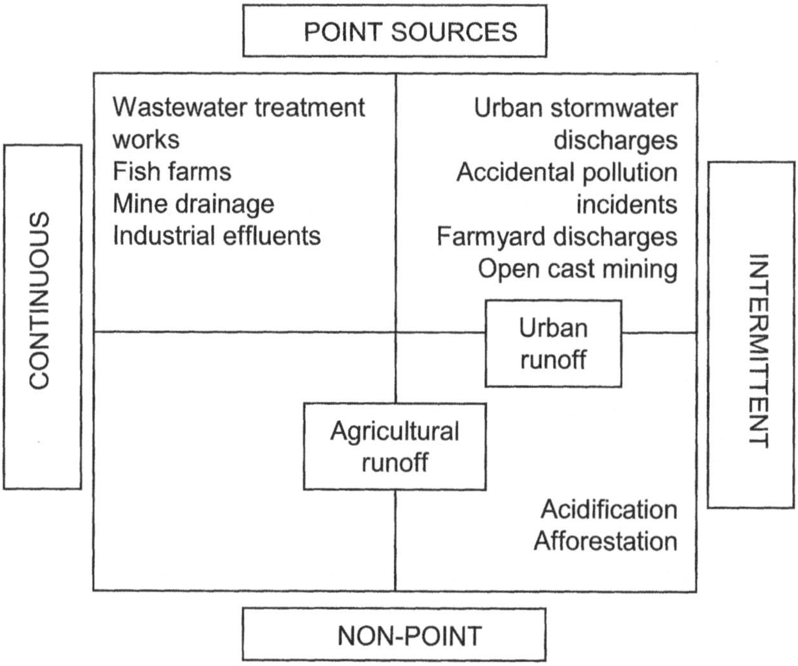

Figure 3.15 Types of pollutant release (adapted from British Ecological Society, 1990)

3.4.2 Context/Severity

The importance and severity of pollutant releases on receiving waters will vary depending on the context, and particularly with respect to the urban catchment, the discharge points and the receiving water.

3.5.2.1 Urban catchment

The nature, type and extent of the drained catchment will influence the characteristics and magnitude of the flow and contaminants discharged.

3.5.2.2 Discharge points

The number, location and type of discharge points will influence the size, duration, frequency and pollutant characteristics of the discharges.

3.5.2.3 Receiving water

The size and type of the receiving water, including its bed sediment, are important in terms of (House *et al*, 1993):

- Transportation of pollutants (e.g. flowing, semi-stagnant and stagnant waters)
- Dilution and/or assimilation of toxic pollutants
- Identification of key pollutants (e.g. biodegradable organic matter for rivers, and nutrients in lakes, coastal areas and the open sea),
- The extent, sensitivity and importance of the local aquatic life.

The release of a pollutant load into the flowing water of a river will transport downstream relatively quickly, interacting with the water column as it progresses, and exposing a significant length of water to contamination for a short period. On the contrary, a shock load discharged to stagnant water in a lake will disperse more slowly and will generally persist for a longer period. Three relevant timescales can therefore be identified:

- Short-term (acute),
- Medium-term (delayed),
- Long-term (chronic or cumulative).

Table 3.11 Leading sources of water quality impairment in the United States

Rivers	**Lakes**	**Estuaries**
Agriculture	Agriculture	Municipal point sources
Municipal point sources	Urban runoff/storm sewers	Urban runoff/storm sewers
Urban runoff/storm sewers	Hydrologic/habitat modification	Agriculture
Resource abstraction	Municipal point sources	Industrial point sources
Industrial point sources	On-site wastewater	Resource extraction

Intermittent discharges are particularly difficult to quantify and regulate because of their nature. Their acute (immediate) impact can only be measured during a spill event, and their chronic (long-term) effects are often difficult to isolate from background pollution. Design standards and performance criteria specifically tailored to intermittent discharges have therefore been developed (see below).

In industrialised countries, the most important pollutant sources tend to be those from the urban water system (urban, point) and agriculture (rural, non-point). This is borne out by Table 3.11, which shows these sources as the main cause of river, lake and estuarine water quality impairment in the USA. The relative rank will change from country to country, with urban sources perhaps overtaking rural ones in areas with weaker point control controls.

3.4.3 Impacts

Impacts on the receiving water can be divided into direct water quality effects (*DO* depletion, nutrient eutrophication, sediment deposition, toxic effects), public health (microbial) issues and aesthetic influences. These are summarised for various determinands and receiving water types in Table 3.12.

3.4.3.1 Dissolved oxygen

The effluents from urban wastewater systems (particularly untreated wastewater and stormwater) contain significant quantities of organic matter, in both particulate and soluble form, which is readily oxidised either biologically or chemically to stable, relatively inert, end products such as carbon dioxide, nitrates, sulphates, and water: an immediate oxygen demand. Decomposition of this organic matter consumes DO leading to reduced concentrations downstream of the pollutant discharge due to the increase in bacterial action. As the resulting oxygen deficit increases so does the rate of oxygen transfer from the atmosphere leading ultimately to a recovery in DO levels. So is formed the well-known oxygen sag curve.

Table 3.12 Qualitative assessment of receiving water impacts of urban discharges (House *et al*, 1993)

Receiving water	**Water quality**				**Public health**	**Aesthetics**	
	Dissolved oxygen	**Nutrients**	**Sediments**	**Toxics**	**Microbials**	**Clarity**	**Sanitary debris**
Streams							
Steep	–	–	–	x	xx	–	xx
Slack	x	–	x	x	xx	–	xx
Rivers							
Small	xx	–	x	x	xx	–	xx
Large	x	–	x	x	xx	x	xx
Estuaries							
Small	x	x	x	x	xx	x	xx
Large	–	–	x	–	xx	x	xx
Lakes							
Shallow	x	xx	x	x	xx	x	xx
Deep	x	x	x	x	xx	x	xx

xx, Probable; x, Possible; –, Unlikely

The DO concentration of a receiving water is a good indicator of its "health". Oxygen is required by all the higher forms of river life and, in the absence of toxic impurities, there is a close correlation between DO and biodiversity. Oxygen is an important limiting factor for fish survival, in particular (Table 3.13). The effect on the biota is a rapid increase in bacterial numbers and can include development of sewage fungus (periphyton) (Gray, 1985). Very low DO zones will lead to elimination of the normal clean water species and replacement with large population densities of tolerant species.

Table 3.13 Oxygen requirements of fish species (Adapted from Gray, 1999)

Characteristic species	Minimum DO concentration (mg/l)	Minimum Saturation (%)	Comment
Trout, bullhead	7–8	100	Fish require much oxygen
Perch, minnow	6–7	<100	Need more oxygen for active life
Roach, pike, chub	3	60–80	Can live for long periods at this level
Carp, tench, bream	< 1	30–40	Can live for short periods at this level

3.4.3.2 Sediments

Sediments and other solids are a ubiquitous component of all urban discharges. These will either be maintained in suspension or, depending on the prevailing velocity fields, be deposited some distance downstream of their discharge.

Inert sediments tend to accumulate on the river bed close to the point of discharge and gradually extend further downstream. The general effect of this is to reduce the diversity of biotic habitats leading to reductions in both species diversity and abundance. There is a selective increase in a few species that can survive the stress imposed by the sediments (Figure 3.16).

Similarly, degradable sediments settle out but, in addition, decompose anaerobically inducing a delayed sediment oxygen demand (*SOD*). Typical undisturbed *SOD* levels are 0.15–2.75 $g/m^2.d$, elevated to 240–1500 $g/m^2.d$ during storm flow conditions (House *et al*, 1993). As the rate of decomposition is slow, the area affected can be extensive. Sensitive benthic organisms are rapidly eliminated but are soon replaced by high population densities of a few species tolerant of the low oxygen silty conditions (Figure 3.16).

The importance and relative magnitude of the delayed oxygen demand compared with the immediate demand depends on the specific circumstances of the discharge and the receiving water. Hvitved-Jacobsen (1986) indicates that in larger rivers immediate consumption dominates, whereas small rivers with flows $< 0.5\ m^3/s$ often show depletion due to delayed consumption.

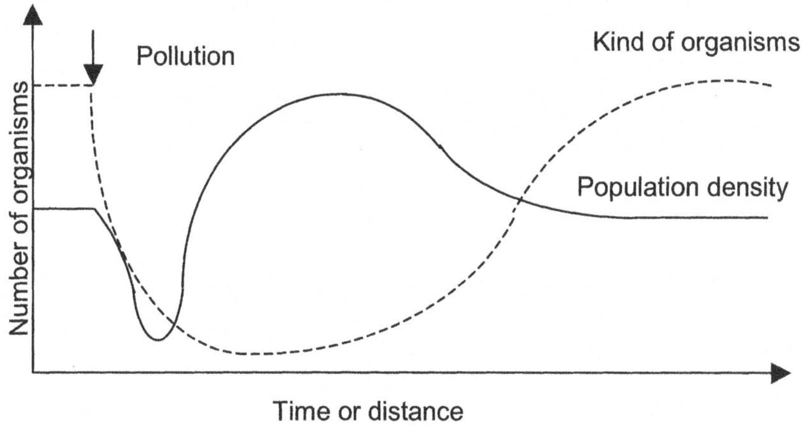

Figure 3.16 Generalised effect of organic sediments on lotic freshwater community structure (UNESCO, 1978)

3.4.3.3 Nutrients

The main sources of nutrients such as nitrogen or phosphorus are agricultural fertilisers, landfill leachates and continuous wastewater discharges. Intermittent discharges are usually a relatively small constituent of the total nutrient load. Sodium tripolyphosphate builders in linear alkylate sulphonate (LAS) detergents are the major source of phosphate in wastewater.

If large quantities of nutrients are discharged into receiving waters, excessive growth of aquatic weeds and algae may occur. The resulting eutrophication can lead to;

- light penetration reduction
- oxygen depletion due to dying algae
- anaerobic conditions in bottom sediments
- aesthetic problems (discoloration, odours)
- toxicity problems due to algal blooms.

Eutrophication is a long-term problem usually associated with shallow stagnant waters such as lakes, estuaries and the coastal zone, but rivers may also be affected. Phosphate is usually the limiting nutrient in freshwater, whereas nitrate is limiting in seawater. Effects on the biota include reduced species diversity, different dominant species emerging and elimination of salmonids.

3.4.3.4 Toxics

Very many toxic heavy metals and synthetic organic and inorganic chemicals can be found in wastewater (both treated and untreated) and stormwater. Toxic substances generally decrease in concentration downstream from their discharge point due to the processes of precipitation and adsorption. Of most concern are those non-biodegradable compounds that are toxic, carcinogenic or mutagenic at very low levels.

Important metal species include arsenic, cadmium, chromium (VI), copper, cyanide, iron, lead, mercury, nickel and zinc. These can exist in particulate, colloidal, and dissolved (labile) phases depending on the prevailing redox and pH conditions (Ellis, 1985). In rivers, metals are predominantly in the particulate phase and will eventually settle out of suspension. This is important because the environmental mobility and bioavailability (and hence toxicity) of metals is highly related to their concentration in solution. Many (particularly the more soluble forms of zinc and copper) are known to have toxic effects on aquatic life and can inhibit biological processes at the WWTP.

Most synthetic organic compounds, such as herbicides and pesticides, become absorbed onto mineral particles. Such compounds in solution can be toxic to a variety of aquatic life at very low concentrations. Some of the more toxic varieties (e.g. chlorinated organics, DDT and PCBs) are no longer used but their residues can still be found in the environment.

The effect of toxic compounds on the receiving water biota is to rapidly reduce species diversity and abundance leading to complete elimination at excessive concentrations. Downstream recovery is generally slower than the loss rate, with tolerant species returning, often at population densities greater than initial levels, due to lack of competition (Figure 3.17).

3.4.3.5 Microbials

Wastewater, in particular, will be contaminated with pathogenic micro-organisms; in proportion to the health of the local community. Practice differs in industrialised countries with regard to disinfecting treated effluents. In most of Europe, wastewater is not routinely disinfected unless discharged to a recognised bathing water. Conventional treatment removes only 80–90% of bacterial pathogens. In the USA, chlorination is widely practised prior to discharge to the receiving water. This will increase removal to > 99.99% (Geldreich, 1991). Discharges from CSOs and SWOs are not normally disinfected. Relatively high concentrations of pathogens may thus be expected from both, and bacterial standards for recreational activity can be violated during even modest storm events.

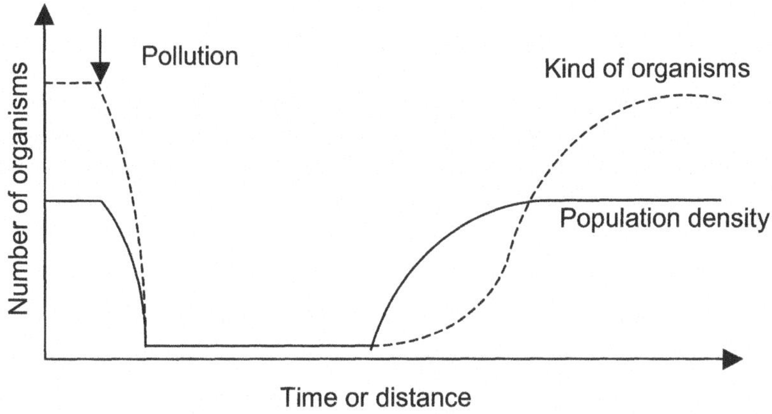

Figure 3.17 Generalised effect of a toxic discharge on lotic freshwater community structure (UNESCO, 1978)

The risk to public health depends on the degree of potential human exposure, and this will be greatest if the receiving water is used for contact recreational purposes. Swimmers are therefore at greatest risk. All pathogens are able to survive for at least short periods in natural waters. Cooler temperatures and the presence of organic matter generally extend this period. Lack of detailed epidemiological studies means actual health risks have not been quantified, and this is an area requiring further study.

3.4.3.6 Aesthetics

In addition to chemical and biological impacts, the appearance of the receiving water is also important. Oddly, the public tends to misperceive rivers of high chemical and biological quality as polluted. Public reaction is based more on the clarity, colour and smell of the water. Solids of obvious sanitary origin in on near to receiving waters, possibly emanating form poorly operating WWTPs or CSOs, are considered to be particularly offensive.

3.4.4 Regulation

3.4.4.1 Basic approaches

There are four basic approaches to the regulation of pollution control: product policy, process standards, uniform emission standards and water quality standards (Mostert *et al.*, 1999). The first of these involves explicit forbidding or allowing the use of specific products, with quality standards for products being set. The second consists of prescribing standards for the environmental performance of production specific processes. In the third approach, identical standards are applied to all emissions in a particular area (usually country). Emission standards are also used in the final approach, but these are usually set in a permit or consent to reflect the quality standard of the receiving water. The advantages and disadvantages of the various approaches are summarised in Table 3.14.

Table 3.14 Regulatory approaches to pollution control (Mostert *et al.*, 1999)

Approach	Advantages	Disadvantages
Product policy	Administrative costs and demands and enforcement can be low (depending on product) Appropriate for diffuse pollution (e.g. pesticides)	Production process often more important than product
Process standards	Standards concerning environmental performance are steps towards ultimate aim of closing substance cycles	Administrative costs and demands on capacity of managers are high Prescribing specific processes may hinder the application of cleaner technologies
Uniform emission standards	Solution for reducing emissions of dangerous substances quickly to zero	Too strict in some cases, too lenient in others May promote end-of-pipe solutions
Water quality standards	Most efficient: pollution reduction efforts concentrated where the need is largest	Difficult to link emissions to water quality: many data needed and good models Administrative costs and demands on capacity of managers are high Not appropriate for pollutants accumulating in the environment; difficult to deal with synergistic effects Unequal treatment of polluters in different basins and between different parts of the basin (upstream-downstream, tributaries) Difficult to ensure minimum level of pollution

The choice of approach depends on a number of factors such as the source of pollution, the type of pollution, the urgency of the problem and the capacity of local administrative structures. No one approach is universally applicable, and indeed, more than one approach may be used simultaneously (Rees & Zabel, 1998).

3.4.4.2 Evolution of European policy

Europe is a good example of the different types of regulatory approach used in water pollution control over the last 25 years. In the early period (1977–1986), several water-related directives were enacted that can be broadly characterised into two types (Somsen, 1990): water-use directives (public health-orientated) and water pollutant directives (orientated to harmonisation of pollution control efforts in the Community). Directives included standards for the quality of waters intended for drinking (at source CEC (1975) and after treatment (CEC, 1980a)), bathing (CEC, 1976a) and for fish and shellfish harvesting (CEC, 1978). These were based on "water quality objectives" with two standards specified: "imperative" (that should be respected in all cases) and/or "guidance" (more stringent levels that Member States should "endeavour" to respect). In contrast, water pollutant directives regulated the permissible levels of discharges of particular pollutants. The two key directives were the emission of dangerous substances to surface and ground water bodies (CEC 1976b; 1980b). Two lists of harmful substances were set: List I (black list) substances for which emission limit or quality standards were set at EU level, and List II (grey list) substances for which Member States were to produce integrated reduction programmes (see Table 3.15).

Following the Maastricht Treaty, a second wave of directives based on the philosophy of "prevention of pollution at source" followed, to account for two particular concerns (Butler *et al.*, 2000): pollution from urban wastewater (CEC, 1991a) and pollution from agricultural runoff (CEC, 1991b). Although the former directive has been positively taken up, although at a high cost, implementation of the latter directive, focused on establishing programmes on "best agricultural practice programs", has been extremely disappointing (CEC, 1991c). This has led to the situation of a general decline in the state of Europe's waters, in many cases because of non-point sources of pollution (such as agriculture, road transport, catchment and hydraulic interventions, etc.), despite considerable reductions in pollution from urban point sources (EEA, 1995).

Table 3.15 List 1 and 2 substances covered by the EU Dangerous Substances Directive (76/464/EEC)

List 1 substances	List 2 substances
Organohalogen compounds and substances that may form such compounds in the aqueous environment	The following metalloids/metals and their compounds: zinc, copper, nickel, chromium, lead, selenium, arsenic, antimony, molybdenum, titanium, tin, barium, beryllium, boron, uranium, vanadium, cobalt, thallium, tellurium, silver
Organophosphorus compounds	Biocides and their derivatives not appearing in List 1
Organotin compounds	Substances that have a deleterious effect on the taste and/or smell of products for human consumption derived form the aquatic environment and compounds liable to give rise to such substances in water
Substances, the carcinogenic activity of which is exhibited in or by the aqueous environment	Toxic or persistent organic compounds of silicon and substances that may give rise to such compounds in water, excluding those that are biologically harmless or are rapidly converted in water to harmless substances
Mercury and its compounds	Inorganic compounds of phosphorus and elemental phosphorus
Cadmium and its compounds	Non-persistent mineral oils and hydrocarbons of petroleum origin
Persistent mineral oils and hydrocarbons of petroleum	Cyanides, fluorides
Persistent synthetic substances	Certain substances that may have an adverse effect on the oxygen balance, particularly ammonia and nitrites

In the early 1990s calls to intensify environmental and public health protection were coming at the same time that high costs of European water regulation were being attacked. In addition, pressure for deregulation and repatriation of water legislation based on the principle of "subsidiarity" was increasing. A long political process has integrated these challenges in the new Water Framework Directive (CEC, 1997), discussed further below.

3.4.4.3 Urban Wastewater Treatment Directive

The most important legislation concerning discharges from the urban water system is currently the Urban Waste Water Treatment Directive (CEC, 1991). This sets out detailed standards for WWTP releases, based on the uniform emission standard approach (see Table 3.16). Absolute minimum concentrations or minimum percentage removals are acceptable. Receiving waters considered to be vulnerable to eutrophication are classified as sensitive, and nitrogen and phosphorus standards have also been established (Table 3.16). Compliance dates for various aspects of the directive range from 31 December 1998 to 31 December 2005.

Table 3.16 Emission limits for wastewater treatment plant discharges under the UWWT Directive (CEC, 1991)

Determinand	**Minimum concentration (mg/l)**	**Minimum reduction (%)**
BOD_5	25	70–90
COD	125	75
Suspended solids	35	90
Total nitrogen*		
10,000–100,000 pe	10	70–80
> 100,000 pe	15	70–80
Total phosphorus*		
10,000–100,000 pe	1	80
> 100,000 pe	2	80

* Only relevant to discharges from population equivalents > 10,000 to sensitive waters

The Directive is less prescriptive for discharges from collection systems. It requires Member States to ensure all urban areas with a population equivalent of 2000 or more are provided with collection systems. Further, the design, construction and maintenance of collecting systems needs be undertaken in accordance with best technical knowledge not entailing excessive costs with respect to:

- volume and characteristics of urban wastewater
- prevention of leaks
- limitation of pollution of receiving waters due to overflows.

The Directive acknowledges the difficulties that arise during periods of unusually heavy rainfall, and allows Member States to decide on their own measures to limit pollution from overflows. These could be based on one of the following:

- dilution rates
- collection system capacity in relation to dry weather flow
- a certain number of overflows per year.

3.4.4.4 Bathing Water Directive

The Bathing Water Quality Directive (CEC, 1976) regulates the bacteriological (and physico-chemical) quality of inland and coastal bathing waters designated by Member States as bathing areas (see Table 3.17). The most relevant measure of wastewater contamination is the faecal coliform standard set at 2000 FC/100 ml.

Table 3.17 EU microbiological bathing water quality standards (CEC, 1976)

Determinand	G (90percentile)	I (95percentile)
Total coliforms/100ml	500	10,000
Faecal coliforms/100ml	100	2,000
Faecal streptococci/100ml	100	—
Salmonella/l	—	0
Enteroviruses PFU/10 l	—	0

In the UK alone there are approximately 500 designated bathing waters. Failure to comply with the standards comes about by a combination of continuous wastewater discharges and intermittent CSO discharges. Hence, control of discharges becomes important where they discharge directly to bathing waters or to estuarine or coastal waters that lead to bathing waters.

3.4.4.5 Water Framework Directive

The Water Framework Directive (WFD) incorporates the main requirements for water management into one single system based on river basins (CEC, 2000). New or re-organised river basin district authorities are to be formed, each with a management plan aimed at achieving the goals of the directive. The key guiding goal of the directive is to achieve "good status" of ground and surface waters, "good" meaning that water meets the standards established in existing water directives and in addition, the new ecological quality standards. A surface water is defined as of good ecological quality if there is only slight departure from the biological community that would be expected in conditions of minimal anthropogenic impact. "Good chemical status" is defined in terms of compliance with all the quality standards established for chemical substances at the European level; a new mechanism for controlling the discharge of dangerous substances, is provided in the Directive to replace the "deadlocked" discharges directive. A combined approach to setting standards is to be used where both emission limit values and river quality standards are to be legally binding.

The WFD is designed to provide an "umbrella" to existing directives as well as incorporating new standards. River basin authorities will designate specific protection zones within their area (i.e. bathing, drinking water or protected natural areas) where the standards in the respective existing EU directives apply, but zones with higher objectives may also be established where more stringent standards must be met. Good ecological and chemical status is the minimum for all waters. The urban wastewater directive and the nitrates directive, are considered as tools to achieve the objectives of the river basin management plans and will be retained. The directives on the quality of surface waters intended for drinking and the fish and shellfish directives will be repealed, as all surface waters will now need to meet "good status".

The directive also sets new rules for groundwater. All direct discharges to ground water are prohibited and a requirement is introduced to monitor ground water bodies so as to detect changes in composition due to non-point pollution and take measures to reverse them.

The initial period of implementation of the directive is 15 years (9 years to prepare plans and 6 years more to achieve specific targets) plus an additional deferment up to two periods of 6 years (Butler *et al.*, 2000).

3.4.5 Intermittent standards

As mentioned above, standards for intermittent discharges are more difficult to derive, and are not specified in the European Water directives. In the UK, intermittent discharges standards have been derived that relate to the receiving water's ability to assimilate pollutants without detriment to legitimate uses of the water. The 'uses' most influenced by intermittent discharges are (NRA, 1993):

- aquatic life
- bathing
- amenity.

3.4.5.1 Aquatic life

Aquatic life intermittent standards are given in the UK *UPM manual* (FWR, 1998). They were derived from laboratory ecotoxicological studies of relevant aquatic species to establish the short-term lethal pollutant concentrations resulting in 50% mortality (LC50). The results enabled relationships to be established between:

- pollutant concentration
- duration of the discharge
- return period of a discharge in which that concentration is exceeded.

Relationships for dissolved oxygen and ammonia are given in Table 3.18.

3.4.5.2 Bathing water

Bathing water intermittent standards are based on limiting the concentration of faecal and total coliforms to 2,000 and 10,000 no/100ml respectively for at least 98.2% of the May to September bathing season, as judged over an average period of 10 years. This requirement has been converted to a surrogate emission standard that limits discharges to no more than 3 per bathing season (NRA, 1993).

Table 3.18 Intermittent standards for dissolved oxygen and ammonia concentration/duration thresholds for sustaining cyprinid fisheries (FWR, 1998)

Return period (months)	DO concentration (mg/l)*		
	1 h	6 h	24 h
1	4.0	5.0	5.5
3	3.5	4.5	5.0
12	3.0	4.0	4.5
	NH_3–N (mg/l)**		
1	0.150	0.075	0.030
3	0.225	0.125	0.050
12	0.250	0.150	0.065

* Applicable when NH_3–N < 0.04 mg/l ** Applicable when DO > 5 mg/l, pH > 7 and T > 5 °C

Table 3.19 Amenity area emission standards (NRA, 1993)

Amenity category	Expected frequency of discharges per year	Standard
High	> 1	6 mm solids separation
	≤ 1	10 mm solids separation
Medium	> 30	6 mm solids separation
	≤ 30	10 mm solids separation
Low	—	Good engineering design

3.4.5.3 Amenity

Receiving waters are classified, in terms of their amenity value, by the amount of public contact, as follows:

- **High**: water used for bathing and water-contact sports, watercourses through parks and picnic sites, shellfish waters
- **Moderate**: water used for boating, watercourses near popular footpaths or through housing developments or town centres
- **Low**: limited public interest.

Amenity intermittent standards are emission based or rely on the provision of good engineering design. They are set in relation to the receiving water amenity-use category and the expected discharge frequency as shown in Table 3.19.

3.5 URBAN WATERBODIES

Most urban areas started their development either near rivers or on the coast. In early civilisations, streams were cleaner, and people depended on them as a source of drinking water and food. However, as civilisations 'developed' streams began to serve more and more as recipients of liquid and solid waste. Today most streams in urban areas have been piped or culverted and been integrated into the sewer network. Lakes and ponds in urban areas initially built to be the places for human rest, recreation and joy, are today mostly polluted, eutrophied, and in need of systematic rehabilitation to return them to their original role of urban amenities. With today's sustainability-conscious concepts in stormwater management, there is renewed hope for the 'reinvention' of urban water bodies, since they are increasingly being acknowledged for what they are, or what they should be: important parts of the urban landscape. This section provides a brief description of some of the problems associated with urban streams, lakes and ponds and provides information on how they could be better integrated into the urban environment of the future.

3.5.1. Streams and rivers

Many watercourses have been degraded as a result of city development. River channels have been confined to narrow corridors, the channels themselves have been canalised and concrete, or other man-made materials, has often replaced the bed of the river. Many urban streams have been converted into closed-conduits sewers. Many now receive both stormwater runoff and raw or dilute sewage from the surrounding area. Additionally, in urban areas in developing and some transitional countries, streams serve as 'convenient' recipients of solid waste.

The historical development of such urban watercourses is presented in Plate 3.2. Starting with early (pre-urban) civilisations, degradation continued through the pre-industrial and industrial period; once steams became a nuisance, most were closed and sewer pipes connected to them. The streams that 'survived' closure were turned into channels which:

- have little or no aesthetic or amenity value,
- support a limited range of eco-systems,
- suffer from limited carrying capacity of sediments,
- develop favourable conditions for the spread of pollution.

Finally we highlight the recent tendency of restoring urban watercourses back to their natural state (or at least as close as possible, given their historical legacy). This process, dubbed "daylighting" (see for example Pinkham, 2000), is beginning to take place as a part of the overall improvement of urban amenities in run-down urban areas in a bid to re-establish an aesthetically-pleasing environment in conjunction with improvements to the natural environment.

One such approach is the re-naturalisation of urban river corridors. At present it is difficult to obtain detailed information on urban river rehabilitation methodologies or schemes. There is a network of River Restoration Centres developing within countries in Europe and also a European Centre for River Restoration. A few rehabilitation schemes have been carried out across Europe and in the USA (Pinkham, 2000). The state of the art is that there is no pan-European framework to assess the suitability of urban rivers for enhancement or to describe the capabilities of relevant rehabilitation techniques should. Additionally, there are no *tools* for assessing the potential for effective river restoration and associated improvements to environmental quality. At the moment, ad-hoc assessments are made in each case. The assessment procedure and the parameters that are considered vary in each case. The methods that are used are normally derived at the time, resulting in repetition and inefficiencies. This ad-hoc approach means there are inconsistencies in which schemes go forward and which are rejected.

Development of the following is still required for watercourses in the urban environment:

- Methods and tools to evaluate the restoration potential of urban streams and rivers as water features or other amenities.
- Innovative methods for retrofitting and rehabilitation of existing systems
- Procedures to integrate economic, social and environmental aspects within the planning and development process
- Approaches to allow greater public involvement and more transparent decision-making.
- Easy to understand tools to aid communication with the general public.

3.5.2. Lakes and ponds

When initially designed as amenity elements in parks and other public places, urban lakes and ponds served the purpose intended, providing a pleasant environment for the public, drawing people to the area, happy to enjoy and appreciate the water feature. However, at the time they were planned, designed and constructed, water quality was not an important issue. So, depending on the means of feeding and maintaining through-flow (amongst other things), the amenity value of these artificial water bodies may be preserved or have become degraded, silted, eutrophied or abandoned.

Even the significant advances in the treatment of wastewater from metropolitan areas (as briefly described above) has had limited effect when the problem was not tackled at catchment level. For example, although over the last two decades wastewater discharges of nitrogen and phosphorus have been greatly reduced in Europe, the fact of eutrophication in urban waterbodies remains.

Further, many urban lakes have significantly elevated levels of heavy metals and hydrocarbons in basal sediments. Indeed, even sediments in park and garden lakes have been found to be contaminated, especially when located near to highways or other pollutant sources. These high pollutant concentrations can lead to secondary pollution of the overlying waterbody; it is increasingly understood that urban lakes are systems with 'memory'.

The study of these systems has been relatively neglected. However, in a recent project, Stoianov *et al.* (2000) attempted to develop a decision-support framework for operational management of water quality in the Serpentine Lake in Hyde Park, London. Here GIS-supported mapping and modelling was used with appropriate indicators to quantify impacts on the water body and its ecosystems.

3.5.3. Conclusions

The condition assessment and potential for restoration or all types of urban water body have been relatively neglected until recent times. Thus we find ourselves at a point where ad-hoc development decisions are being made (and remade) with little reference material, methodologies and tools being in the public domain. Of course, the basic understanding of urban and natural systems and relevant modelling tools are in the literature (e.g. Butler & Maksimović, 1999b; Chapra, 1997), but these are not yet specific enough for routine use in this application. Further work is definitely required in this area.

On a brighter note, increasing introduction of sustainable urban drainage practices that rely on application of source, site and regional control measures bodes well for urban waterbodies because of the significant role they have to play. New ones will be built and old ones retrofitted and rehabilitated. However, wise management remains a challenge.

3.6 TOOLS

3.6.1 Informatic support

The entities involved in planning, development, and running modern urban water systems such as water companies, municipalities and other interest groups are increasingly relying on strong informatic support in order to be reliable, efficient, financially viable and transparent to water users which, in the end, cover their running and investment costs. Some of the major roles for informatic technology are access to external (global and regional) data, acquisition and processing of in-house 'technical data' (including surface and underground assets, corporate business data (Ray, 1998)), analysis, planning, design, operational and business management, training and communication. Basic features of these groups are presented in the Table 3.20. It should be mentioned that, traditionally, these tools have been applied separately for various single tasks. However, integrated management requires that special attention is paid to the following:

- Data quality, reliability and quantification of uncertainty,
- Interactions of individual components of urban water systems
- The corporate concept

Mastering the implementation of these concepts enables water utilities to improve the quality of their day-to-day business operation. In addition, it should enable a change of concept with respect to managing available resources and improving overall efficiency in the face of unfavourable natural episodes such as floods, droughts and socio-economical instabilities.

As an example of this approach used in practice, Anglian Water, a private water company in the UK, operates a corporate GIS-based concept (Ray, 1998). This attempts to maximise benefits by integration of all business systems within the company. It is based on an 'asset driven' (customer focused) approach, rather than a 'process driven' system, in which high value is given to data; indeed, data is becoming one of the most valuable assets of the company. The data structure in this system is presented in Fig. 3.18.

Table 3.20 Groups of informatic tools applied in urban water management

Activity	Informatic product	Reliability, Uncertainty, Sensitivity & Remarks
Access to external sources of data, knowledge, Internet based data acquisition, international communication	General purpose hardware and software. Corporate "concept"	Easy start, low initial cost, training needed, reliability dominated by the quality of internal communication
Acquisition and processing of local data (climate, hydrology, assets, water availability, demand, losses, interactions)	Reliable ("appropriate") sensors, customised data acquisition systems, customised - target oriented data processing software. Data quality assurance. Emerging technologies	Data Reliability and quality critical, Internet based data transfer could reduce the cost of data transfer. Emerging sensor and communication technologies could significantly reduce costs
Analysis, planning, design, operational and business management, development of interfaces for analysis of interactions of subsystems and interactions with the environment and with society. Analysis of sustainability.	Software packages and GIS based or GIS centred integration for: catchment based resources (water quantity and quality) analysis, including flood & droughts. Treatment processes modelling. Network analysis and optimisation Rehabilitation scheduling.	Reliability of the results strongly dependent on data quality. Data exchange i.e. share of a common system (corporate GIS) can reduce total cost. Integrated modelling requires development of new interfaces GIS centred approach.
Training of: top level decision makers, managers and planers, specialists, operators, administrative staff, users.	Specialised educational version of software training modules. Distant learning modules.	Corporate approach needed. Regular check and retraining needed. Feedback analysis.

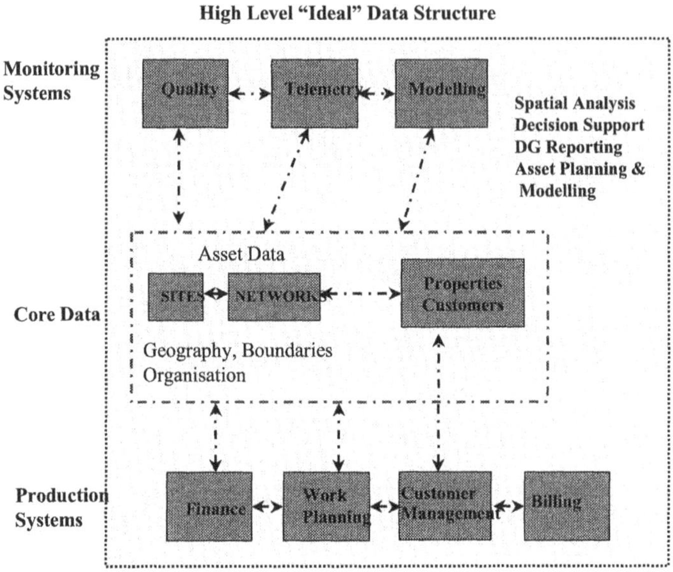

Figure 3.18 Data structure in an asset driven system (Ray, 1988)

3.6.2 Data

Providing that such a system is made operational, it is possible to carry out technical analyses such as distribution or sewer system network analysis, network optimisation, rehabilitation planning and so on, based on common sources of data. However, the quality of the system is strongly dependent on the reliability of data. Verification of data quality involves also analysis of possible sources of errors and quantification of uncertainty. Sources of errors include sensors, data transmission and acquisition system, and data processing. As an example, Stoianov *et al.* (2000) develop an approach to fault detection and error quantification as presented in Fig. 3.18. It is argued here that the systems of the future should always include methods for securing high quality data and quantifying uncertainty, so that the uncertainty in the decisions made or the riskiness of the solution adopted can be quantified.

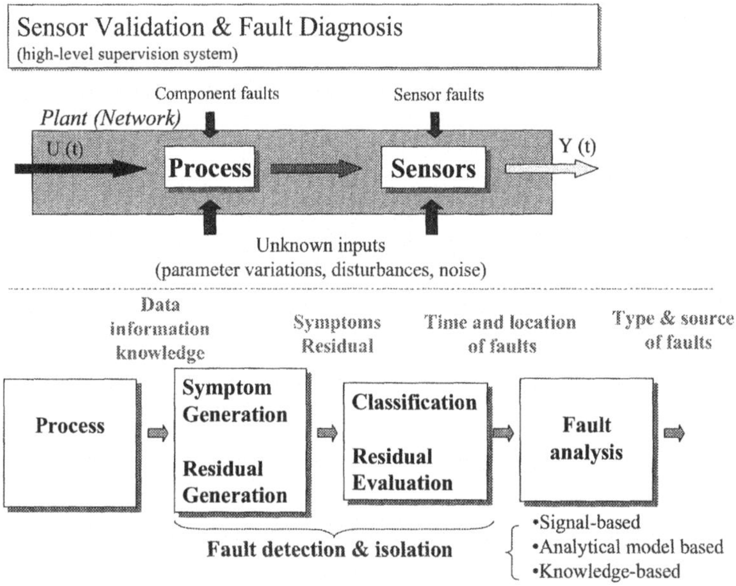

Figure 3.19 System for sensor's fault detection and quantification of uncertainty in water supply systems (Stoianov *et al.*, 2000)

3.7 CONCLUSIONS

This chapter has presented the main interactions of the urban water sub-systems with the environment and their implications. Attention has been focussed on the many processes involved and the sometimes subtle, but mostly blatant, interplay of infrastructure and environment. We have seen the major driving force behind infrastructure provision to be human development, which requires larger and larger volumes of clean water, generates ever-increasing quantities of wastewater and covers the earth surface with impervious urban sprawl. If we are to cope with a scenario of even greater population growth, it is important to acknowledge and exploit the nature of our urban water systems in a more integrated fashion. We can only hope to achieve this if these interactions are understood *and* included in the design, analysis and operational management of systems in practice. We offer this chapter as one small step in the right direction.

3.8 REFERENCES

Ainola, L., T. Koppel, A. Vassiljev (2000) Complex Approach to the Water Network Model Calibration and the Leakage Distribution, *Hydraulic Software VIII*, WIT PRESS: Southampton, Boston, 91–100.

Ainger, C.M., Armstrong, R.J. and Butler, D. (1997) *Dry Weather Flow in Sewers,* Report R177, CIRIA, London.

Anderson, G., Bishop, B., Misstear, B. and White, M. (1996) *Reliability of Sewers in Environmentally Sensitive Areas*, Report PR44, CIRIA, London.

Balmforth, D.J., Saul, A.J., and Clifforde, I.T. (1994) *Guide to the design of combined sewer overflow structures*, Report FR 0488, Foundation for Water Research.

Barrios J. E. (2000) How to Deal with Wastewater in Developing Countries: Some Experiences from Mexico, *UNEP–IETC Workshop of Sustainable Wastewater and Stormwater Management*, Rio de Janeiro March, 184-201.

Beck, M.B., Chen, J., Saul, A.J. & Butler, D. (1994) Urban drainage in the 21st century: assessment of new technology on the basis of global material flows. *Water Science and Technology*, **30** (2), 1–12.

British Ecological Society (1990) *River Water Quality*, Field Studies Council, Shrewsbury.

Butler, D. & Davies, J.W. (2000) *Urban Drainage*, E. & F.N. Spon, London.

Butler, D., Kallis, G. & Mills, K. (2000) Implications of the New EU Water Framework Directive, *CIWEM Millennium Conference*, Leeds, UK.

Butler, D. & Maksimović, Č. (1999a) Urban water management – Challenges for the third millennium, *Progress in Environmental Science*, **1** (3), 213–235.

Butler D. & Maksimović, Č. (1999b) Developments in Urban Drainage Modelling, *Water Science and Technology*, **39** (9).

Butler, D. & Parkinson, J. (1997) Towards sustainable urban drainage. *Water Science and Technology*, **35** (9), 53–63.

Butler, D. & Payne, J.A. (1995) Septic tanks: problems and practice. *Building and Environment*, **30** (3), 419–425.

Butler, D. & Smith, S.R. (In press) Wastewater Treatment - Septic Tank Systems. In *Encyclopaedia of Environmental Microbiology* (Ed G. Britton), John Wiley & Sons, New York.

BS EN 752-4:1998 *Drains and Sewer Systems Outside Build*ings. Part 4: Hydraulic Design and Environmental Considerations. European Standard.

CEC (1975). *Directive concerning the quality of surface waters intended for the abstraction of drinking water* (75/440/EEC).

CEC (1976) *Directive concerning Quality of Bathing Wa*ter, 76/160/EEC.

CEC (1976b) *Directive concerning pollution caused by dangerous substances discharged into the aquatic envir*onment (76/464/EEC).

CEC (1978) *Directive concerning the quality of fish and shellfish waters* (78/659/EEC and 79/923/EEC).

CEC (1980a) *Directive concerning quality of water for human consumption* (80/778/EEC).

CEC (1980b) *Directive concerning the protection of groundwater against pollution caused by certain dangerous substances* (80/778/EEC).

CEC (1986) *Directive on the protection of the environment, and in particular of soil, when sewage sludge is used in agricul*ture (86/278/EEC).

CEC (1991a) *Directive concerning Urban Waste Water Treatment*, (91/271/EEC).

CEC (1991b) *Directive concerning protection of water against pollution by nitrates from agriculture* (91/276/EEC).

CEC (1991c) *Implementation of Council Directive 91/271/EEC of 21 May 1991 concerning Urban Waste Water Treatment*, CR-16-98-198-EN-C, Luxembourg, 1999.

CEC (2000) *Directive establishing a framework for Community action in the field of water policy* (2000/60/EC).

Chapra, S. (1997) *Surface Water Quality Modeling*, McGraw-Hill.

Chilton J. Editor (1997) Groundwater in the Urban Environment, *Proc. of the XXVII IAHR Congress on Groundwater in the Urban Environ*ment, Nottingham, UK, September.

Craun, G.F. (1981) *J. Ameican. Water Works Association* , **73**, 360–369.

Davis, R.D. (19960 The impact of EU and UK environmental pressures on the future of sludge treatment and disposal, J. *Chartered Institution of Water & Environmental Engineering*, **10**, Feb., 65–69.

Dixon, A., Butler, D. & Fewkes, A. (1999) Water saving potential of domestic water re-use systems using grey water and rainwater in combination, *Water Science & Technology*, **39** (5), 25–32.

Djujić, A., Č. Maksimović, M. Ivetić, D. Milovanović (2000) Lijevče Polje Aquifer under the Threat of Environmental Degradation, *Proc.CNG, Italian National Council of Geologists*, Rome, August 2000.

DoE (1996) *Code of Practice for Agricultural Use of Sewage Sludge, Department of the Environment*, London.

Ellis, J.B. (Ed.) (1985) *Urban Drainage and Receiving Water Impacts*, Pergamon Press, Oxford.

Ellis, J.B. (1986) Pollutional aspects of urban runoff. In: *Urban Runoff Pollution* (Eds. Torno, H.C, Marsalek, J. and Desbordes, M.), NATO ASI Series . Environment, , Springer Verlag, Berlin, 1-38.

Engelhardt, M.O., P. J. Skipworth, D.A. Savic, A.J. Saul, and G.A. Walters, (2000), Rehabilitation Strategies for Water Distribution Networks: A Literature Review with a UK Perspective, *Urban Water*, **2** (2), 153–170.

European Environment Agency (1995) *Europe's Environment: the Dobris Assessment*, Copenhagen.

Fiddes, D. and Simmonds, N. (1981) Infiltration – do we have to live with it? *The Public Health Engineer*, **9** (1), 11–13.

Foster, S., A. Lawerence, B. Morris (1998) *Ground Water in Urban Development*, World Bank Technological Paper No. 390.

FWR (1998) *Urban Pollution Management Manual*, 2nd Edn., Foundation for Water Research, FR/CL0009.

Geiger W. F., (1998) Principles of Integrated Water Management for the Revival of Old Industrial Areas, In *Environmentally Devastated Areas in River Basins in Eastern Europe*, Edited by A.G. Buekens and V. V. Dragalov, ASI Series, **45**, Springer, 57–103.

Geldreich, E.E. (1991) Microbial water quality concerns for water supply use, *Environmental Toxicology and Water Quality*, **6**, 209–223.

Gray, N.F. (1985) Heterotrophic slimes in flowing waters, *Biological Reviews of the Cambridge Philosophical Society*, **60**, 499–548.

Gray, N.F. (1999*) Water technology. An Introduction for Scientists and Engineers.* Arnold, London.

Green, M., Shelef, G. and Messing, A. (1985*)* Using the sewerage system main conduits for biological treatment. *Water Research*, **19** (8), 1023–1028.

Harrison, P. & Pearce, F. (2001) *AAAS Atlas of Population and Environment*, Univ California Press, 216 pp.

Herrmann, T. and Schmida, U. (1999) Rainwater utilisation in Germany: efficiency, dimensioning, hydraulic and environmental aspects, *Urban Water*, **1**, 307–316.

Ho, B.S.W. & Tam, T.-Y. (1998) Giardia and Cryptosporidium in sewage-contaminated river waters, *Water Research*, **32**, 2860–2864.

House, M.A., Ellis, J.B., Herricks, E.E., Hvitved-Jacobsen, T., Seager, J., Lijklema, L., Aalderink, H. & Clifforde, I.T. (1993). Urban drainage – impacts on receiving water quality, *Water Science and Technology*, **27** (12), 117–158.

House of Commons Environment, Transport and Regional Affairs Committee (1998) Second Report: *Sewage Treatment and Disposal*. HC266-I. The Stationery Office, London.

Hvitved-Jacobsen, T. (1986) Conventional pollutant impacts on receiving waters. In *Urban Runoff Pollution,* (Ed., Torno, H.C., Marsalek, J. & Desbordes, M., NATO ASI Series, ,Springer-Verlag, Berlin, 345–378.

Inman, J. (1975) Civil engineering aspects of sewage treatment works de*sign*, *Proceedings of the Institution of Civil Engineers*, Part 1, **58**, May, 195–204, discussion, 669–672.

Jennings-Wrey J.(2000) *Modelling turbidity in Water Distribution Networks*, MSc thesis, Department of Civil and Environmental Engineering, Imperial College, London.

Larsen, T.A. & Guyer, W. (1996*)* Separate management of anthropogenic nutrient solutions (human urine). *Water Science and Technology*, **34** (3–4), 87–94.

Larsen, T.A. & Guyer, W. (1997) Sustainable urban water management – technological implications, *EAWAG News*, 44E, Dec., 12–14. Check WST paper – special 35/9

Maksimović, Č., F. Calomino, J. Snoxell, (editors), (1996) *Water Supply Systems – New Technologies*, ASI Series, Springer,

Maksimović, Č., N. Carmi (1999) GIS supported analysis of pressure dependent vulnerability of distribution networks to leakage, In *Water Industry Systems: Modelling and Optimization,* (Ed D.Savic & G.Walter), Vol. 1, Research Studies Press, Baldock, 1999, pp.85–96

Maksimović, Č., G. Ho (2001) *Sustainable Wastewater and Stormwater Management*, UNEP – IETC, Osaka, Japan, Training Material.

Mostert, E., van Beek, E., Bouman, N.W.M., Hey, E., Savenije, H.H.G. & Thissen, W.A.H. (1999) River basin management and planning. In *River Basin Management, Proc. Int. Workshop* (Ed. E. Mostert), The Hague, October, 1999. UNESCO IHP-V Technical Documents in Hydrology No. 31, 24–55.

National Research Council NRC (2000) *Watershed Management for Potable Water Supply,* National Academy Press.

Niemczynowicz, J. (1997) Water profession and Agenda 21, *Proc. 6th IRNES Conference*, London, September.

NRA (1993) *General Guidance Note for Preparatory Work for AMP2* (Version 2), Oct.

Ongley, E.D. (1996) *Control of Water Pollution from Agriculture*. FAO Irrigation and Drainage Paper 55, Food and Agriculture Organization of the United Nations, Rome.

Otterpohl, R., Grottker, M. & Lange, J. (1997) Sustainable water and waste management in urban areas, *Water Science and Technology*, **35** (9), 121–133.

Pierrepont, P. (1998) The 'ADAS Matrix': the food and water industry in agreement. *Wastes Management*, December 1998, 28–29.

Pinkham, R., (2000) *Daylighting: New life for Buried Streams*, Rocky Mountain Institute, http://www.rmi.org/sitepages/pid172.asp.

Pokrajac, D., (1999) Interrelation of wastewater and groundwater management in the city of Bijeljina in Bosnia, *Urban Water*, **1** (3), 243–255.

Porto, M.(1999) *Sustainable Urban Water Supplies The Metropolitan Region of Sao Paulo, Brazil*, Keynote lecture at Stockholm Water Symposium.

Postel, S. (1996) *Dividing the waters: food, security, ecosystem health, and the new politics of scarcity*. Paper 132. Worldwatch Institute.

Ray, C. (1998), *UK Corporate GIS, Successes and lessons learned*, Lecture notes of the course: GIS in Urban Water, Imperial College, London.

Rees, Y. & Zabel, T. (1998) Regulation and enforcement of discharges to water. In Water Resources Management in Europe. Vol. II: *Selected Issues in Water Resources Management in Europe* (Ed. F.N. Correia), Balkema, Rotterdam.

Royal Commission on Environmental Pollution (1996) Nineteenth Report: *Sustainable Use of* Soil, Cm 3165, HMSO, London.

Saegrov S., G.K. Hansen, A. König, J. Røstum, W. Schilling, J.Vatn,(2000) Water network management – Forecasting network-rehabilitation needs by probabilistic methods, *Proc. IWA Congress*, Paris.

Sanders and Yevjevich (1996) *Urban Water Demand in Water Supply Systems*, ASI series, Springer Verlag, Ed. by Č. Maksimović, Calomino, F., and Snoxell, 7–17.

Saul, A.J. (1998) CSO state of the art review: a UK perspective. 4th *International Conference on Developments in Urban Drainage Modelling*, **2**, London, September, 617–626.

Savić, D.A., G.A. Walters, M. Randall-Smith and R.M. Atkinson (2000), Large Water Distribution Systems Design through Genetic Algorithm Optimisation, *ASCE 2000 Joint Conference on Water Resources Engineering and Water Resources Planning and Management*, July–August, Minneapolis, USA,10.

Scandura, J.E. & Sobsey, M.D. (1997) Viral and bacterial contamination of groundwater from on-site sewage treatment systems, *Water Science and Technology*. **35**, 141–146.

Shiklomanov, L.A. (1993) World fresh water resources. In *Water in Crisis* (Ed P.H. Gleick), Oxford University Press, pp. 13–24.

Smith, S.R. (1994) *Agricultural Recycling of Sewage Sludge and the Environment*, CAB International, London.

Somsen, H. (1990) EC Water Directives, *Water Law*, November.

Stanley, G.D. (1975) *Design Flows in Foul Sewerage Systems*, DOE. Project Report No.2.

Stilinovic, B. Plenkovic-Moraj, A. Zutic-Maloseja, Z. and Zafran, J. (1995) The first microbiological results of research in the Plitvice lakes national park in Croatia after its liberation, *Periodicum Biologorum*, **97**, 359–364.

Stoianov, I., Č. Maksimović, N. Graham (1999) *Management of Uncertainty and Data Accuracy for On-Line Monitoring and Burst Detection in Water Networks*, Internal Report Department of Civil & Environmental Engineering, Imperial College, London.

Stoianov, I., Chapra, S. & Č. Maksimović (2000) A framework linking urban park land use with pond water quality, *Urban Water*, **2** (1), 47–62.

Threlfall, J.L., Crabtree, R.W., and Hyde, J. (1991) *Sewer quality archive data analysis*, Report FR 0203, Foundation for Water Research, Marlow.

Twort, A.C., Ratnavaka, D.D., Brandt, M.J. & MacDonald (Eds) (2000) *Water Supply*, 5th Ed edition, Butterworth-Heinemann.

UNESCO (1978) *Water Quality Surveys*. Studies in Hydrobiology: 23, UNESCO and WHO, Geneva.

UNESCO (2001) *Proceedings of the Workshop on Urban Development and Freshwater Resources: - Small Coastal Historical Cities*, Kotor, Montenegro, Yugoslavia (In press)

US Bureau of the Census (1990) *Statistical abstract of the United States 1990: The national data book,* 110th Edition, US Department of Commerce, Washington DC.

US EPA (1992) *Technical Support Document for the Reduction of Pathogens and Vector Attraction in Sewage Sludge,* EPA 822/R-93-004, Office of Water, Washington DC.

van Leeuwen, M., Butler, A.P., te Stroet, C.B.M. and Tompkins, J.A. (2000). Stochastic determination of well capture zones conditioned on regular grids of transmissivity measurements. *Water Resources Research*, **36** (4), 949–957.

Veenstra, S. & Alaerts, G.J. (1996) Technology selection for pollution control. *Workshop on Sustainable Municipal Waste Water treatment Systems*, Leusden, the Netherlands, November, 17–40

White, M., Johnson, H., Anderson, G. and Misstear, B. (1997) *Control of Infiltration to Sewers*, Report R175, CIRIA, London.

WHO (1995) *Protection of water sources* (Local authorities, health and environment briefing pamphlet series; 4), World Health Organization, Geneva.

Yevjevich, V., Sanders, (1996) Availability and Selection of Sources of Water Supply Systems In *Water Supply Systems,* ASI series, Springer Verlag, Ed. by Č. Maksimović, Calomino, F., and Snoxell, pp. 25–42.

4

Infrastructure integration issues

Pierre-Alain Roche, François Valiron, René Coulomb and Daniel Villessot

4.1 INTRODUCTION

Many examples show that the concept of urban development in the 20th Century did not always properly integrate water both as regards the benefits that it brings and the risks that it represents. The short-lasting character of these policies and outcome of them are developed in Chapters 1–3, which take stock of these difficulties. There is, however, no lack of positive, recent or old experience, and many innovative works suggest possible normal practices that will be adopted by the end of the next 20 or 30 years (Roche, 2000). Technological changes in process are contributing to finding these solutions, provided that they allow for the complexity of urban systems, that matters of cost, maintenance, operation, training and institutional development are integrated in the concept of them and they are in keeping with the overall treatment of economic, social and political stakes (IAURIF, 1997).

 Frontiers in Urban Water Management: Deadlock or Hope
Edited by Čedo Maksimović and José Albert Tejada-Guibert. ISBN: 1 900222 76 0

4.2 INTEGRATING AND ENHANCING WATER IN URBAN DEVELOPMENT

4.2.1 Urban development tools and methods

4.2.1.2 A town planning crisis?

By stating "that the age of creators is over", many commentators have pronounced the end, or at least a crisis in urban policies. We are far from the time when "the prince and his architect", or later the town planner, used to design towns as a whole and organised them as he pleased. There is no choice but to accept that the main part of urban development, particularly in developing countries, comes from unplanned expansion, barely controlled by the public institutions. The importance of population growth combines with a huge rural exodus in a context of weakening unqualified manufacturing jobs, to induce development of districts with no infrastructure and no services, in the most disadvantageous places grouping families together with no jobs and no resources.

4.2.1.2 Long-lasting urban framework: integration of various scales

Urban complexes need, however, to be organised on extremely varied territorial scales. They tend to be structured now more and more in the multiple polar form, indeed like archipelagos, as the only way to face up to the need for each one to live in a space scale allowing daily life (housing, work, leisure, etc.). Apart from some highly polluting or hazardous manufacturing work, the time for large-scale zoning in the functionalist logic of progressive town planning is doubtlessly over, as the Athens Charter proposed it. It should not be thought that the overall scales of towns should be given up for all that.

In fact non-interventionism and anarchic development of districts offering poor services do not offer a long lasting outlook. Modern urbanisation cannot do without strategic orientation documents (master plans, urban plans, etc.) that allow planning of essential infrastructure on a wide enough scale. But this should also be adaptable and flexible enough to allow a diversified urban framework to be set up on a block or district scale. Enhancing the historical framework and diversity of towns, sometimes built centuries ago, becomes vital for development depending on the identity of these "urban complexes".

This exponential geographical extension of towns beyond municipal boundaries requires institutional adaptations. In 1950 Budapest absorbed 7 towns and 16 villages. A pertinent question is whether appropriate institutions for managing urbanisation of towns should be coordinated with respect to water. Table 4.1 gives two examples of these structures, which vary from one city to another according to local institutions and also the speed of urban development. In Mexico in particular, the Federal District should come to a compromise with the State of Mexico, which these days involves nearly half the total population of the city, and with the "Cuenca del Valle de Mexico" (Mexico Valley Basin) for water management. In the same way Greater Paris, which only used to involve 3 departments, today is spread over 8 (2 original departments having divided into 6 new ones) and from now on coordination has to be located on the scale of the Ile de France area (Table 4.1).

Table 4.1 Two examples of town planning and water management structures (Cairo and Mexico)

Cairo	The town groups together **Cairo** and several towns, including **Helouan** to the South covering 911 km^2 of which about 60 are urban. - It is spread over 3 provinces with **governors**: Cairo, Guiza, Qualubia.	- **General Organization for Physical Planning** (GOPP) for urban planning and development - 3 Governors for local implementation	- **General Organization for Greater Cairo Water Supply** (GOGCWS) - **General Organization for Greater Cairo Sewerage and Sanitary Draining** (GOGCSSD) for water.	**Ministries**: - Local Government - Finance - Health - Public Works and supervision by the **President of the Republic.**
Mexico	**Mexico Federal District covering** 1,300 km^2. The capital consists of 12 authorities, 4 towns, and a central department; **suburbs in the State of Mexico** with 35 towns since 1990 forming the **metropolitan area** covering 4,600 km^2 in all.	- **Federal District** with a **"regent "** appointed by the Governor - **State of Mexico** with an elected **governor,** both responsible for urban planning with general powers for water in their areas.	**Water and drainage in the District** (DGCO) **and floods** (CGRUPE) coordinated by the "district department". - Responsibility for water and drainage in the State of Mexico are spread over the 35 towns.	- Water coordination through the **National Water Commission** (CNA) dependent on the Federal State. - By the State of Mexico in 35 towns. -- Valle de Mexico **Consejo de Cuenca** in process of being set up for water and the Environment.

4.2.1.3 Tools for town policy, urban planning, regional development and water planning in France

In the various fields of town policy, urban planning, regional development and water planning France has a quite recent set of laws, summarised in Table 4.2. The general trends of these new requirements are:

- Flexibility and adaptability, which have led to legal-type provisions being transferred to planning documents, and likewise taking everything that was not of a strategic nature from planning documents, to include it in generally 5–6 years' validity programming documents, easier to review and adapt.

These programming documents are generally drafted in the form of contract agreements between the State and local authorities of various territorial levels.

- The integration of sector policies for an integrated, overall and strategic outlook, directed towards long-lasting development, to provide coherence of policies expressed on various pertinent area scales which cannot all be equivalent to the specific jurisdiction of a community.

Table 4.2 Planning and programming tools for water and land use in France

Scale of topic	**Territorial Development**	**Water**
Planning tools		
National	Law for sustainable development -LOADDT- 1998 sectorial public utilities schemes (State) Law for the solidarity and town renewing – LSRU 2000	1964 Water Act 1992 Water Act 1995 Natural hazard prevention law 2001 Water Act (in discussion)
Region or basin	Regional public services schemes (Region) Territorial Development Directives - DTA (State)	Water use directory scheme SDAGE (Basin water committee/State)
Urban area	Urban coherence planning - SCT (municipality)	Water master plan / zoning (municipality) Natural hazard preventive plans (State)
Programming tools		
Region or basin	6 year Regional contracts (State/Region/County)	6 year programme (Water Agency)
Urban area	Urban contracts (state/region/municipality) Municipal programs (municipality)	Water contracts (Water agency/municipality)

4.2.1.4 Two key factors: population forecasts and urban density

The predominance of the urban fact in future years is clear: each day an additional 160,000 people live in urban areas. Urban concentration outlooks are huge: in 1950 there were 3 cities with more than 10 million inhabitants (New York, Tokyo and London); today there are 21 of which 17 are in developing countries, and there should be more than 50 in 2025 (Figure 4.1).

Today urban areas are expected to take 90% of future population growth, estimated at between 1.5 and 4 thousand million in the next 50 years, with the bulk in the next 25 years. Population rate increases stay more or less stable in various continents (Figure 4.2). They are greater in Africa (5% per year) and the Arabian Peninsula (the only significant reduction noted, but from record levels: 6.5% in 1975–1980, 4% now comprising a relatively small population) than in the rest of the world (2.8 % on average), as shown in Figure 4.3.

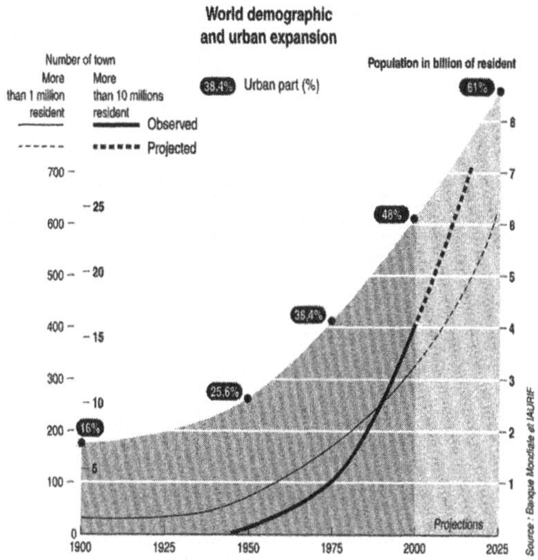

Figure 4.1 Population growth in the world and urban development (World Bank, 1999)

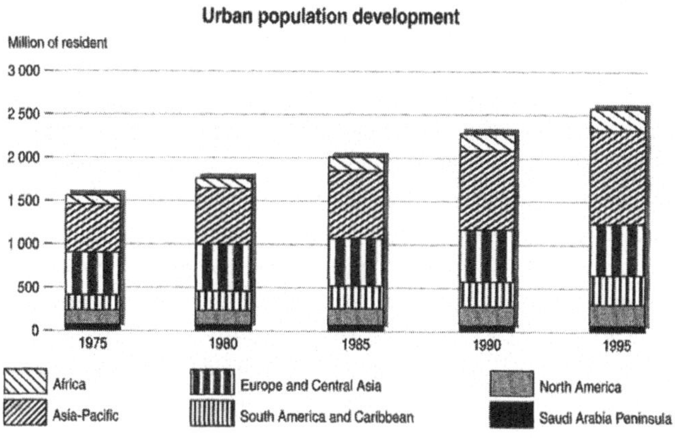

Figure 4.2 Urban population changes (UNESCO, 1997)

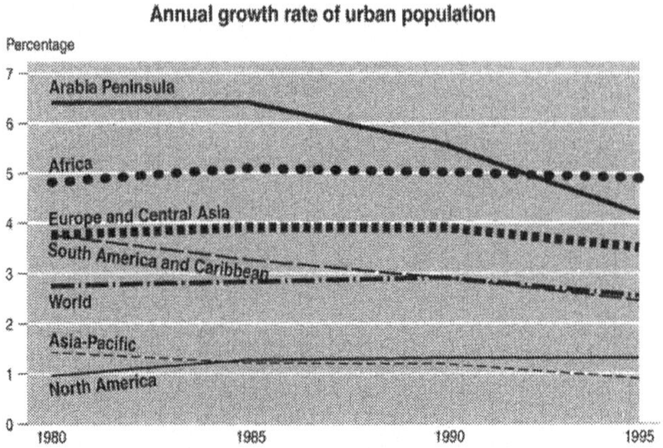

Figure 4.3 Urban population rate of growth (UNESCO, 1997)

Population forecasts are still one of the basic difficulties of plans or master plans. Besides the overall uncertainty of migratory flows, they are faced with the complexity of population displacements and activities inside towns.

Urban density varies enormously particularly according to the proportion of individual dwellings. In many countries owning a house is, or has become, the symbol of access to a stable social status. The heart of old towns has frequently lost population in rich countries as well as in poor countries, with the resulting impoverishment of the city core because of the outward displacement of the well-to-do social strata and middle classes of the population (Lima, etc).

So in many towns heavy migration from the centre to the outskirts, started between the world wars, became more noticeable after 1950. Moreover, outlying areas accommodated a high proportion of many arrivals. This trend had been very rarely anticipated in town planning documents: they planned for greater growth to the outside while keeping a reduced growth rate in the centre, but the downtown areas of Paris, London, Madrid, Munich, Boston, Mexico, Buenos Aires, Cairo, etc., lost an appreciable proportion of their population (10–25%). Traditional suburbs often have even been deserted for the benefit of more distant outlying areas as in Paris, with the growth of the second outskirts, and in Madrid and Boston.

Between 1980 and 1994 the population of the 1,320 km^2 Federal District in Mexico dropped by several hundred thousand people while the towns in the metropolitan area (4,600 km^2) grew at the same time from 5 to 7 million. The Federal District represents only 55% of the total population of 15.8 million people (Figure 4.4). Today, however, a return to the centre particularly in Harlem, New York or in the heart of Paris is noted.

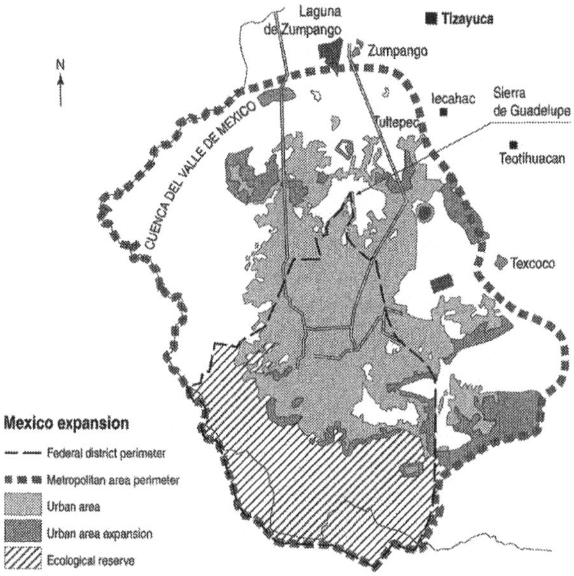

Figure 4.4 The growth of Mexico

Often the keen interest of people in individual houses with gardens was not anticipated, So districts with public housing and also some housing estates for very poor people, these days require concentrated restructuring work, going as far as the demolition of some prominent large buildings 20 or 30 years after they were built, symbolically an admission of failure: the urban model which did not survive longer than a generation, is an obvious setback, while the town rebuilds itself for centuries by progressively adapting.

Density itself depends not only on housing requirements, which establishes the number of housing units per hectare, but also on the number of occupants of them. This number depending on habits and life style varies between 3, in developed countries, and 5 and even 10 in developing countries.

So using the equivalent space, surcharges for public utilities (roads, water, electricity) and also increased transportation needs vary but can be significant. In California, where it is expected that the population will triple between now and 2040, this is an extra 1.5 million hectares of urbanisation to be considered at the present 7 houses per hectare density. The city of Phoenix in Arizona is over 1,500 km^2 for just over 1 million inhabitants, while all the islands of Singapore are hardly 650 km^2 and accommodate 3 million people.

From this should it be deduced that, under any circumstances, towns should be subject to maximum concentrations to reduce structuring infrastructure costs? Surely not. In many cities in "strong government" countries like Hanoi, Shanghaï or only a short time ago Leningrad, the choice was to accommodate many of the new arrivals in existing housing, and the resulting quality of life was sometimes unbearable.

The new Singapore Concept Plan, while applying a strict effective planning in a space shortage situation, provides new facilities to accommodate an extra 1 million people. This no doubt occurs because there really is no alternative and the concern for multiple ethnic integration by means of a policy of central allocation of housing is a major political issue. In 1819 Singapore created by England, became the owner of this large island of 620 km², with its archipelago of 40 small islands at the southern point of Peninsular Malaysia, on the shipping route to China and Japan, and the Pacific. The deep water port of Keppel Harbour, protected by 2 islands to the South, has prospered very quickly: 200 inhabitants in 1820, 10,000 five years later, 80,000 in 1860 and 4 million today as shown in the following table (Figure 4.5).

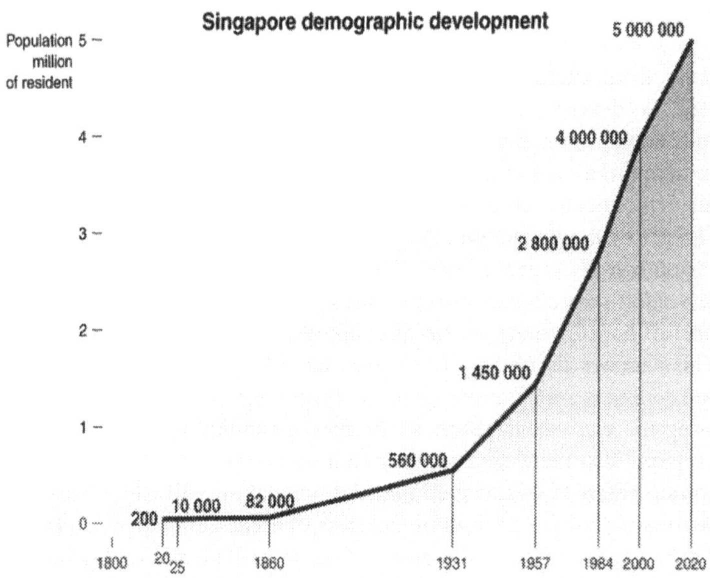

Figure 4.5 Growth of Singapore's population

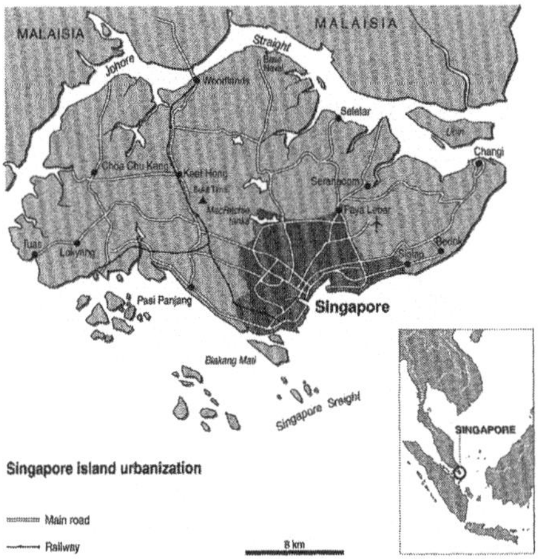

Figure 4.6 Singapore's urban development

The Selat Johor strait to the North, now crossed by a road and railway bridge, had made it possible for a well sheltered naval base to be built. Natural resources were poor; not enough water for an urban complex with only three small reservoirs; water that has to be imported today from Malaysia. But this site also opens onto the great air routes and a hard-working population, 70% of which are of Chinese origin, has the essential elements complemented by intelligent management. This well administered "City-State " became independent in 1946 relies on a banking system and industrialization directed towards new technologies with a cheaper labour force than Japan or Europe. The GNP of $30,000 per person, even if unevenly distributed, attracts immigrants who have to be housed and made to work. With only 5% forest and 8.5% land under cultivation, the only solution is to concentrate the urbanised 520 km², none the less reserving the land required for the work of the million new townspeople expected between 2000 and 2020 (Figure 4.6).

New towns (or satellite towns) were created in London, invented by town planners who wanted to prevent the spread of buildings in the country coming from towns, fleeing the centres that had become stifling and too noisy, or to protect country people from being attracted by the urban magnet. This concept spread in multiple forms first in Paris, and to some extent everywhere in the world today. These days it is found in Brasilia, Buenos Aires, Cairo, New Delhi or even in Seoul. The metropolitan area in Seoul, spread over 605 km^2 with over 12 million inhabitants, covers over 11,600 km^2 and comprises 5 subsidiary areas with nearly 20 satellite towns and 5 million people, or more than a quarter of the population of Seoul itself. It was possible to contain the growth of Seoul, as the influx of immigrants was, however, concentrated in the outside area. The urban area concept takes its pace from that of the town (Figure 4.7).

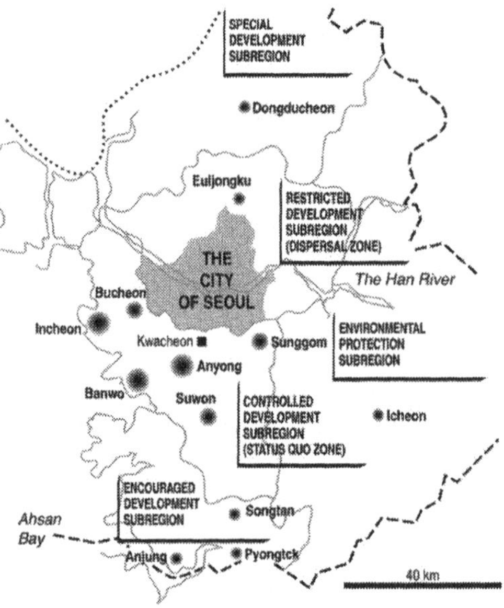

Figure 4.7 Seoul and its satellite towns

These satellite towns can be real ghettos for poor people, or even for the rich. Sao Paulo, which has experienced development (a large group of high-rise tower blocks for the well-to-do or middle classes), today has an increase in island satellite towns, highly protected for security reasons; segregationist town planning in all its excesses.

Many of these satellite towns, generally chosen by town planners based on their own particular values, have led to significant surcharges, on account of their water service and drainage, which could have been avoided if water had been one of the selection criteria.

4.2.1.5 "Spontaneous" districts

Unplanned town growth accommodating most of the new inhabitants in urban complexes in developing countries, as (Philippe Haeringer, 1998) analysed in detail, comes from a phenomenon appreciably different from those that created urban development, even very spectacular ones, of the 19th and first half of the 20th century. These responded to the attractiveness of manufacturing jobs in industries experiencing a high rate of expansion. Today against a background of rural production being of decreasing importance, it responds to a major population imbalance linked to the drop in infant mortality rates, and it inexorably keeps on going even in situations of economic crisis, leaving no hope of jobs nor income. Today urban communities, far from wanting this influx, see it as a source of trouble and imbalance, which they try to hold back both the good and the bad. According to UNCHS (1996) it is estimated that there are 600 million people living in these sectors under permanent threat to life and health and over 100 million homeless. Some of these districts even grow 10–20% per year (UNEP, 1999).

This spontaneous urbanisation, generally with no legal backing, carried out by self-building, usually takes place under the worst conditions, generally occupying the most marginal land (the floors of swampy valleys, areas exposed to flooding, hillsides subject to runoff, gullying, mudslides, etc.). These therefore are the most costly to provide with public utilities (pipe length, ground instability, depth of foundations, size of road systems required for public transport etc.). And at the same time, they are these zones in which community investments are the lowest. Being by default the places for accommodating great poverty, there is little possibility of the inhabitants taking responsibility. So most often it is a dead end.

Under the best conditions, rarely achieved unfortunately, the first area has running water, gradual housing programmes and main roads. Sometimes the rest of this area will also have running water but most often it is drinking fountains. Beyond this the area where immigrants crowd together under the worst conditions is found. Controlling urbanisation is much more difficult in these fast-growing towns; each day the demand to be satisfied is larger and the land area destined to it is bigger. Now, in addition to technical problems to be solved, these towns generally do not have enough financial resources to provide infrastructure for the sites in advance. Figure 4.8 is a diagram of this urbanisation process (Valiron, 1997). The heart of the settlement is extended in a star shape generally along the main roads by differential extension areas, one accommodating a fairly dense settlement, individual in strips or small buildings intended for the middle class, the other consisting of low cost housing plots, comprising parcels for permanent housing, complemented by denser housing in small buildings along the main outlines of the service. Lastly various business areas are also planned. Organised controlled urbanisation is ineffective and does not manage to take on spontaneous urbanisation linked to migratory pressures and population explosion.

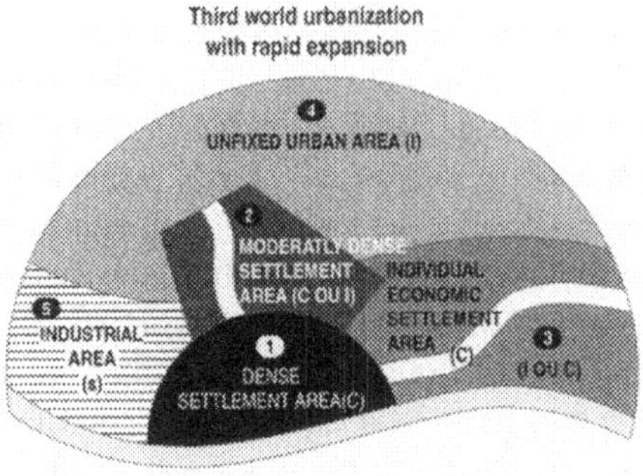

Figure 4.8 Expansion process of towns

In many megapolises of the world, the future of these towns is viewed as apocalyptic. Lack of rehabilitation concepts and actions to remediate the situation sustain this pessimistic view, often shared by their inhabitants. In the following paragraphs we shall see that significant efforts were made to get out of this vicious circle, particularly to create vital infrastructure, by enhancing population organisation capacities to the maximum. One should be aware that the community, whether willingly or unwillingly, shall afterwards come to formalize its status as an urban unit and equip these settlements which were colonized by pioneers. This is an additional reason for not denying the evidence of them, and quite the opposite, mobilising the resources needed for gradual infrastructure, through successive appropriate stages. The stake for the next 30 years really is facing up to this influx.

The case of Ouagadougou shows that, in districts in full extension, it is possible to organise a water supply, minimal for sure but effective, and to control risks from excreta, waste water and refuse. Ouagadougou, which had fewer than 200,000 inhabitants 50 years ago, now has over a million. Sales of water to managers of drinking fountains at independent stations serve 40% of the population and only 18% of the water distributed. The managers who first were private individuals selected by the water company, rely on decentralized political authorities at district level, particularly the least favoured. Another part of the needs of 20% of the population without running water or these fountains is covered by deliveries for resale bringing water to individuals (Figure 4.9). The transport and resale system well organised by the authorities, complies with typical contracts preventing sudden rises in price. The authorities in parallel set up a system for refuse collection in these districts and equipment for plot excreta removal for which the figure opposite provides the data on the infrastructure. In 1998, only 25% of central districts in the city were not on mains drainage compared with 75% in the peripheral areas as shown in Figures 4.10 and 4.11.

Figure 4.9 Sanitation in Ouagadougou

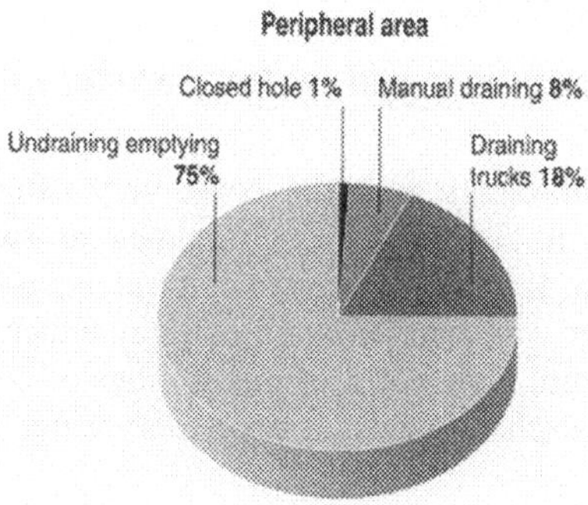

Figure 4.10 Sanitation in the periphery of Ouagadougou

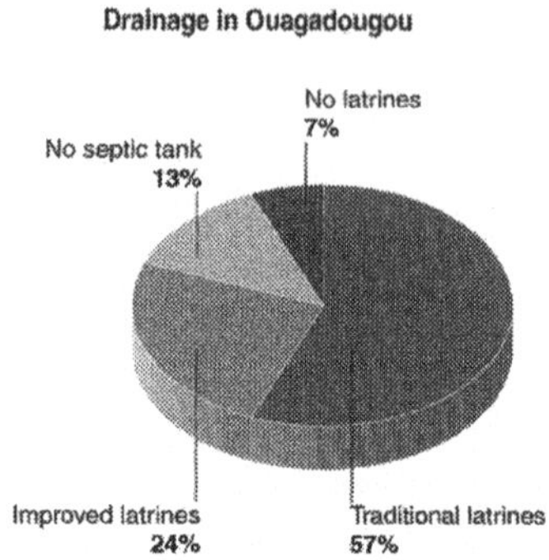

Figure 4.11 Sanitation in the periphery of Ouagadougou

4.2.2 Water's complex role in the city

In this general crisis of growth of large towns dominated by housing, employment and transport stakes, the water question can seem accessory. And unfortunately it is very often that in a subsidiary way that it is dealt with in town planning plans. All the same this is ignoring the central part that water plays in development and urban development.

4.2.2.1 The social, economic and cultural stake; accessible water in towns

To begin with, rivers or banks near the coast very often were the human installations as shown in table 4.3, which relates to twenty or so towns mentioned in this section. In many towns the cultural, economic and social part of this closeness to water was vital, and still is today either because it explains development of towns from their ports, or possibilities of connecting with river traffic, or as a place of exchange even if poor fresh water resources slowed it down a bit.

Table 4.3 Geographical position of cities mentioned and their relationships with water

Continental or coastal cities (bold)			
Lake	Main water course	Average water course	Small surface water course
Brasilia Mexico	Bordeaux Budapest Hanoi London Paris Cairo Delhi Seul **Shanghai** Saint Petersburg	**Boston** **Jakarta** Munich **New York** **Osaka**	**Casablanca** Ouagadougou **Marseille** Madrid Mexico Nimes Singapore

4.2.2.2 Water the "urban art" material

The usual town, which took centuries to build, enhanced this closeness to water made one of the major "urban art" tools from it. During periods of prestige town planning, gardens, parks and fountains were placed in the middle of public areas. It was only in the second half of the 19th century and the beginning of the 20th that the sight and pleasure of water was ignored by town planners. Water and town planning also created then parks and gardens, the forerunners of which were the builders of Babylon, and then the Arabs. Madrid, with the El Retiro park (Figure 4.12) and the three beautiful fountains in the old city, has been certainly inspired by this tradition by which what was created at Granada are still admired today.

So capital cities also built palaces with parks and fountains for royal residences; Versailles is an example of successful incorporation between landscape and architecture where water has been brought by very technical feats at exorbitant cost. This connection with water as an essential resource is still engraved in stone today in many towns where many streets or squares recall an ancient well or fountain. Longchamp palace in Marseilles marks the arrival of water from la Durance in 1849 (Figure 4.13).

Figure 4.12 The Retiro fountain in Madrid

Figure 4.13 Longchamp palace in Marseilles

Rivers integrated into the urban landscape took advantage of the sight offered by navigation and transport (the London Docks on the Thames, the Garonne waterfront in Bordeaux, etc.). In Paris the Seine was made the link for its prestigious architecture. St Petersburg, a city built on a system of diverticula of the Neva and canals, is renowned for its successful integration between the city and water sine the 18th century.

This enhancement of towns by water for the last 30 or so years has again become a concern for town planners with the rise in power of the defenders of the environment and ecology as well as that of the water specialists for controlling stormwater discharge.

So the green swathes linked to water appeared in 1976 in London, Budapest as in Paris or Osaka. In this city the *Green Master Plan* planned to create there green reservations on the hills and along the rivers and waterfront, capable of changing the park area and green reservation from 5 to 15 m^2 per inhabitant and the green reserve from 40 to 50 m^2.

Water plans and swimming areas in parallel on certain rivers were implemented a little everywhere, particularly in Munich, Marseilles and Boston. Finally hydraulic engineers organised new retention basins in the urbanisation area reducing much of the pollution while offering recreation areas.

4.2.2.3 *Water and public hygiene*

The concern for hygiene developed in the 19th century particularly when doctors acknowledged the major part of water in public health. Did not Louis Pasteur used to say justifiably: "we drink 90% of our illnesses"? Having safe water distribution through drinking water supplies and stormwater and waste water drainage through an underground system since then has been one of the symbols of modern cities. Treating these waters firstly discharged simply downstream, only took place subsequently. The Osaka example displays the link between improving the running water service and reducing the number of people affected by sickness connected with water (Figure 4.14).

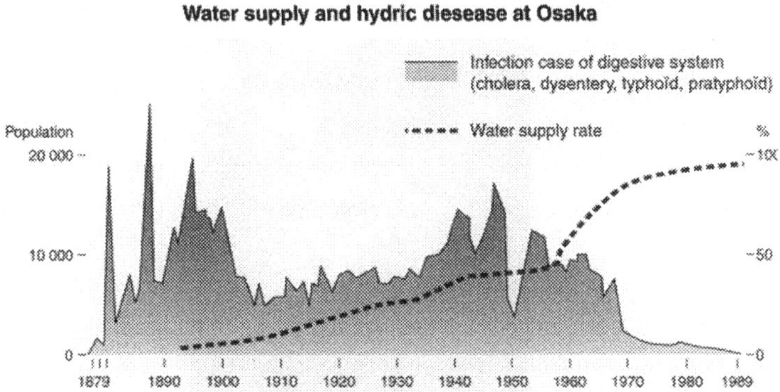

Figure 4.14 Water supply and water-related illnesses in Osaka

Eliminating cholera cases in Marseilles was linked to chlorination of the canal water (undertaken between 1942 and 1950). A very serious warning was given about 1935 where 2,600 fatal cases were attributed to the distribution of raw water polluted during transportation through open channels (Figure 4.15).

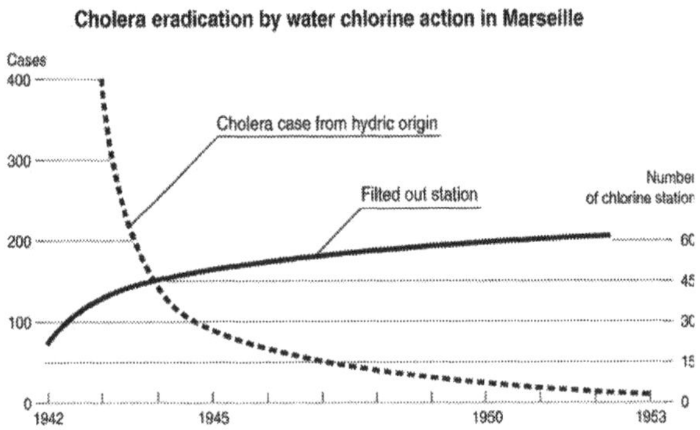

Figure 4.15 Cholera stamped out by chlorination of water in Marseilles

Analogous occurrences in Munich link development of water mains to changes in death rates. Between 1870 and 1995 in Buenos Aires the death rate was halved while the service was greatly increased (Figure 4.16).

Past and present-day epidemics are linked to water infrastructure in cities and gaps noted in water infrastructures are found again 50 years or more later (Figure 4.17).

Figure 4.16 Water porterage

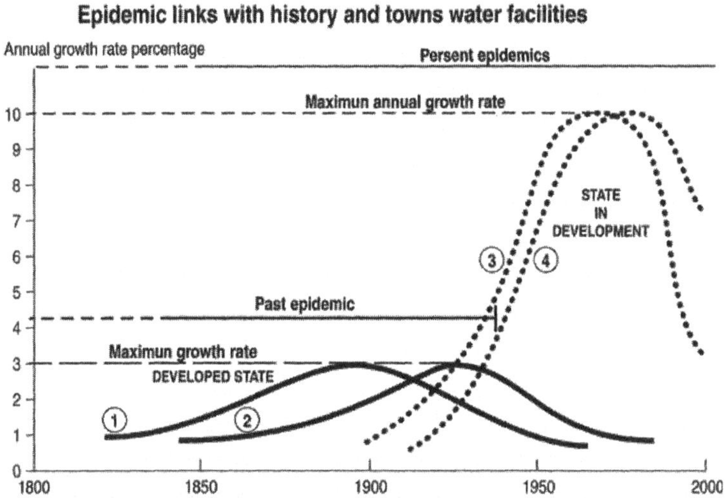

1	Paris	2	Marseilles	3	Mexico	4	Delhi
	London		Madrid		Casablanca		Shanghai
	Munich		Buenos Aires		Cairo		Ouagadougou
	Boston		Saint Petersburg		Jakarta		
			Budapest		Seoul		
			Osaka				

Figure 4.17 Links of epidemics with history and the water infrastructure

4.2.2.4 The influence of the existing structures

The cities where we are working today have large systems, sometimes very deteriorated, sometimes built in areas liable to flooding. Urbanisation there was developed in unsuitable districts; the structure of the old city was not appropriate for the intrusion of "modern comfort". So many likely reasons for dreaming of a new modernist Utopia, an optimised City, rebuilt to get rid of its obsolete residue from the past. We have seen too much of the damage from this ideological view of town planning, too much qualitative poverty of functionalist undertakings to abandon this history today, this diversity. So this is the influence and cultural and social value of the existing structures from which we will try to build long-lasting cities.

In the past planners and town planners remodelled towns to adapt them to new requirements, particularly those from roads and traffic. But, however talented as they were, many of them to achieve those aims destroyed certain buildings that deserved to be preserved: in Paris Baron Haussmann succeeded in giving the city its present appearance with wide boulevards and outlooks which some consider too expensive in cultural terms. For sure, the task was not easy but the lesson struck home. So M. Beaudoin, a French town planner, was able to save Tunis from radical work which would have razed a large part of the Souk district to the ground which is one of the charms of this city and made it part of its heritage and soul.

4.2.3 Water quality requirements

The quality of water required for supplying mankind, industry, agriculture, or even swimming and fish farming, is not the same. The most demanding requirement relates to drinking water and certain industrial uses; requirements are less strict for washing streets and drainage of certain discharges. There are standards for determining the limits of them. They were drafted by international bodies like the World Health Organization (WHO), multinationals like through the European Union or the USA, or by the appropriate national government agencies.

Natural environments are generally defined in terms of their sensitivity, and categorised accordingly into zones that are susceptible to eutrophication (uncontrolled algae growth due to the presence of excess nutrients), shellfish farms (fishing or fish breeding zones), swimming areas (fresh or coastal waters), and raw water intakes used to produce drinking water. A great many countries have developed rather complex regulations to protect these areas (Table 4.4).

Table 4.4 Water quality standards

	Country	Potable Water	Irrigation	Waste Water	Fisheries	Bathing Water
1	European Community	X	X	X	X	X
2	France	X	X	X	X	X
3	UK	X	—	X	X	X
4	Germany	X	X		—	X
5	Spain	X	X	X	X	X
6	Hungary (1)	X	X	(to be provided)	—	X
7	Czech Republic	X	X	X	—	X
8	WHO references	*				
9	United States	X	—	X	—	—
10	Canada	X	—	X	—	X
11	Argentina	X	X		—	X
12	Philippines	X	X	X	X	—
13	Indonesia	X	X	X	X	—
14	Malaysia (2)	X	—	X	X	—
15	Brazil	X				

"X" specific regulations exist and were sent to us
"—" no specific regulation
* Web site
1 A new set of regulations must be adopted this year
2 There are no specific regulations but simply "guidelines"

For example, the EEC standard for quality of water intended for human consumption of 1998 (98/83 EC) contains 53 parameters (microbiological, chemical, indicators, radioactivity). Each one is allocated a "100 parametric value" which is the minimum constraint level to be respected. Table 4.5 shows the disparity between standards for organic micropollutants:

Table 4.5 Various standards for micropollutants

(Level in µg/l)	USA	WHO	EEC
Cyanide	200	70	50
Nickel	100	20	20
Lindane	0.2	2	0.1*
Methoxychloride	40	20	
Trichlorethylene	5	70	10†
Xylene	10,000	500	

*Total pesticides 0.5 µg/l

†10 µg/l is for tetrachroethylene and trichlorethylene

These parameter values are reviewed often enough, often in the sense of more constraining levels. For lead, for example, the EEC reduced the content from 50 µg/l to 10 µg/l, while falling into line with the WHO figures. The approach generally is based on health considerations or toxicity, where the carcinogenic effects are taken into account. For cancer, for example, the guideline value is evaluated with the probability for a man at 1 in 100,000, but experimental data are rare and hardly reliable. This explains the marked differences by using a more or less strong precaution principle.

The European Union also drafted "Directives" from 1975 to 1980, covering a very wide field, for example the one for surface water intended for the production of water for drinks, swimming, fish-farming and shellfish breeding and lastly drinking water. It also adopted the list of hazardous substances then, secondly, a Directive on the conditions to be complied with for municipal water treatment and one for agricultural origin nitrates. Finally, it adopted a Framework Directive grouping together all requirements and providing for implementation in each catchment basin including the Directive in the project for industrial water treatment. This Framework Directive recommends integrated management relying on coordination of measures, an integrated approach and a management plan per catchment area with the public participating and recovery of operating costs thanks to applying the "polluter pays" principle in particular.

These requirements show the need for cities to worry, not only about water supply problems, waste water and stormwater drainage but also other uses of water, in appropriate urban areas.

The scarcity of fresh water on the Atlantic coast of Morocco, combined with population pressure concentrating on several large towns in Kenitra to the north and south of Casablanca, led ONEP to make mobilisation of the required water resources common, mobilised by dams, from Sebou in the north, Bou Regreg and Fouarat in the centre, and Oum Er Rbia in the south. The resources master plan not only allows for town water requirements but also those of agriculture and hydraulic electricity generation. Casablanca would take about 55% of the urban water requirements. It provides for the construction of new dams on the Sebou and heightening of some others, and fixes dates taking account of present population growth and its requirements. In parallel, the Casablanca drainage master plan provided for part of the water treated in a big purification plant in Sidi Bennousi to be used for industrial purposes and above all for developing the park areas in greater Casablanca and irrigation programmed by the urban development plan.

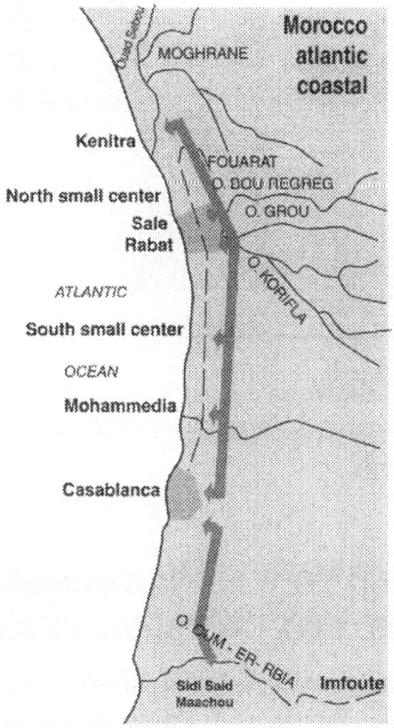

Figure 4.18 The Atlantic coast in Morocco

The Jakarta example shows the effects of land use. Growth of Jakarta led to the urban space being enlarged. The city of 26 km^2 at the beginning of the century has become a metropolitan area of 6,800 km^2 at the dawning of the 21st century with over 700 urbanised in Jakarta itself and over 100 in Tanger and and Bekasi to the East and West and in Bogor to the South. This enlarged area called "Jubotabek" had 11.5 million inhabitants in 1980 and nearly 20 million in the year 2000, including 60% in Jakarta itself. Urban growth in this originally agricultural area caused many upsets, deforestation, erosion, flooding risks, salt water intrusion, ground subsidence with pumping intensification. This led to the integrated management of water resources for all uses being attacked, made difficult by the lagoons of the appropriate structure despite the strong powers of the area capital, the DKI (Figure 4.19).

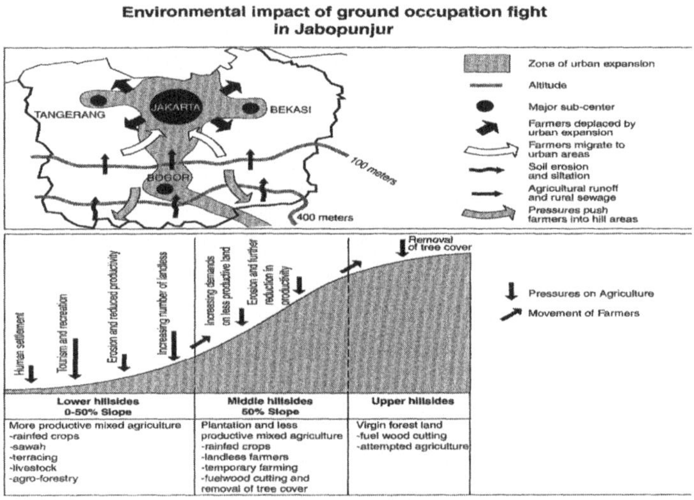

Figure 4.19 Environmental impacts of ground occupancy conflicts in Jabopunjur (greater Jakarta planning)

4.3 GETTING USED TO WATERCOURSES OVERFLOWING AND CONTROLLING URBAN RUNOFF

Flooding, whether linked to watercourses overflowing, runoff or stormwater drainage systems being saturated, is nevertheless at the heart of urban development problems. Without returning to flood control outside the urban area dealt with in Chapter 2, measures to be taken in towns recommend having the resources for forecasting floods and warning systems. Moreover a balance should be found between the technical and regulatory resources mentioned here and the insurance systems considered in Chapter 8.

4.3.1 What to do in floodplains

4.3.1.1 Do not develop new urbanisation in exposed areas

In a sustainable development outlook the most immediate logic consists in preserving, restoring, even creating flood relief areas rather than hindering the natural meandering of river beds where they run through urban complexes. At the same time this solution enables damp areas of great ecological value to be preserved. The tools for preserving these areas are mainly statutory regulations banning construction. The land value of these areas is often favourable to construction (many transport infrastructures, ease of construction, etc.) and makes the withholding of construction permits especially difficult economically and politically. None the less public authorities, at national or municipal level, have from now on understood the stakes and at last established very strict requirements.

In France, "risk prevention plans" (RPPs), set by the 1995 law, introduce construction limits in areas defined as liable to flooding thanks to a survey of historical floods, followed up by statistical hydrological analysis and generally detailed hydraulic modelling. These requirement plans drawn up by the State give rise to public enquiries but, frequently, only make use of too little of inhabitants' knowledge of old floods. Zoning established in this way singles out areas where construction is not permitted and those where it may be permitted in return for special precautions. Existing properties may then be insured, and there is no variation in insurance premiums according to risk ; this is a significant defect in this device not making individual decision makers sufficiently aware of their responsibilities (Bourrelier *et al.*, 1999). In the end it is the community which pays by way of solidarity (Roche and Bourrelier, 1998) in the event of disaster.

4.3.1.2 Upgrading spaces by developing them for leisure

One of the solutions reconciling floods and urban occupancy for the best consists in using areas liable to flooding for urban parks or sports and leisure facilities (Dégardin and Gaide, 1999). These in fact are readily compatible with the occasional presence of water, and besides need much space for the kind of natural structure of the landscape in valley floors (large meadows, afforestation, etc.). Of course it is advisable to make sure that the facilities in no way obstruct flood discharge and they should be managed and maintained with great care.

The drainage system of the new town of Vitrolles, near Marseilles, comprises 13 reservoirs for restricting the Laladière river discharge to 21 m^3/s in flood, preventing flooding in the Marseille-Provence airport district and the towns of Marignane and L'Estoublet. One of them, Lionant (Figure 4.20), of 1.4 hectares which can store 40,000 m^3, was developed as a cycle track and football field. The Frescoule one (1.5 hectares and 25,000 m^3) is used for a sports field and public park for a group of 900 apartments (Figure 4.21).

Figure 4.20 Playground in Vitrolles

Figure 4.21 Landscaped storage reservoir in Vitrolles

In Seoul, which in 1965 already had over 5 million inhabitants, floods used to cause considerable damage despite dykes, etc., which led to an overall project being launched for training the river Han and protecting surrounding areas, comprising several dams upstream and creating flood zones on the two banks for use as flood relief areas to reduce the effects on the international airport facilities in particular. As Figure 4.22 shows, these areas of 700 hectares were developed in large public parks created for the Olympic Games in 1986. Thanks to the dams upstream, especially Champil and Chai of Pelleng in the city which in fact is a lake for pleasure boats, the water level nowadays in front of the two parks hardly fluctuates except at the time of very severe floods during which the parks are under water.

Figure 4.22 Public park in an area liable to flooding in Seoul

These examples show the part which one could take advantage of the diversity of landscapes, water and areas of more or less high dampness to create an atmosphere in urban parks in areas liable to flooding. Educational development may also be designed there to disclose the ecological richness of damp areas to large populations. Conversely large busy areas, especially business and industrial ones also should clearly be banned.

4.3.1.3 Constructing or rehabilitating by reducing vulnerability

As seen previously construction in areas exposed to risks should not be permitted except in areas already built up. It is advisable in fact to stop urban development on valley floors liable to flooding. This, however, may be necessary for complementing urban infrastructure already existing in these areas. In already developed floodplains in fact, "dovetailed" construction and rehabilitation of frequently deteriorated housing is necessary for urban continuity and the dynamism of districts considered.

These buildings should then be constructed ensuring the safety of the riverbank people and minimising the district's vulnerability. Flood risks generally lead to the first inhabitable levels being built above reference flood levels. Architectural problems of this heightening requiring the use of piles should be solved. Construction or reconstruction in areas liable to flooding also means appropriate urban requirements for reminders of the closeness of water and risks, and thus contributing to permanent preventive information. Developers could then rely on landscape features (the mirror effect of water, structuring part of tree lines).

4.3.1.4 Protecting sensitive equipment

One of the significant stakes is preserving safety in flood situations. This means that much sensitive equipment has to be reinstalled under cover. For this purpose an accurate inventory is required, complemented with hydraulic modelling enabling top quality representation of a flood event developing. Making utilities secure (electricity supply, telephone and gas) should originate in the establishment of a hierarchy of acceptable risks and of course be covered by special forecasting and warning procedures.

Osaka, one of the big cities in Japan, at the seaside, suffers from flooding linked to typhoons and heavy rain which accompanies them, reduced the frequency and repercussions of them thanks to the main guidelines to be complied with for operating the Biwa lake which collects a great part of the surplus water. This enables floods to be prevented and a reserve for low water to be stored. The system is complex as it also involves management of many dams on rivers flowing into the lake (Figure 4.23).

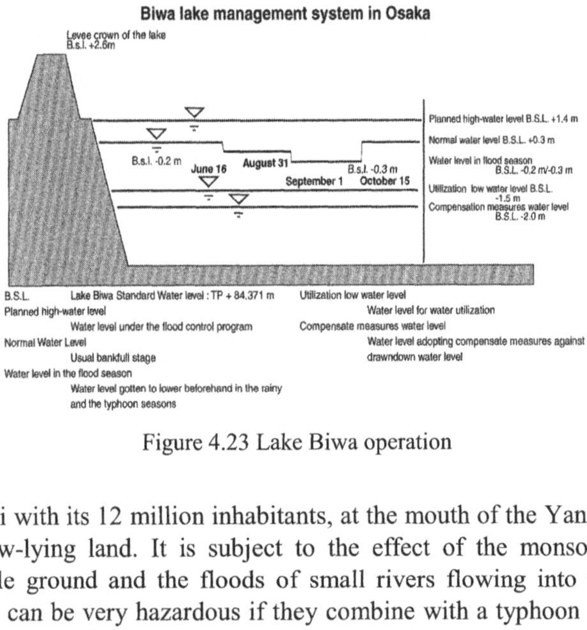

Figure 4.23 Lake Biwa operation

Shanghai with its 12 million inhabitants, at the mouth of the Yangtze, is built on very low-lying land. It is subject to the effect of the monsoon rains on impermeable ground and the floods of small rivers flowing into the Tai Hui lake. These can be very hazardous if they combine with a typhoon despite very significant dykes. In this case the warning and forecasting system enables monsoon stormwater to be stored in the lake and the rise in water levels to be greatly reduced also (Figure 4.24).

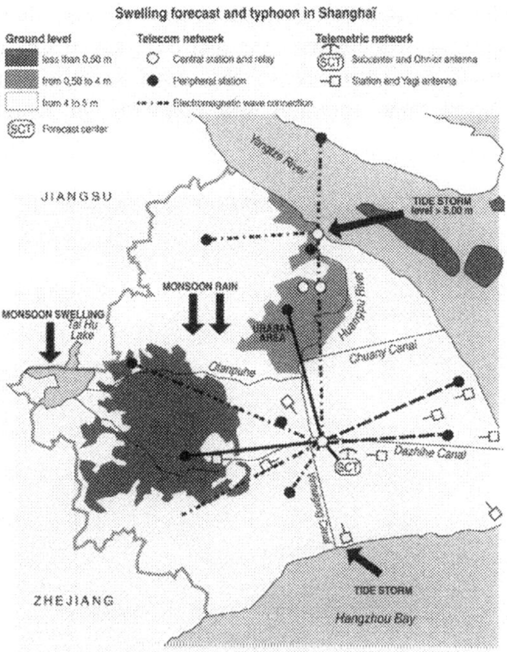

Figure 4.24 Forecasting floods and typhoons in Shanghai

Arrangements implemented in other metropolises like London could have been shown, relating to the Thames and its tributaries and tidal effets or multi-purpose dam control (flood reduction and raising low flow marks in Paris and Madrid).

4.3.1.5 Constructing roads suitable for occasional floods

Of course roads form part of these sensitive infrastructures. It is a question of acting in such a way that in communications crisis situations, especially for emergency requirements, they can be maintained at minimum level.

But it is also a great advantage to design roads that accept being submerged occasionally. In Osaka the city has benefited from flood control provided by controlling the level of the Biwa lake to develop vehicle traffic on the riverbanks by opening it to the city (Figure 4.25). Roads, docks and parking areas can be developed on low banks flooded by five or ten year floods without heightening (roads on the banks of the Seine in Paris).

Recent developments systematically include public gardens and pedestrian footpaths or cycle paths. Car traffic on the banks in fact can obstruct pedestrian access to the water deteriorating riverbank quality. So care should be taken that there is visual opening between the river and city, which was ignored in some developments. Road infrastructure of this kind greatly contributes to the possibilities of spare time activities and approval for landscaping densely populated areas.

Figure 4.25 Osaka; roads on rivers banks

4.3.2 Controlling urban runoff

Developers and town planners have a very special responsibility as regards urban runoff. Design has to be provided at the same time on plot scale and that of public infrastructures (GRAIE, 1998). The aim of course is to arrange it so that upstream areas do not contribute by being impermeable to aggravate downstream flood situations. There are very many techniques nowadays for facing up to these questions. By relying on town planning or drainage regulations, action can be taken requirements for construction of buildings, or the resources for retaining water in road systems or with mini basins.

4.3.2.1 Statutory regulations for town planning

Public communities can require builders, in new complexes, either to take construction measures themselves that will not aggravate the imperviousness coefficient, or to pay them a tax for re-establishing the situation. In parallel with the compensation structures linked to plots required by these regulations new complex runoff leaching structures were developed (Valiron and Tabuchi, 1992).

They were required by drainage regulations which, as in Bordeaux since 1984, only permitted discharge in systems which runoff excess after leaching in appropriate structures. This is the case for the Fontaudin water basin in Pessac, which was designed to be inserted in the surroundings and also reducing stormwater pollution (Figure 4.26).

Figure 4.26 The Fontaudin's basin

Bordeaux, built on low ground in a loop of the Gironde with its tides, is dominated by its suburb built on plateau sites. This situation is the origin of strong sudden floods linked to the impermeable ground. This is why the Bordeaux Urban District (C.U.B.) was a precursor in France for stormwater control by adopting since 1982 compensating measures for urbanisation enabling discharges not to be increased in the area involved. The arrangements reviewed in 1988 in the SOP (Soil Occupancy Plan) return to authorise only at the outlet of urbanised plots the only discharge allowable by the downstream system. The developers should therefore compensate the impermeability caused by their buildings by undertaking infrastructure works, approved by the CUB, for storing, delaying and restricting stormwater flow. To date, over 500 compensating structures, individual as well as community ones, have been built. These developments call on a dozen odd different kinds (storage basins of all kinds, triangular section stone filled ditches, reservoir structures, porous parking surfaces (Figure 4.27) drainage ditches, infiltration blocks and roof terraces).

Figure 4.27 Porous parking areas

Since 1960, Munich has been the inventor and refiner of these compensation techniques on site. The city in fact used to want to preserve the Isar, which crosses it, and strengthen its part as a green lung (Figure 4.28). For this purpose, town planners and water specialists combined their efforts and creativity to set out stormwater runoff reduction as far downstream as possible and as near as possible to built plots thanks to all possible techniques which they experimented with and developed. These resources very widely used in the city and suburbs were complemented by reduction storage set out in the drainage system (140,000 m^3) and in storage reservoirs (380,000 m^3) like the Hirschgarten one which, with 90,000 m^3 capacity, is the biggest in Europe.

Figure 4.28 The Isar

4.3.2.2 *Building measures limiting runoff*

Thc problem can be considered at plot level. Roof reservoir devices and individual reservoir systems allow the quantity of runoff water to be limited. Systems for recovering this stormwater can also supply uses other than for drinking water. It is advisable in this case to pay great attention to hygienic storage conditions.

Using terraces of houses, covered with 5–10 cm of gravel enables stormwater to be stored and drained off taking advantage of the slowdown on account of storing and gravel. Steps taken in Aix on 26 hectares occupied by 900 apartments, with 20 hectares made impervious and 7 hectares of terracing, showed a 30–70% reduction in discharge following rain and restoration spreading over 6 kilometres on average.

4.3.2.3 *Local technical resources*

In particular the following may be mentioned (Azzout *et al.*, 1994):

- Porous pavements:

  There are very many porous surfaces nowadays both for roads and parking areas. They generally consist of a layer of bituminous concrete (for traffic) resting on crushed material used for storing infiltrated stormwater. They reduce stormwater runoff and pollution very appreciably. This reservoir can be replaced by any porous material. It is isolated from the soil with a waterproof film, a geotextile for example. A branch drain picks up water restored by storage. Such surfacing (or parking areas) moreover is applied for discharges which is their main role, a significant effect for reducing stormwater pollution. Many initiatives have shown in France and elsewhere (Figure 4.29).

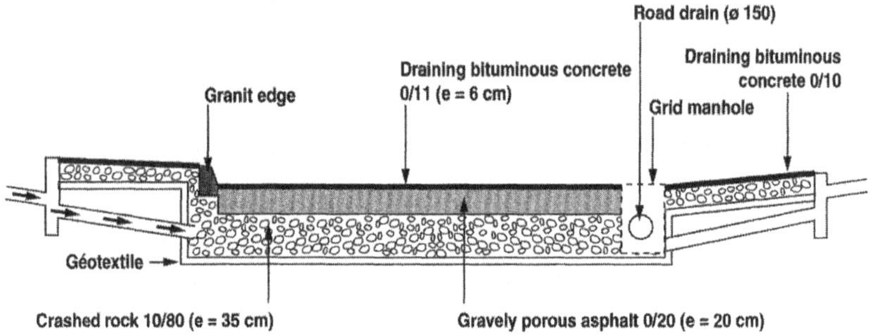

Figure 4.29 Porous pavements

- Absorbing wells, infiltration or injection wells (Figures 4.30 and 4.31):
  It is however advisable to pay great attention to health conditions especially as regards injection wells, and to prevent groundwater pollution, particularly by running pollution control before discharge into the surroundings.

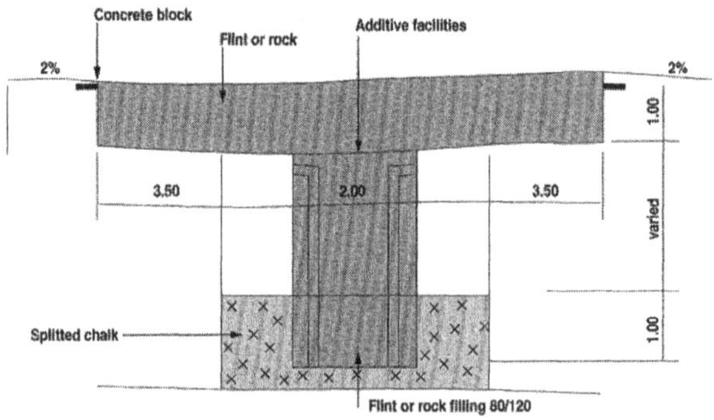

Figure 4.30 Infiltration wells

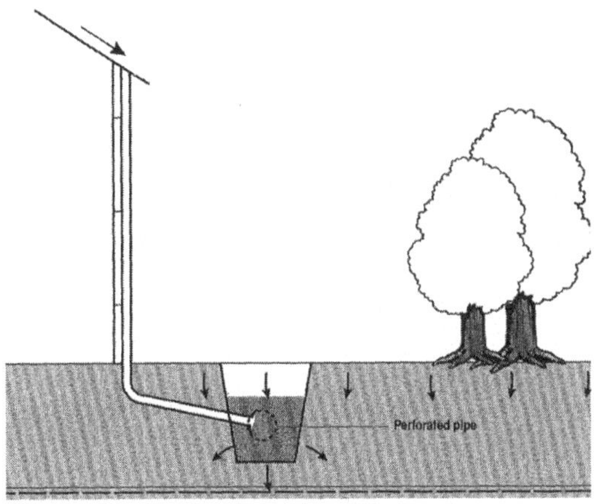

Figure 4.31 Infiltrating pipes

- Ditches, swales and water meadows:
  Open ditches also are outstanding options for storage but in a context where it is possible to check the urban surroundings quality. Filled with gravel they store or slow down drainage down ditches for impervious surface areas or by infiltrating stored water for permeable ones. They can in fact become cesspits, especially if waste water is discharged, or refuse dumped into them (Figure 4.32).

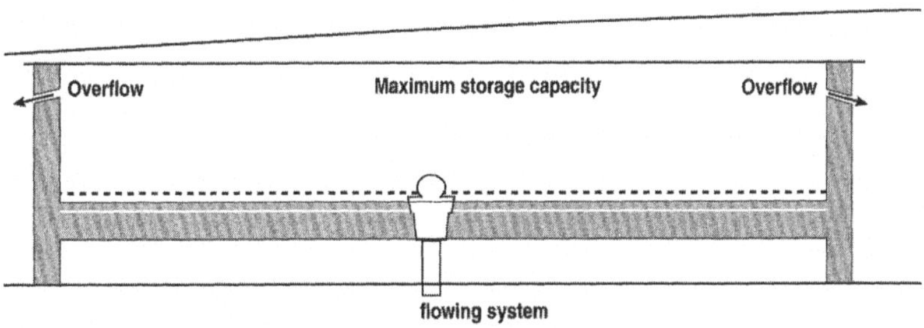

Figure 4.32 Infiltration ditches

Water swales (Figure 4.33) replace stormwater systems with wide shallow ditches on green catchment areas. Drainage or porous pavements consisting of a layer of drainage bituminous concrete for traffic on top of a drainage layer resting on crushed material used for storing infiltrated stormwater that brought from roofs or ditches, this layer, replaceable by other storage materials, plastic for example, is insulated from the surface area with a waterproof film, geotextile for example.

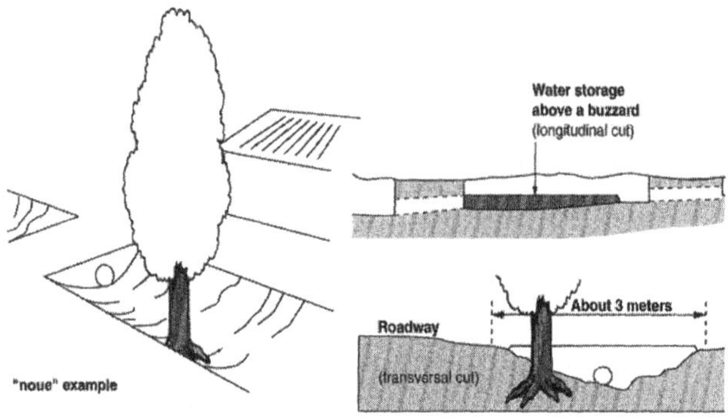

Figure 4.33 Swales

4.3.2.4 Detention basins and their integration into the urban scene

One outcome of urbanisation is making land impervious, which greatly increases stormwater runoff and reduces underlying groundwater supply. This increase in discharge to be drained off is all the more annoying when it occurs on peripheral surface areas located in such a way that to drain them areas already built up and provided with mains drainage designed for local needs, unable to cope with new loads, have to be crossed (STU, 1994). Detention basins are the symbol of the urban integration of big cleaning infrastructures and very many examples have shown the possibilities of reconciling significant storage volumes with a very wide range of operations. Sports facilities, pleasure pools, squares and meeting places accordingly create opportunities for temporary water storage.

The Bordeaux example clearly shows the extra works to be undertaken: for draining stormwater from small hills in the suburbs which have to cross the lower built up part of the city, old drains and storage upstream had to be reinforced to prevent flooding during storms. Figure 4.34 shows solutions used, in diagram form, combining reservoirs and compensating techniques. The same situation is found especially in Budapest, with very fast building on the heights of the right bank of the Danube, and in Limoges with building on plateaux. Many retention basins were resorted to and the use of compensation techniques was developed on a massive scale, to reduce pipe diameters and pumped volumes, in the low area as well as on plateaux. The 25 reservoirs built store 1,200,000 m^3 and pumping is limited to 100 m^3/s during floods or tides in the Garonne. Many of these reservoirs specially designed for this purpose play a part in reducing stormwater pollution.

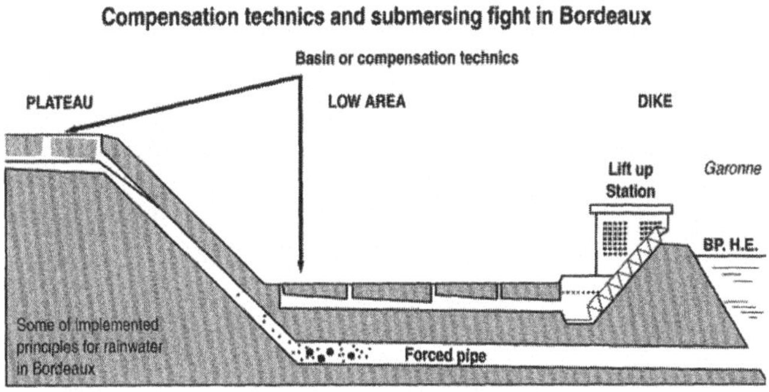

Figure 4.34 Bordeaux reservoir diagram

4.3.2 5 Adapting urban management to the risks of flooding

It is necessary to have infrastructures for controlling as much water as possible upstream, but it is also advisable to adapt roads and urban buildings to be flooded under exceptional circumstances. It is preferable to integrate the possibility in the design of all infrastructure, rather than refuse to accept the chance of an exceptional torrential rain event.

The 1988 disaster that occurred in Nimes, France, illustrates this need. The city was seriously damaged as a result of rains with a 50 year return period. Nine deaths in the city centre, 95% of small businesses destroyed, 2000 houses made uninhabitable, 45,000 damaged, 6000 vehicles destroyed, 700 million dollars damage and some districts under 3.5 metres of water and mud; the damage was out of all proportion even for a rare event. Inadequate clear openings in hydraulic structures caused water surges in sloping streets. These quickly became blocked by stationary car wrecks piling up, and street furniture torn from fixings (road signs, bus shelters, decorations, etc.). A little thought into making parking conditions suitable, or the buildings, would have made it possible to minimise the cost and social consequences, clearly in addition to making the hydraulic infrastructure suitable to reduce the frequency of the event.

4.4 DRINKING WATER SUPPLY

4.4.1 Demand and resource

4.4.1.1 Predicting water needs

Urban water requirements vary considerably from one country to another (Valiron, 1995). Domestic requirements are only on the order of two or three dozen litres per person per day (the Vision for the 21st Century (Commission Mondiale, 2000) estimates a minimum of 40 litres per person per day). However, these figures are considerably different for Western Europe (Table 4.6):

Table 4.6 Domestic and total consumption in Europe (Centre d'information, 1998)

Country	Domestic Consumption* (litres/person/day)	Total Consumption * (litres/person/day)
Switzerland	264	402
Italy	220	293
Norway	199	300
Denmark	194	291
Sweden	175	350
Luxembourg	171	259
Netherlands	159	195
Spain	158	217
Finland	156	288
France	147	211
Germany	146	196
Austria	131	271
Great Britain	132	267
Belgium	108	166

* "Total consumption" includes "domestic consumption" (home water use) and community consumption venues such as schools, hospitals, street cleaning, work environments, and businesses.

These figures are significantly lower than those for the United States. However, they represent the water consumption of developed countries and, as such, can be considered indicative of consumption levels that may eventually be reached in most of the world's cities.

Indeed, these amounts are already being exceeded in some of the wealthier neighborhoods in large cities such as Buenos Aires. Moreover, the outlying areas of many large metropolises are often not connected to the public distribution system and ways must still be found to provide them with a minimum of basic services. Water consumption rates equivalent to those of Europe will only be reached once these disadvantaged communities have been connected to the public distribution system and their inhabitants have acquired household appliances and sanitation facilities similar to those of European cities. Thus, although household appliances are being designed to consume less and less water, we must still prepare for a continuing rise in water demand in the great megalopolises and their sprawling suburbs, and in the small and medium-sized cities of the Third World, the megalopolises of the future, particularly in Africa and Asia (Cosgrove and Rijsberman, 2000).

However, it has been difficult to predict water needs reliably enough to avoid grossly over- or undersizing water projects. And, in any case, before costly projects are undertaken to develop new water resources, existing resources must be put to their maximum good use.

4.4.1.2 Steps to avoid investing in new resources

Water is lost during all phases of water management, some during production, but most as water is being transported and distributed. In some cases, as much water is lost as is actually distributed (50% loss rate) and even in the best of systems it is impossible to avoid losses completely. A good distribution network has an 80% effective water yield. The best systems, like those of Paris where leaks may be easily repaired in easy-access sewage networks, have yields of 90% or more.

In order to limit investments in new resource development and transport and distribution systems (water mains, pumping facilities, reservoirs, distribution networks) the following steps must be taken to address the problem of water loss:

- Distribution systems must be carefully maintained, protected against corrosion and replaced when old, with careful attention paid to leak-prone areas.

- Water levels must be measured during production, at specific points in the water mains and distribution system and, if possible, at the connection point to each building (although some countries, such as Great Britain, often apply a flat rate for water service, most have concluded that effective water management can only be achieved through systematic metering of all water customers. Because metering makes each customer financially responsible for the water he actually uses, it is also an effective tool for limiting demand).

However much national and international authorities may view the job of water management in terms of simply limiting the development of new resources, this is its true mission: to promote water conservation through finding ways to curb individual consumption.

Obviously, leaks in internal plumbing networks must be avoided, and if water is correctly priced building occupants will also see this as being in their interest. The water is paid for in any case, providing the necessary means to manage and maintain the distribution system.

A simple economic incentive can lead to self-limits on the demand for water for personal grooming, household appliances, landscape irrigation, swimming pools, washing the car, etc., while still preserving the freedom of the user to choose the level of comfort he desires. Demand has gone down to some degree in all countries in which sufficient water prices have provided this incentive and made water users responsible for their choices.

In fact, the results of a recent study by the OECD (OECD, 1999) show that the setting of fee schedules can be an effective "demand management" tool under certain circumstances.

One may legitimately ask whether tying water demand to price levels and/or rate structures (that is "price elasticity of demand") can provide a strong enough incentive to change behaviour. There are indeed several factors that may prevent fee schedules from having an inhibitory effect on water consumption:

- It may not be possible to price water high enough to be economically "relevant" to the customer. For a person with a high income, for example, a modest price increase will not take much of a slice out of the monthly budget.
- Consumers need a minimum amount of water for basic subsistence and will pay whatever they must to obtain this basic amount.
- The more complex a rate system is, the more difficult it will be for consumers to grasp its total significance and make the appropriate choices.
- Consumers may have insufficient financial and technical knowledge of existing water conservation and effluent reduction techniques.

Despite these limitations, price changes have been shown to have some effect on demand under the right conditions.

In recent years, domestic water prices have been increased to some degree in certain OECD countries, particularly Denmark, France, Hungary, the Czech Republic, and Luxembourg. In each of these countries in which fees have gone up substantially in real terms for various reasons (in three of them, reduced subsidies played a major role), household water consumption has gone down.

Correlation between price increases and lower consumption rates are harder to establish in countries where only modest rate increases have occurred. In England and Wales, for example, total recovery of economic costs has been a reality for many years now, but because relatively few houses have meters (and are thus without this means of making the price – water volume connection) domestic consumption has continued to rise.

The results of the OECD study on the effects of metering in individual homes showed a significant reduction in demand after meter installation. The effects on peak-time demand are often much greater than those on mean demand if the rate schedule incorporates seasonal variations. For some European public water distribution services, the elasticity value (ratio of new price to old price, multiplied by the difference between the new and old volumes consumed) is as follows:

- For 216 communities in eastern France:	- 0.22
- For the Gironde department (France):	- 0.17
- For 282 Swedish communities:	- 0.20

This proves that an economic incentive can be used to control demand.

The OECD study confirms the results of a recent Italian study (Instituto per studie ricerche sui servizi idrici, 1998) which showed that, above a basic minimum cost, price increases do lead to lower water use, with consumers monitoring their own consumption. But the elasticity value is also dependent on how rapidly prices increase, on the socio-economic characteristics of the populations involved, on climate, and on the initial level of consumption. For this reason, there are wide disparities in the figures for different countries, making it impossible for the time being to draw any general conclusions about price elasticity that can be used to reliably predict water demand.

There is an extreme lack of uniformity among nations and this underscores the importance of finding leaks and repairing water mains first (IWSA, 1999). Until a distribution system's technical yield is known, it is impossible to determine whether it is even necessary to plan new investments.

Although sections 4.4.1.1 and 4.4.1.2 may do little more than underscore the difficulty of planning new facility construction without risking over- or under-investment, we nevertheless wish to emphasize the need for prior technical and economic studies as a step to limit investments (and the corresponding interest payments) in cases in which they will become useful only at a later date. The following should be borne in mind in particular:

- Water intake and treatment facilities can be built in modules and sections that can be rapidly completed (in under two years) once the decision is made to do so.
- Distribution systems must be modelled mathematically to determine necessary reinforcements in terms of both water mains and booster stations, so that distribution may be improved if demand increases or the network is extended.

Issues concerning water resources and their uses

Planning which water resources to develop in order to meet future drinking water needs is as complex and difficult as trying to predict demand. The following questions must be answered:

- What imperatives must be addressed to enable a system to withstand crisis situations (occasional pollution, flooding, exceptional drought conditions, electrical power failures)?
- Should a single water resource system be designed to allow spontaneous or intentional water transfers between urban consumers and other types of users (in particular, agricultural)?
- Should non-conventional urban water resources (brackish water, treated wastewater) be developed for specific urban uses in order to alleviate pressure on high quality water sources?

What level of safety should be sought in system design?

Continuity of service in crisis situations is a major public health issue when it comes to drinking water; it is also important to the local economy. In developed countries in which quantity demand is stable, a higher percentage of investments are now being directed toward the creation of interconnections, increased storage capacity, emergency wells in preserved aquifers, etc., all of which contribute to limiting the risk of service interruptions. Although sophisticated tools (warning stations, computer modelling of stream and aquifer pollution propagation, complex network simulations, etc.) are often used in emergency planning for dense population zones, there is no recognised method for analysing the overall operational safety of a drinking water system and making economically optimal investment decisions.

In particular, the notion of an acceptable failure risk under extreme drought conditions is rarely addressed in design studies, although it is a systematic feature of irrigation system design.

Nevertheless, cities in Mediterranean or arid climates have a certain amount of flexibility during very dry periods, because often up to 50% of consumption goes to maintaining landscaped areas. This type of crisis can therefore usually be weathered through local restrictions on watering lawns and washing cars, or through more or less equitable rationing (rotating water service interruptions). In southern California, the Metropolitan Water District also uses rate systems to control consumption by offering low-priced schedules for "interruptible" service, similar to the systems used by electrical power suppliers. In addition, there is often a surcharge on water provided during droughts.

In many cities in which the public water systems are not reliable, users adapt by using private wells in addition to water from the public network (Beirut, Manila, etc.). But in these cases the aquifer is often overused, compromising its future.

Should water resources be used for multiple uses?

The second question – water transfers between urban and agricultural users – goes back to the universally recognised concept of integrated management of water resources. The need for such exchanges can be determined by geography – drainage basins – or by the existing infrastructure. In semi-arid regions the existence of large water transport systems makes it possible to "share" water on a regional or even national scale (as in South Africa, Israel, the southwest US, Ukraine, Spain, etc.). In these countries, increased depletion of natural resources is becoming an issue because of water supply or geopolitical concerns (in the Middle East), or environmental impact. However, agriculture is usually the biggest consumer (generally more than two thirds of water resources in current use). This means that the main alternative resource for meeting increased urban demand is diversion from agricultural use. Such transfers can most often be made without serious impact on agricultural production, as long as they include financial compensations to allow farmers to modernise their collective or individual irrigation equipment and improve water efficiency. Transfers of this kind have been occurring on a voluntary basis for several years in the western United States and Chile, where water rights may be traded. Spain is currently implementing legislation to allow similar mechanisms.

In any case, the growth of new megalopolises in the southern hemisphere is leading or will lead to the need to develop new water resources that will probably be located at some distance from the urban areas. For 70 of the world's cities that have populations of more than one million, the next resource will be located more than 50 kilometres away. On this scale, water development projects are no longer "municipal" endeavours but regional undertakings. The water use of rural populations and farmers in the areas in question must be included in system planning.

What contribution non-conventional water resources can make?

On the theoretical level, this option may appear as a panacea. However, it usually requires a second distribution system for non-potable water in buildings and/or the city, and public health officials are hesitant to embrace this. New quality standards developed by public health authorities or the World Health Organization for recycled wastewater used to irrigate crops, water landscaped areas, and recharge aquifers will require treatments at costs that are often too high to pass on to users (with the exception, perhaps, of industry). "Primary network" users must pay for the "secondary network", which actually makes some economic sense. For these reasons, the number of concrete instances of recycled wastewater use remains rather low. But the following successful experiences may mark the beginning of a growing trend in coming years:

- In Paris, a distribution system for raw river water has been in use for a long time for watering parks and gardens and washing down gutters.
- In Kuwait, a raw water ("second class water") distribution network is used to distribute brackish water. In Gibraltar, sea water is distributed in this way.
- In Japan, "grey water" from domestic use (sinks, bathtubs, etc.) is treated in some buildings and used to flush toilets.
- On the Arabian Peninsula, many landscaped areas, golf courses, etc., are now irrigated with treated wastewater.
- In southern California, distribution systems are being developed to distribute wastewater treated with advanced techniques (including membranes) for industrial use, and to recharge aquifers subject to sea water intrusion.
- Finally, a single case of direct recycling: one quarter of the drinking water needs of the city of Windhoek in Namibia have been met for the past 25 years by treated wastewater. Intermediary lagooning, a polishing treatment using

activated carbon, and ozone disinfection are used. We have much to learn from this unique experiment:

- It is possible only through the enforcement of strict urban planning restrictions (industries are located in their own special areas; their effluents are collected separately and are not recycled).
- There has been no discernible effect on public health in the neighbourhoods that receive the recycled water mixed with fresh water.

4.4.2 Satisfying new regulations and reducing by-products

New drinking water quality regulations are stricter (see Table 4.4), requiring water treatment facilities and distributors to optimise and control the entire treatment process. Therefore, research in recent years has focused on developing technical know-how in basic treatment process controls, taking into account the entire treatment process as a whole.

Important advances in process controls for traditional treatments such as coagulation, flocculation, sedimentation, and filtration have been achieved through new developments in computer modeling tools. These powerful tools make it possible to optimise hydrodynamic aspects that were previously poorly integrated into process controls. Such approaches allow treatment managers to rigorously control the conditions for injecting and mixing chemicals and the doses applied, in order to achieve the desired treated water quality while minimizing the formation of undesirable disinfection by-products. Treatment costs are also optimised.

Thus the ozonation phase can now be better controlled to achieve desired levels of parasite inactivation for bacteria such as *Giardia* and *Cryptosporidium* while at the same time limiting the formation of by-products such as bromates. In the same way, sedimentation, filtration, and disinfection facilities are currently being designed or modified on the basis of these now essential criteria. Highly advanced, innovative sensors are also being used to further automate treatment facilities, thus achieving greater reliability and facilitating efforts to comply with new regulations. Neural computer networks are being used in drinking water treatment to control certain processes in real time, thereby safeguarding and maintaining optimum quality. Research in the area of treatment techniques and technologies has led to the development of interesting alternatives to assist in meeting new water quality regulations. Such alternatives include membrane filtration (micro-, ultra-, nano-filtration), oxidation, and UV light technologies, which have the added advantage of reducing the formation of disinfection byproducts. These technologies can be implemented alongside or in combination with conventional processes.

All of this recent progress has been documented and consolidated with existing know-how in written treatment guidelines and good practice manuals that contribute to the development and implementation of concrete approaches to ensuring quality at drinking water production plants.

4.4.2.1 Further refining the quality of treated water

Regulations concerning the quality of distributed water sometimes call for the development of new treatments to complement traditional coagulation-flocculation-sedimentation and sand filtration treatments. These traditional methods do not eliminate contaminants such as pesticides or chlorinated solvents, making it necessary to apply technologies that combine oxidation and activated carbon adsorption. Activated carbon adsorption is the best method for eliminating pesticides or organic micro-pollutants and should always be used downstream of any ozone oxidation treatment. Activated carbon may be introduced in powder form during sedimentation, or in combination with a membrane filter such as ultra-filtration. Activated carbon may also be used in the form of granules in a filtering and adsorption bed. Using models that have been developed to control these processes, facilities are now sized to better meet the targeted objectives. Carbon powder dosage and the frequency with which the activated carbon granules are regenerated are now optimised, thus meeting the dual objective of controlling quality and keeping down production costs.

4.4.2.2 Use of membranes in drinking water treatment

The use of membrane processes has expanded considerably in recent years. The most recent applications in drinking water involve micro-filtration (MF) and ultra-filtration (UF), the low-pressure membrane processes. Since 1988 when the first UF membrane plant was commissioned in France (Anselme *et al*, 1990), (Anselme *et al*, 1993), UF technology has boomed: there are now more than 70 plants worldwide.

Clarification and disinfection

Both MF and UF processes filter out mineral, organic, and biological particles such as algae. Both ensure a high level of removal for bacteria and cysts such as *Giardia* and *Cryptosporidium* without the use of disinfectants. It should be noted that, unlike MF, UF technology also removes viruses, owing to its tighter pore size (approximately 0.01 μm). UF has been found to exceed all current and anticipated disinfection regulations, regardless of variations in resource quality. In comparison, the efficiency and reliability of conventional clarification and disinfection processes are limited, and depend on resource quality variations and plant operating conditions.

UF can be used on its own for treating drinking water if the organic matter content of the feed water is not too high. But for surface water with high organic contents, UF is often not efficient enough to ensure the required water quality. To broaden its scope of application, UF can be used in combination with conventional treatments (adsorption, coagulation, etc.), making it possible to meet additional quality criteria as discussed below.

Micropollutants and organic matter

Treatments applying coagulation / sedimentation steps prior to UF may be considered in particular for surface water with fairly high organic contents, because the UF membranes cannot be operated directly on the resource because of the likelihood of fouling. This combined treatment may be also considered when upgrading conventional treatment plants to meet current and future regulations (regarding turbidity, *Cryptosporidium*, *Giardia*, disinfection by-products (DBPs) formation, etc.), replacing sand filters with UF membranes.

Innovative drinking water treatment process combining PAC and UF technology, has been successfully applied on very large scale: first at the Vigneux (55,000 m^3/day) and Apié (28,000 m^3/day) plants, then at Kopper (10,000 m^3/day), and now at Lausanne (65,000 m^3/day). Combined coagulation-sedimentation/UF process is being applied on an industrial scale in the United States in the San Antonio, Texas, project. This 34,000 m^3/day capacity production plant was commissioned in January 2000. A second phase leading to a doubling of capacity is under consideration at this stage of the project. The process combines powdered activated carbon adsorption (PAC) with UF membranes. For removing organic matter and trace compounds, this combined process is a viable alternative to more conventional methods such as granular activated carbon filtration (GAC) or ozonation.

The membrane provides a physical barrier preventing the passage of the PAC (Baudin & Anselme, 1995), thereby retaining the organic compounds adsorbed by the PAC, compounds which otherwise would the membrane trap not all. These compounds include organic matter, pesticides, compounds affecting taste and odour, precursors of disinfection by-products, etc. This process combines the advantages of UF and PAC adsorption. The main advantages of this process over conventional treatment methods are that:

- It limits the competition effect between organic matter and trace organic compounds.
- It is flexible: the dosage and type of PAC can be adapted to variations in influent water quality and spikes of particular pollutants.
- The process helps transfer the organic molecules and diffuse them towards the PAC.
- There is no breakthrough, in comparison with the conventional GAC process.

Nano-filtration (NF)

Nano-filtration is used mainly in the US for groundwater containing very low levels of suspended solids but high levels of total hardness, color, and precursors for forming trihalomethane and other disinfection by-products. These types of water are characterized by Total Dissolved Solids (TDS) concentrations of approximately 2 or 3 g/L and may contain 5–20 mg/L of Total Organic Carbon (TOC). The plants have demonstrated that nano-filtration is suitable not only for removing hardness but also for eliminating organic matter. Most membranes used in NF eliminate organic compounds with molecular weights above 400 daltons.

Nano-filtration provides a very high degree of overall organic matter elimination, but it generally needs to be associated with oxidation and/or adsorption to ensure total pesticide removal. An alternative to this polishing treatment is to combine adsorption with UF. The process discussed above in 4.4.2.2 makes it possible to eliminate a high proportion of organic matter. In addition to its lower cost, this process has two other advantages over nano-filtration: it can totally eliminate pesticides, and it does not retain salts. Furthermore, it can be applied directly in the treatment of surface water.

Integrated multi-objective membrane systems: UF pre-treatment followed by NF/RO

Recently, Thin Film Composite (TFC) membranes have been developed that operate at very low pressures (<10 bar), making it possible to substantially reduce energy consumption. But as these membranes are more sensitive to fouling, use of an efficient and reliable pre-treatment is critical for deriving the expected energy benefits. Raw water pre-treatment is essential with any NF or Reverse Osmosis (RO) system to avoid membrane fouling that can seriously affect plant capacity, degrade water quality, and shorten membrane life through exposure to harmful contaminants and/or cleaning chemicals.

The resulting Integrated Membrane Systems (IMS) combine the advantages of MF or UF for particle removal with the selectivity of NF or RO, thereby minimizing NF/RO fouling while controlling DBPs and Synthetic Organic Compounds (SOCs). Tight NF membranes remove over 95% of Trihalomethanes (THMs) and Halo Acetic Acids (HAAs) and 95% of total hardness.

Currently, pre-treatment is based on one or more conventional water treatment operations consisting of a disinfection step followed by thorough removal of the residual disinfectant, flocculation, deep-bed filtration, and addition of an anti-scalant chemical. These conventional pre-treatment processes are not always effective, resulting in a decline in membrane performance and deterioration of water quality. Thus the use of membranes for pre-treatment appears to be a very attractive alternative and could provide significant advantages, such as reliable protection of the RO membrane from foulants, elimination of flocculation chemicals, and smoother system operation. Feed-water pre-treatment is of particular interest in the case of polluted raw water sources such as river water, pond drainage streams, and irrigation streams.

Denitrification

Combination of a biological reactor with UF or a membrane bioreactor (MBR) process associates a biological reactor to eliminate nitrates, to remove solids and disinfect. The combined process is an MBR process with a hollow fiber membrane in which PAC is added to also adsorb SOCs.

The Douchy UF plant (France), which has been in operation since 1989 (Anselme *et al*, 1993), is an example of an industrial application of this process in drinking water production. The MBR makes it possible to treat the nitrates while retaining the advantages of UF to reduce turbidity and disinfect.

Conclusion

The use of membrane technologies in drinking water treatment systems depends on two factors: the quality of the source water and the quality objective for the treated water. The quality of the effluent produced by the different types of membranes will depend upon whether the influent is:

- sea water or brackish water,
- groundwater with a light organic matter load, or
- surface water, e.g. river water, with a heavy organic matter load but varying mineral content.

Ultra-filtration is perfectly suited to clarifying groundwater, mainly for particle elimination (suspended solids or micro-organisms). Coupled with PAC or a clarification step, UF can be used equally well to treat groundwater contaminated by micro-pollutants or to treat surface water with a heavy organic matter load.

Nano-filtration or low pressure RO can be used to treat highly mineralised surface water requiring softening. It eliminates organic matter and hardness in a single operation. Moreover, while reverse osmosis is necessary to make sea water or very brackish water drinkable, nano-filtration is sufficient for softening and partly demineralising brackish water. However, for use on surface water, both processes require intensive pre-treatments.

4.4.3 Use of desalination techniques

Various technologies are used today to remove salt from water. These technologies, reviewed below, may be divided into two main categories: distillation and membrane filtration. In distillation, the two processes used are multi-stage flash distillation (MSF) and multiple-effect distillation (MED), without or with a vapour compression distillation cycle: Mechanical Vapour Compression Distillation

(MVC). The membrane processes include reverse osmosis (RO), nano-filtration (NF), and electrodialysis (ED).

4.4.3.1 Distillation

Distillation is the oldest method used to remove salt from sea water. The principle is the same for all processes: sea water is boiled under certain pressure and temperature conditions and the evaporated water is condensed in chamber called a cell or stage. The process most often used today is MSF, although MED appears to be a more competitive alternative for small and medium-sized facilities.

4.4.3.2 MSF

This process consists of circulating sea water through a series of chambers set at increasingly lower pressures and temperatures to avoid having to supply additional heat at the entrance to each chamber. An external heat source (boiler) generates steam that is used to bring the water to a temperature of approximately 110 °C by the time it enters the first stage; the temperature is then gradually reduced to approximately 40 °C by the time it reaches the final stage. In the upper end of each chamber, the steam that is generated condenses on the heat exchangers through which the pre-treated seawater circulates. It is this seawater that gradually heats up across the different chambers until it reaches 110 °C as it then enters the first chamber. The water condensed in each chamber is collected in troughs and a pump is used to extract the brine in the last chamber. To limit the risks of clogging and scaling, precipitation inhibitor and anti-foaming products are added upstream of the heat exchangers.

4.4.3.3 MED

This process is based on the same principles of condensation and evaporation in a series of chambers (i.e. effects or stages). Unlike with MSF, the sea water is sprayed on tube or plate heat exchangers in which the steam circulates. With the heat exchange, the steam inside the tubes condenses into potable water, while the water sprayed on the outside is transformed into steam. This new steam is then introduced into the exchanger of the next chamber, and so on. Adding a thermal compression system may optimise energy efficiency. This system compresses the steam in the last chamber through a thermal compressor powered by external steam. The steam generated powers the exchanger of the first stage. When there is no external steam source, a mechanical compressor may be used in place of the thermal compressor. In this case, the process is referred to as mechanical vapour compression (MVC). The problem of scale formation is now under control, thanks to the development of more advanced inhibitors and to operating temperatures generally limited to 65 °C in the hottest stage.

4.4.3.4 Reverse Osmosis

Choosing which process and membrane to use depends on the type of water to be treated and the quality level to be achieved. The vast array of membranes available on the market today makes it possible to make choices on the basis of the pollutant(s) to be treated. Reverse osmosis is the only membrane technique that will be discussed for desalinating seawater. This process makes use of semi-permeable membranes through which the water passes. All solutes are rejected, with the exception of few small-sized molecules. To obtain economically viable yields of purified water from salted water, the osmotic pressure of the feed water solution must be at least doubled. Thus, for brackish water containing a few grams of salt per litre, pressures of between 5 and 30 bars must be applied. For seawater, pressures may reach as high as 50 to 80 bars.

4.4.3.5 Distillation versus Membrane Treatments

MSF and RO are the main technologies used for desalination. According to the latest Wanggnick survey (Wanggnick, 1998), of the 20,000,000 m^3 of water produced worldwide by desalination units, 19,000,000 m^3 are treated with RO and MSF. However, it should be noted that since 1990 the total installed capacity for osmosis has increased by more than 50%, compared to less than 30% for MSF (Figure 4.35). But RO can be used for all types of water, while MSF is only used for removing salt from seawater. Although mainly used to treat brackish water (50%), reverse osmosis has become more competitive in recent years for desalination of seawater because of the ongoing decline in membrane prices and the energy they consume. Nevertheless, membranes are very sensitive to clogging, requiring complex pre-treatments when the water to be treated contains a lot of suspended solids. However, it remains difficult to compare the costs of these two processes, because each situation has its own requirements based on water quality, energy and chemical costs, etc. Costs for the two processes do seem to be coming together. At capacities of 20,000 m^3/day, the cost of treated water in recent projects has been on the order of 0.8 US$/$m^3$, for an investment cost of about 1,000 US$/$m^3$ per day installed, regardless of the process used.

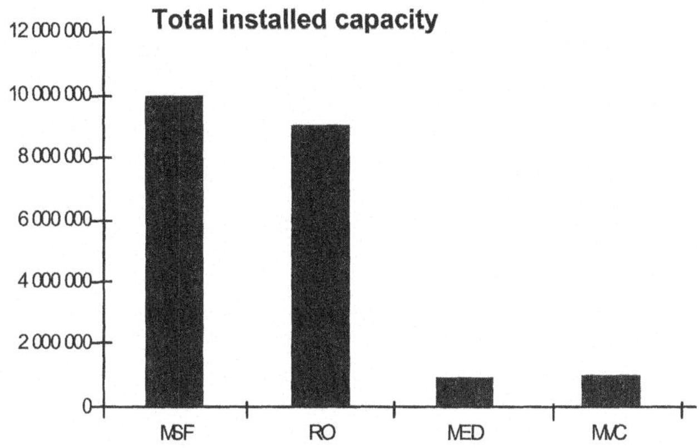

Figure 4.35 Installed capacity of various desalination processes

Energy Consumption in Desalination Processes

Unit capacity has little influence over how much energy desalination processes consume. The same process can consume different amounts of energy due to variations in equipment efficiency (Table 4.7 and Figure 4.36). High efficiencies result in reduced consumption, but at increased capital costs. An economic evaluation should be made to determine the optimum solution as a function of local conditions.

Table 4.7 Energy consumption of various desalination processes (Wade, 1993)

Process	Vapour consumption kJ/kg	Electricity consumption KWh/m^3	Primary energy consumption	
			Single process kJ/kg	Co-generation process kJ/kg
MSF	290.6	3.9	380.9	175.8
MED*	193.8	2.9	368.9	120.2
MVC	—	8.0	80	—
RO**	—	6.5	65	—

* With thermo-compressor

** With turbine for energy recovery

The conversion factor for converting electricity into primary energy (fuel) is 10.

85% efficiency for the boiler/turbine and 36% efficiency for co-generation.

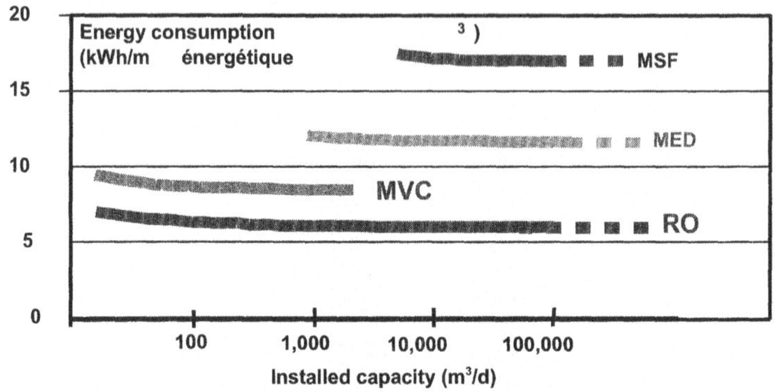

Figure 4.36 Variation of energy consumption with the installed capacity

4.4.4 Treating the wastewater and wastes from drinking water production

Treating natural water to produce drinking water generates wastewater that must be treated before it is released into the natural environment. Stricter environmental requirements have led to the development of new advances in the area of wastewater treatment techniques. Thus, backwash water from sand filters is now treated and then recycled back into the water production plant process and the suspended solids are treated and disposed of. Sludge produced by treatment plants is thickened and then discharged into sewers, or dewatered and sent to landfills. Intensive treatments (advanced separation equipment: rapid settling tanks plus filter presses or centrifuges) or extensive treatments (lagooning) may be used, depending on local conditions. Research and development are still underway to control and recycle wastewater with high salt contents resulting from membrane treatments such as nano-filtration and reverse osmosis. This research focuses on integrating the notion of re-use, and the wastewater in question is no longer considered as a waste. Dewatered sludges can now be re-used as construction materials or for road embankments.

4.4.5 Diversity of water distribution systems

4.4.5.1 Introduction

Given the immense diversity of urban water distribution systems throughout the world, it is difficult to define common key issues. However, above and beyond obvious distinctions between developed and developing nations, each with its own unique situation, several basic universal concepts should be addressed:

Viewing the system as a whole: In order to gain an accurate picture of a water distribution system, one must consider both the existing physical structures and the methods used to operate and develop the system.

Service to the end user: The final goal is always to provide adequate service to the user, who should pay for this service to the best of his abilities.

Longevity: Two issues must be addressed with regard to water supply:

Water sources must be of sufficient quantity and their quality must be protected. System maintenance and replacements must be conducted in such a way as to safeguard continuity of service.

Finally, a solution must be found to the problem of **financing** the considerable efforts necessary for developing urban drinking water supply systems across the globe.

4.4.5.2 Water distribution systems in developed nations

Most urban water systems in developed nations, and in a certain number of developing nations, have been around for some time. Current laws, regulations, and existing institutional frameworks relating to water have fully and explicitly defined the role of all the players, addressing obligations to water users, controls imposed by local authorities, rate systems, and financing mechanisms.

- The water resources in current use are subject to both quality and quantity controls. They supply water to all inhabitants and satisfy administrative, commercial, and industrial needs (for those without their own resources). Barring accidents, the service provided is continuous and of high quality (drinking water delivered 24 hours a day at an adequate pressure, even during peak consumption times).
- In most cases, these systems employ a metering system. Macro metering is used to measure the volumes of water produced and delivered to the distribution network inlet. The practice of metering individual consumers is common in Europe[(1)] and is spreading in the United States, where it is replacing the flat rate systems that are still widely used.
- Water prices have increased significantly over the past 10 to 15 years, now reaching between 2 and 4 euros per cubic metre. At the same time, per capita consumption for all types of uses has stabilized or declined, and is now generally between 200 and 300 L/person/day.
- Water distribution systems have a certain number of common characteristics, concerning which the following key points may be made (these do not include resource management and treatment issues, which will be discussed elsewhere):

[(1)] With the exception of the United Kingdom

- ***Transport - storage:*** Pipes for transporting water to storage reservoirs at distribution heads are sized to handle the average load on peak days. The storage reservoirs are designed with total capacities of about 50% of average daily production, allowing them to function as output regulators (keeping hourly flow fluctuations from impacting the transport and production systems)[(2)], pressure regulators, and emergency or fire-fighting reserves.
- ***Distribution*** systems share the following characteristics:

  Sizing: Peak flows and fire-fighting reserves must be taken into account, while still avoiding over-sizing, which may lead to low velocities that can contribute to deterioration in water quality.

  Equipment/maintenance: Proven materials must be selected for network extensions and new construction (ductile iron, polyvinylchloride (PVC), High Density Polyethylene (HDPE), as well as gradual refurbishment/replacement of the oldest mains, paying special attention to the interaction between soil and materials in order to implement adequate safety measures when necessary.

  Working pressure: This is regulated within a specific range that is often laid down by a water authority regulation. The minimum limit should be enough to supply pressure to buildings several stories high without overpressure; maximum limits (often 5 to 6 bars) should be set with a view to avoiding wear and tear on plumbing. Depending on topography, other measures may be needed to obtain the appropriate pressure, such as creating pressure zones, adding pressure regulators or, in some cases, installing variable speed pumping systems and/or booster stations at the end of the line.

  Quality of distributed water: Changes that occur in water quality between the treatment plant and the user's tap, due in particular to interactions between water and pipe materials, are receiving greater attention, leading to the development of appropriate corrective measures.

  Service connections: Replacement of old lead service connections is a subject of growing concern. New service connections are generally made of polyethylene with approved characteristics for drinking water.

- Investment needs for urban water distribution systems in developed nations are specifically identified. Low population growth and a trend toward lower per person consumption make for low investments in system reinforcement

---

[(2)] The ratio of peak flow rate to average flow rate often exceeds 2 to 1; at night this ratio often exceeds 4 or 5 to 1.

or extension. However, major investments must often be made to satisfy two main objectives with regard to water distribution networks:

- ***Making distribution networks more secure*** in order to prevent accidents (pollution of water resources, bursts of primary water mains, prolonged interruptions in power supply).
- ***Replacing*** existing physical structures, which are very old in most cases (more than 50 years).

- In the area of facility management, the main concerns of water distributors include the following:
    - Ongoing efforts are being made to ***reduce the non-revenue water ratio*** (Roche 2000) by reducing physical losses and installing more meters. The non-revenue water rate is generally less than 30% and is often as low as 20%, and the goal will usually be to maintain the low rate already achieved, rather than to significantly improve it.
    - Efforts are underway to optimise technical management of facilities using ***centralized management systems*** that incorporate decision-making software aids and simulated or real-time modeling in order to improve safety and reduce operating costs (Bos & Jarrige, 1990).
    - To improve management of existing physical structures, tools are being developed that make use of the full potential of ***geographic information systems***.

4.4.5.3 Water distribution systems in the urban environment of developing countries

In developing countries, urban water distribution systems run the gamut from situations comparable to those of developed ones, to cases in which only a tiny fraction of the population has easy access to adequate water from a source of acceptable quality. The difficulties are aggravated by rapid population growth that continues to increase each year, increasing the number of inhabitants that do not have access to the most basic urban services. These situations result partly from a lack of legislation, regulations, and appropriate institutions, but they also stem from insufficient human and financial resources on the part of the entities in charge of the water sector. Water prices are often too low and bill collection is sometimes poorly enforced, making it impossible to finance the investments that would be required.

Extreme variations are also encountered with regard to water resources, from situations of abundance (agglomerations located on the banks of large rivers) to critical situations in which already insufficient water resources are overtaxed and/or are of questionable quality due to urban or industrial pollution or the risk of salt intrusion. Water table depletion is extremely serious in many cases. Associated with these extreme variations in water resource quantity and quality are situations that are just as extreme with regard to consumption levels. Where major investments have been made to develop water distribution systems for nearby abundant resources, per capita consumption rates are abnormally high, due to both physical losses within the network and user waste. By contrast, where there are insufficient water resources and low investments in drinking water supply systems, per capita consumption rates are low, and lack of water often leads to interruptions in service for those who do have access to the public distribution system.

For the great majority of existing urban drinking water supply systems, the biggest issue is the effort to extend water services to everyone. The percentage of inhabitants who have direct or even indirect access to the public water distribution system is rarely 100% and in extreme cases can be lower than 50%. In addition, only a fraction of the population receiving water services has water delivered to a service connection; the rest receive it through various other means (most often, public standpipes, water vendors, etc.). Those who do not have water service must turn to other sources of usually unregulated quality (private wells, streams, etc.).

The great majority of systems have macro-meters at production point outlets. However, these are far from accurate and reliable. In many cases, customer water bills are still based on a flat rate system. In cases in which all consumers are metered, the metering system is often poorly managed, leading to under-counting and, consequently, to major under billing. The ratio[(1)] of non-revenue water is not always well known due to uncertainties with regard to macro metering, among other things. Its very definition poses a problem in the case of flat-rate billing since the volumes delivered must be estimated. However, typically, the ratio is about 50% and the two main factors that contribute to non- revenue water (physical losses on the network and under metering – or underestimation of the volume of water delivered to users[(2)]) are often equally to blame.

As regards technical issues, urban water systems in developing nations generally share the following characteristics:

- Water resource quality and quantity are not always controlled, and in some situations there is severe shortage. Proven, traditional treatment methods are generally used.
- The "storage" function is often partly fulfilled by water users, either because the distribution systems have been designed that way or because users who receive intermittent service store the water they need.

[(1)] ratio $(V_D - V_B)/V_B$ V_D = Volume delivered to the distribution system
V_B = Volume billed to the customers

[(2)] This also includes certain water uses that may be provided without charge: fire service, street cleaning, etc.

- Networks are often not large enough and sometimes do not serve outlying areas.
- Unsuitable pipe materials are sometimes used, such as steel that is insufficiently protected against corrosion, and asbestos–cement, which experience has shown will weaken with exposure to frequently corrosive water, and deteriorate over time.
- A number of factors (storage methods, insufficient network sizing, major physical losses and wastage by users) help to smooth down hourly production peaks. [(3)]
- Working pressures are often too low. When there is no pressure (intermittent service) the quality of distributed water can be affected by possible pollution intrusion in the system.
- Finally, available financial resources generally are not such as to ensure adequate maintenance and replacement of the existing physical structures.

Efforts must first focus on management so that existing systems are put to their best possible use. Reducing non-revenue water is a priority (UNESCO, 1999) in this regard because it will allow the distributing entity both to increase its financial resources and to direct volumes of lost or wasted water to under-supplied users. However, the investment needs are immense because it is often necessary to ensure all of the following at the same time:

- Development of new water resources (which are sometimes far off and costly).
- Refurbishment of deteriorating existing structures.
- Reinforcement of production and transport infrastructures to address current problems and provide service to new inhabitants.
- Extension of distribution systems to inhabitants who are not currently receiving water services and to new arrivals from population growth.
- Improvement of service quality.

The enormity of these needs calls for responses that are tailored to each situation (Mathys & Chambolle, 1999). To meet the urban drinking water supply challenge for the 21st century, innovative solutions must be made widely available, so that consumers will have services that meet their needs and expectations. Some of these solutions are already being implemented, such as social policies for offering service connections to poorer consumers, pay standpipes whose management is delegated to standpipe managers, or even special solutions for underprivileged neighbourhoods. These responses must take the big picture into account. They must be clearly defined and involve all

[(3)] It is not uncommon to observe peak time/average time ratios of below 1.15 or 1.10.

participants: government bodies, water managers and, in particular, water users. Water cannot be provided free of charge; payment is a necessary condition for avoiding waste. Contrary perhaps to what is often said, most users are well aware of this. It is widely known that those who are deprived of access to the public distribution system often pay a premium per cubic meter for the small amount of water they consume. Thus the future user must be involved in defining and implementing these solutions – with regard to both the required investments and their later management. Only then is it possible to ensure both a willingness and an ability to pay and to provide long-lasting service. However, all imaginable sources of financing must be mobilised. One measure is of particularly vital importance: to introduce rate schedules and well designed metering systems that discourage waste by users already on the system and make them stakeholders in the refurbishment and development of the drinking water supply systems from which they already benefit.

4.4.6 Quality – Operations – Maintenance

4.4.6.1 Quality of distributed water

The public drinking water system must guarantee the consumer high quality water at the point at which it enters his private system, that is, at his meter, if all of the water provided is metered, as is preferably the case. However, more and more often, water quality is measured at the consumers' tap, and therefore reflects any changes in quality that occur through contact with the buildings' internal plumbing system.

Water intended for human consumption must comply with the standards set by each nation and, in Europe, by the European Union (member countries may adopt differing standards, but only if they are stricter). The standards incorporate recommendations by the World Health Organization (WHO, 1999) which, to ensure maximum safety, are based on studies that include extrapolations from scientific data using analytical methods that can now be used to detect compounds in even trace amounts. Even the conditions of analysis are extremely important, since the leaching of metals from water pipes (lead, copper, etc.), for example, can invalidate comparisons between values obtained from water samples taken as soon as the tap is turned on and values from samples taken after the tap has run long enough to discharge all of the water that has been sitting in the internal plumbing network.

The most important thing is to rid water of pathogens (bacteria, viruses), that is, to disinfect it. For safety reasons, even water from non polluted sources of exceptional quality must be disinfected. In some cases one may wish to avoid the by-products of disinfection techniques using, in particular, chlorine or chlorine derivatives, because of their effect on taste or for health reasons. However, even in these cases a good disinfection technique must be used. If ozone is used, the water should still be lightly chlorinated to eliminate any bacterial regrowth in the distribution system.

In recent years, the pathogenic nature of certain protozoa (*Giardia*, *Cryptosporidium*) has been made known, and it has been necessary to thoroughly filter water before disinfecting it, since disinfection alone may not eliminate these parasites. Removing these pathogens can even require supplementary water treatments.

Thus, in addition to having access to approved testing laboratories, distribution system operators must have their own laboratories for conducting the most current analyses (especially bacteriological ones). If possible, a central research and analysis laboratory should also be available to all the operations of a single company for more advanced analyses of certain parameters. The individual operators' laboratories and central laboratories must be calibrated so that their results may be validly compared. They should also undergo a quality certification process.

4.4.6.2 Operation

What consumers want above all is uninterrupted service, that is, they want quality water at an appropriate pressure on a continuous basis. Service interruptions to make necessary repairs or for other actions must be very limited in number and of short duration and, if possible, they must be planned in advance. In particular, the water facility electrical power supply must be very secure, because water cannot be treated or delivered without electricity; it is indispensable for nearly all facility operations.

Although water treatment plants are susceptible to plant upsets and must be closely monitored, distribution facilities are generally very stable and, more and more often, automated. Nevertheless, they still require skilled and trained personnel to operate them, because quality service (supplying water of good quality with the appropriate pressure in the system) cannot be provided otherwise.

Distribution systems must be systematically checked, and observations should be made during repair interventions or in response to customer complaints. These observations should include:

- Pipes: age, bursts and their causes, and the condition of internal and external liners.
- Other structures: incidents related to construction work and fittings.

System output must also be monitored. The operator must certify the quality of the service he provides in accordance with ISO 9000 and ISO 14000 standards.

4.4.6.3 Water metering — customer management

Users of water services (including sanitation services if the cost of this service appears on the bill, as it often does) are very aware of the administrative and financial end of service management. They want to be listened to and informed of any new conditions that may arise, whether these be financial or technical, and of any change affecting water quality.

Private companies providing delegated management services on behalf of local authorities pay close attention to customer management. The practice of remote meter reading will continue to grow, saving labour costs and creating fewer disturbances at customers' residences but, in most cases, the meter reader will continue to serve as a link with the customer, advising him about excessive water consumption, how to protect meters against frost, etc.

In the future, the time between meter reading and payment collection will grow shorter, and the meter reader may even be able to draw up the bill, collect payment, and record the resulting transaction himself.

The customer needs someone to go to, if possible the same person every time, who can provide immediate information and help solve technical problems (adding a new service connection, repairing leaks in the public distribution system or, at the customer's request, in private plumbing, etc.). This person should have knowledge of the customer's billing status and be able to modify payment conditions (automatic payment, payment at the service office, split payments or instalment plan).

When a customer cannot pay his bills due to financial insolvency, he should seek financial assistance through local social services rather than receive water without paying. However, for bills relating to major leak repairs at the customer's home, it is best to accept partial payment and defer instalments over time.

Service administration involves paying close attention to bill collection; reminders must be sent to negligent or non-paying customers. Although companies may, as a last resort, try to obtain payment by threatening to cut off water service, this should be done only in the most extreme cases, and even then it may be best simply to reduce the supply of water to the service connection.

4.4.6.4 Maintenance

As emphasised in section 4.4.1.2, facilities must be carefully maintained and even refurbished from time to time.

On the production end, this involves upkeep of underground intakes, which inevitably age, leading to electrochemical and bacterial pipe corrosion, and mechanical, chemical, and biological clogging of the well. Before wells can be refurbished, their characteristics must be thoroughly known, and additional hydrological and chemical studies must be conducted.

Groundwater sources may become permanently or temporarily polluted, and care should be taken to protect them (protection perimeters for to the immediate, adjacent, and remote surroundings). Care should also be taken to avoid over exploiting them, as this may alter flows between stream and aquifer and, in coastal areas, create the risk of salt contamination.

When drinking water is produced from surface water, one must prepare for instances of incidental temporary pollution (if necessary, by installing warning stations far enough upstream of the water intakes). It may be necessary to stop pumping the raw water and/or change the dosage of treatment chemicals. To guarantee water quality, rigorous preventive maintenance methods must be applied and recorded clearly in operators' files.

As a general rule, all facilities should:

- Establish a preventive maintenance schedule
- Maintain a maintenance and repair shop
- Organise an emergency repair system with radio equipment for on-call staff
- Plan for possible incidents or accidents by developing precise written procedures for preventing and managing crises.

4.5 WASTEWATER AND STORMWATER COLLECTION

4.5.1 System design

New regulations on wastewater treatment plant are increasingly demanding in terms of reliability and safety, the aim being to improve protection of the receiving environment not only from the effects of chronic pollution but also from the consequences of pollution surges due to rain or accidental industrial discharges. Also, the treatment waste management costs are rising fast and permanent outlets for waste elimination or utilisation need to be established, with tight control measures and total traceability. For all these reasons, what is required is an overall management of the sewer system and its three components:

- collection network,
- treatment system
- elimination or utilisation (recycling) of waste and biosolids.

Optimum functioning of a wastewater treatment system therefore now means replacing the direct-to-sewer system with increasingly selective and increasingly closely monitored collection systems, so as to regulate the pollution loads reaching the treatment plants. Of course it will be easier to apply new sanitation designs to new town networks than in existing towns that often still have combined systems.

In many cases it is difficult to change combined systems in the old centres of major towns, owing to traffic disturbance, the cost of the pipe laying work and the numerous underground pipes and cables already present. Because of this, one often encounters mixed systems with combined networks in the city centre and separate networks in outlying districts.

Given the difficulty of achieving an ideal collection system with constant flow rates and pollution loads, with no danger to the treatment system, regulation systems are needed to temporarily hold back pollution before it is conveyed the treatment plant. Temporary storage of outflow from combined systems, equalisation basins at treatment plant inlets and mobile weirs.

In regions with regular, light rain, comparative analysis of investment and operating costs may lead decision makers to choose combined systems provided the treatment plant has sufficient capacity. But in Mediterranean regions, for example, where rain often comes in the form of heavy storms, separate networks are probably preferable.

4.5.1.1 Combined systems

Combined systems generally collect in a wide-diameter sewer (Valiron & Affholder, 1996):

- domestic wastewater,
- industrial wastewater (pre-treated or not),
- surface runoff,
- water from street cleaning.

These sewers, in stone or concrete (prefabricated or cast *in situ*), are subject to much wear and tear:

- erosion of floor,
- deposits of grit and bulky waste entering from surface drains,
- settled organic solids in sections with slight slope, with attendant risk of H_2S production,
- load increase due to heavy storms, which in extreme cases may lift manhole covers and possibly cause traffic accidents,
- discharge into the environment in wet weather,
- danger for operating staff, who in most cases have to enter the sewer for maintenance (toxic atmosphere, risk of sudden rise in water level, deterioration of handrails),
- variations in pollution load owing to changes in the hydraulic regime, so that the network stores water in dry weather and releases it in wet weather.

Most of these problems can be avoided when new networks are being designed, by making sure to lay self-cleaning sewers with high velocity gradients in the bed. Collector slope and profile must be designed to limit mineral and fermentable deposits as far as possible. For existing networks with mild slope, special measures can be taken to reduce these problems:

- create gully holes with settling compartments to trap grit and bulky waste at the surface before they are carried down into the sewer;
- create grit extraction sumps or traps for solids upstream of slight slope sections;
- use motorised mobile weirs to momentarily increase network load so as to keep within the network the additional flow due to light rain.

Similarly, procedures for access to structures must be established, taking the risks to staff into account. The air quality must be controlled before any inspection. Remote monitoring centres responsible for safety in the event of sudden rain or an incident in the water supply system (reservoir emptying, broken pipes etc.) must be alerted to the presence of operating staff in the sewer network. In collectors with access points and with particularly heavy grit deposits, it may be necessary to introduce mechanised cleaning procedures to re-establish hydraulic capacity. Such procedures may use mobile equipment (sluice boats, hydraulic cleaning with hydrojet trucks, winches, hydrojet valves) or Hydrass valves. Besides the problem of horizontal movement of grit in the sewer, there is the problem of vertical extraction; this can be solved using hydrojet or vacuum trucks or simply mechanical shovels or buckets.

4.5.1.2 Separate systems

In these systems stormwater and street cleaning water are collected in a separate network from domestic and industrial wastewater. These systems, too, may be subject to a number of malfunctions:

- partial obstruction by accumulated solid matter or restaurant grease;
- roots penetrating joints;
- crushing or breakage by traffic on the road above and/or earth movements;
- incoming connections;
- penetration of clear water into poorly sealed deep sewers that are temporarily below the groundwater table;
- illicit stormwater connections made by private individuals or firms to solve stormwater evacuation problems on their land, or because there is no stormwater network nearby;
- discharge of wastewater into the environment in wet weather;
- variations in flow rate and pollution load owing to storage/release events similar to those found with combined systems;
- quality of effluent collected in industrial areas.

Most of these anomalies can be diminished if mains and connection pipes are laid under good technical conditions in terms of pipe bed, joint quality, regular slope and branch connections. There are precautions engineers must obviously take in order to provide the services with as long a lifespan as possible: applying quality charters agreed with pipe laying contractors, and strict control of the work (testing backfill compaction, CCTV inspection, air- and water-tightness checks).

4.5.1.3 Stormwater networks

Urban spread and the increasing proportion of impermeable ground in large towns are greatly increasing flow rate surges during storms, creating a risk of flooding that can endanger both property and lives. As with combined networks, stormwater conduits are generally of large diameter and are subject to:

- silting,
- entry of bulky waste via gully inlets,
- surcharge of storm sewers in regions where rainfall is very intense.

The risks can be reduced locally in outlying areas by building stormwater basins or underground tanks, but it is in the interests of local authorities to encourage or oblige private individuals to ensure that runoff water can infiltrate on their land, before turning to much more costly alternatives such as porous roadways or reservoir roadways.

Studies of annual quantities of suspended solids discharged into the environment show, taking the example of Marseille (Table 4.8), that:

- the treatment plant eliminates over 80% of suspended solids flow from combined and sanitary sewer networks. If one takes into account the flow from streams at low water, this percentage falls to 70%;
- the stormwater network collects 47% of suspended solids output volume during the year and 44% taking wet weather into account.

Table 4.8 Suspended solids in the Marseilles stormwater network

Type Of System		Volume ($10^6 \times m^3$)	Suspended solids	
			Entering (tonnes)	At outlet (tonnes)
	Combined	66	11,900	
	Sanitary	36	11,000	5,150
Dry Weather	Stormwater	78	2,150	2,150
	Total	180	25,050	7,300
	Combined	7	800	700
	Sanitary	0	0	0
Wet Weather	Stormwater	18	2,400	2,400
	Total	25	3,200	3,200

In many cases there is no alternative to treating stormwater. Despite the efforts already made in equipping collecting networks and treatment plant, the suspended solids elimination yield at treatment plant outlets is, as a yearly average, slightly over 50% of the flow produced, for example, by the city of Marseille and its catchment. This shows how much work still has to be done in order to treat stormwater and overflow from combined systems.

Additionally, there is the problem of the pollution surge that occurs when it rains, given that urban stormwater can be up to ten times more polluted than dry weather wastewater owing to wash-off from roads, pollution gathered during percolation through the soil, and the re-suspension of deposits accumulated in the sewers. These sudden increases in flow discharged into streams can occur within a few hours wipe out the benefits to the environment of all the measures taken through the year to protect it. In all cases of particularly vulnerable natural environments, stormwater must be treated. New technical solutions (lamellar settling with sludge recirculation) are now available to treat stormwater by physico-chemical means, in very compact facilities. This problem of overall management of dry weather and wet weather pollution flows has revived the debate over the comparative advantages of combined and separate sewer systems.

Which way are planners now moving? Certainly, separate networks are more expensive to build, especially if stormwater must be treated to protect the environment from pollution surges during wet weather. But the technical solutions for treating wastewater (eliminating organic pollution biologically) and stormwater (eliminating solid pollution physico-chemically) are very different. Of course it is very hard to give a general answer to the question, as one must take account of such regional particularities as:

- intensity of precipitation,
- vulnerability of the receiving environment,
- existing infrastructures.

4.5.1.4 Vacuum and pressure networks

Sewer systems under vacuum and pressure offer an alternative to traditional gravity-flow wastewater collection systems. Their use should be considered when one or more of the following conditions exist:

- Insufficient slope of the terrain
- groundwater table close to surface
- Low population density
- Subsoil not suitable for trench work
- Intermittent wastewater production (campgrounds, seasonal activities, etc.)
- Significant environmental constraints
- Possibility of major road repair costs
- Subsoil obstructed by existing networks

Pressure systems are manifold systems in which each sub-branch is equipped with a discharge or injection device located in a central main. Several variations exist, which differ primarily according to the installation method and type of injection device used (air ejector or pump).

Vacuum systems have a central tank equipped with a vacuum pump. The tank communicates directly with a sealed collection network. Household wastewater flows into cisterns at atmospheric pressure. When a cistern is full, a valve opens automatically and the water is sucked into the system and transported to the central tank.

The choice of which technique to use should be based on a technical and economic comparison. An important advantage that systems under vacuum or under pressure have over gravity-flow systems is ease of installation. Capital and operations costs are highly dependent on the number and density of residences receiving service.

4.5.2 Private connections

Private connections to dwellings are designed to collect domestic wastewater and convey it to the public sewers. There are some basic rules for installing branch connections, e.g.:

- they must be capable of resisting the pressure when the network is full to natural ground level;
- they must have an inspection hole beside the highway, as close as possible to the façade of the connected property, giving maintenance staff easy access;
- the inner diameter of 0.2–0.3 metres, straight routing and longitudinal slope of 3% are designed to facilitate evacuation of solids and avoid the formation of deposits;
- the junction with the public sewer should be at an angle of 45° to 60° facilitates wastewater inflow.

Two types of junction device can be installed to avoid backflow of unpleasant odours:

- disconnector siphon with water seal,
- direct passage device with odour-tight seal. In practice, this device creates fewer problems of obstruction and hence maintenance because it has no low point liable to get clogged.

Private connection quality control has also become obligatory for network management, to protect the public network and minimise subsequent maintenance operations. When a connection is requested, and when inspections are made, the service managing the system must also make sure the private systems are properly separate, to avoid stormwater entering the public sanitary sewer.

These compliance checks generally take place in two stages. First, any sanitary sewers that are receiving stormwater (gutters, gully gratings) are identified by injecting smoke into the separate networks and finding where it re-emerges. The work team assess the area affected so as to class the anomalies in order of priority. The next step is to make the necessary administrative approach to users whose installations are not in compliance (demand that the installation be brought into compliance, assessment of the type of work needed, reminder letter, etc.). Lastly, periodic information campaigns must be organised to remind users that the networks are not intended for evacuating chemicals, pharmaceutical products or, above all, used motor oil.

4.5.3 Industrial connections

It is common practice for industrial effluent to be conveyed into the public sanitary sewers. In France, for non-domestic effluent, industrial facilities are not obliged to be connected to the main sewer and therefore have to have the authorisation of the community. The Mayor or wastewater treatment company sign a special waste discharge agreement in which the network manager lays down particular conditions (such as pre-treatment on site, retention of toxic substances at source, flow regulation basin or self-inspection by the company which seeks connection to the public network). The provisions adopted in France can easily be applied in other countries, as they have a number of advantages:

- economic advantages, especially for smaller firms, because collective wastewater treatment is much cheaper than individual treatment. Above a certain business size, this advantage is obviously slight or nil.
- better control of discharge quality for the network manager where *pre-treatment* facilities are required.
- responsibility is shifted to the company in the event of non-compliant discharge having an impact on the public treatment plant and hence the environment.

Connections for industrial discharge can cause various malfunctions if they are not closely monitored:

- collector sewers can be deteriorated by free acids or grease that coats the walls of the public network as it cools down
- the biological treatment process can be disrupted since it is vulnerable to toxic or chemical substances or variations in organic pollution load
- the quality of the treatment sludge can be affected, and levels of organic matter or mineral trace elements may make agricultural use of the biosolids inadvisable
- sewer maintenance staff who have to enter accessible parts of the network may be endangered.

In a word, industrial connections need to be tightly controlled by the network manager. The nature of this control depends mainly on the type of industry. Control is exercised through the authorisation granted after thorough investigation of the company's activities, the uses water is put to and the types of pollutant that may find their way into the effluent. In some cases authorisation may be conditional on technical specifications designed to render the industrial effluent comparable to domestic effluent. This generally means making *pre-treatment* obligatory to adjust pH, retain toxic substances at source, remove grit and grease, or settle the effluent before discharge. A special connection agreement must be appended, defining the company's technical and financial commitments towards the network manager, who will then have to regularly check that the effluent is in compliance with the specifications.

Obviously effluent from food industry facilities will not need such strict controls as leather tanning or surface treatment plants. Controlling industrial discharge at source is probably the only way to make businesses more aware of the consequences of uncontrolled discharges, make sure they pre-treat or even self-inspect, and ensure that the network manager is updated with any inspection results.

4.5.4 Pumping stations

Pumping stations are needed for built-up areas at lower altitudes than the gravity-feed networks. Delivery conduit length and static discharge head vary, sometimes very widely, according to ground topography. Design defects can expose pumping stations to:

- inflow of bulky waste in the case of combined systems, clogging pipes or suction screens
- grit deposits on basin beds or grease on surface
- anaerobic fermentation leading to formation of hydrogen sulphide and evil smells at the point of discharge
- electrical or electro-mechanical breakdowns.

Most of these problems can be solved at the design stage, by correct sizing of tanks and pumping capacity, choice of pump and delivery conduit routing.

In combined networks, installing bar screens at the tank inlet can be a way of protecting the pumps from obstruction. Basin bed geometry can also make it easier to recover grit and solid sludge. Mechanical stirrers installed in the large tanks to stir up the water can be an elegant way of reducing the problem of grit and grease accumulation. As regards automatic controllers, installing ultrasound head measurement sensors (not in contact with the effluent) can be far more reliable in the long run than the usual mercury ball level controller which, being in contact with surface grease, requires regular maintenance.

Lastly, the risk of H_2S formation can be assessed at the outset to allow the engineers to prevent sewer corrosion further downstream. The conditions that favour H_2S formation are now well known (prolonged retention time in delivery conduits, pumping, septic capacity of the effluent, insufficient effluent velocity in the conduits). Similarly, the recommended measures are now thoroughly mastered and solve the odour problem. So when retention time in the delivery conduits is long, the engineer must install an effluent treatment system that uses either:

- aeration by air injection
- addition of an oxidizing reagent
- addition of an iron based reagent (ferrous sulphate or ferric chloride) to precipitate out the iron sulphide.

4.5.5 Quality, maintenance and operation

Sewer systems must be operated in such a way as to maintain continuous flow, prevent discharge into the environment and, in short, improve the network's collection rate. Maintenance quality can be assessed by a few indicators, e.g.:

- the number of pipe obstructions per kilometre of network
- the number of emergency interventions on branch pipes
- the rate of collection or volumes of untreated wastewater discharged into the environment
- the number of justified complaints concerning the state of gullies or the presence of evil smells.

Some tools have now become indispensable to network operators for controlling the quality of their services:

- a centralized control system recording all remote monitoring data from the field:
    - state of the lift pumps
    - flow rate or head in the main collectors or at the main storm overflows
    - storm gauging of the area
- computerised map with associated data base allowing easy updating of structures; making plan extracts or visualizing longitudinal profiles
- high service management quality compared to standards such as ISO 9000 or ISO 14000, making it possible to standardise procedures, identify cases of non-compliance, take corrective action and manage quality indicators.

These increasingly indispensable operating aids are designed to help optimise the methods and efficiency of preventive sewer cleaning. Preventive maintenance of 10–20% of the network's length per year can keep down the incidence of obstructions to 0.5 obstructions per km of sewer per year. Similarly, taking account of pipe slope and the past chronology of incidents focuses the network manager's attention on the "problem" parts of the network and allow him to take diagnostic action and hence make proposals for work required, or modify the preventive cleaning operations.

Lastly, it is essential for decision makers to have diagnostic studies run if they are to program network rehabilitation, extension work or new connections. These analyses can be carried out by outside consultants or on an ongoing basis by the network operator.

4.6 WASTEWATER TREATMENT

4.6.1 General

Wastewater treatment methods are designed and implemented in such a way as to provide the maximum protection to the environment that will receive the resulting discharge, keeping within regulatory limits. They are chosen in accordance with the specific characteristics of the natural surroundings and their ability to absorb limited residual pollution flows downstream from the treatment plant.

In some cases secondary wastewater treatments are used if the water is to be recycled for industry, where it has many applications for which potable water is not required (cooling water or even process water in some cases). Recycled wastewater may also be used for aquifer recharge, particularly in coastal areas where salt water intrusion may be a problem or in urban areas that need to protect against runoff intrusion. Finally, it may be used to irrigate crops and landscaped areas, both public and private. Likely candidates to receive this specially treated wastewater are, of course, regions with limited raw water resources. A good case study can be provided by the West Basin project (Levine & Lazarova, 1999) (U.S.A.).

The West Basin Research and Development Program was organised to develop safe methods for recycling wastewater in order to operate facilities more efficiently and cost-effectively. The project has four main partners: the West Basin Municipal Water District (WBMWD), UCLA, UWS, and LYONNAISE DES EAUX (CIRSEE). All research activities are in the applied research category, and results will be used to improve technical and analytical methods for wastewater recycling.

Research activities fall into these major areas:

- high purity water production using reverse osmosis and other membrane-based technologies (ultrafiltration, nanofiltration)
- membrane cleaning
- water infiltration modelling
- corrosion assessments
- filter and ultra violet disinfection performance evaluations
- water quality monitoring

For this last activity, the quality of the re-used water will be monitored at three different sites in the Los Angeles region and the final effluent qualities from each treatment plant will be compared to those of the region's natural resources. The effectiveness of membrane filtration and reverse osmosis techniques will be compared to the effectiveness of clarification after lime softening and reverse osmosis. In addition to general treatment parameters (TOC, nitrogen, etc.), new quality parameters proposed by CIRSEE will also be monitored (trace organics, boron isotopes linked to anthropogenic pollution, and stannous organics for sanitary tracing).

The research project is still ongoing, but some conclusions can be drawn from now. Wastewater treatment methods have evolved considerably in recent decades, becoming safer, more effective, and better suited to urban and natural environments.

The disposal of solid waste (sand, screenings, grease, and residual sludge), the final product of the wastewater treatment process, remains a challenge that must be fully addressed by owners, development engineers, and city planners before treatment plant operators will be able to provide a full set of high-quality services.

4.6.2 Toward safer treatment methods

Current treatment methods are generally designed to eliminate pollution linked to suspended organic matter in the first phase, and are often accompanied by efforts at partial or maximum removal of the nitrogen and phosphorus, particularly if the waste is to be discharged into sensitive areas or recycled for industry or water table recharge.

Combined systems that collect rain water must be designed to handle the resulting surges in hydraulic and organic loads, and incoming flows must be limited if one is to avoid irreversibly disrupting the systems' hydraulic and biological functions; an overload coefficient – often equal to three times the dry weather flow rate – must be applied.

Some areas, such as seaside or winter resorts, face the problem of extreme variations in seasonal populations. Overloads on the order of ten times the off-season value are quite common, calling for the use of special treatment methods. To ensure high quality effluent in the presence of highly variable pollution loading rates, the treatments used in such areas must be very flexible and extremely safe. Conventional biological methods using activated sludge to treat domestic wastewater are precluded because of their long response times, which make it too difficult to adjust to variable flow rates and thus remain within effluent standards. Physico-chemical treatments that adjust more rapidly to incoming loading rates have been implemented alone or as a complement to existing biological processes, depending on the effluent quality that must be attained.

In many cases, the multiplicity and complexity of the processes involved, and the need to continuously monitor them, have led to the development of new solutions in the form of plant design (parallel systems so that one line may be maintained without shutting down the other), structure size (lower hydraulic and applied volume loads) or, more recently, continuous sensors and analysers. These sensors and analysers are reliable, require less maintenance, and can be used to automate various adjustments and control processes, such as aeration or sludge circuits, in real time.

Such an integrated monitoring is under development within a research CIRSEE programme (Caulet *et al*, 2000).

The C3A research programme (Capteur, Contrôle, Commande en Assainissement – Sensor, Control, and Command in Sanitation) signed between Lyonnaise des Eaux, Thomson-CSF Services Industrie, Degrémont, and SERES has been organised to develop a global management system for wastewater treatment plants to improve:

- real-time adjustment of wastewater treatment plants based on variations in plant influent
- operator assistance in diagnosing problems with a view to preventing malfunctions

Several operations are involved, including:

- development and adjustment of a sensor for identifying and monitoring effluent quality at the plant
- inclusion of a test data acquisition, processing, and transfer medium
- data processing (fuzzy logic tools, neural network) for extracting and using data in models and automated systems
- dynamic simulation for correcting possible malfunctions.

The project is being implemented for the first time at the SARO (Syndicat d'assainissement de la région d'Orgeval) treatment plant in Morainvilliers, France. It is open to other communities to assist them in meeting the new regulation concerning self-monitoring for sanitation systems.

4.6.3 Better treatment performances

As discussed above, recycling options and standards for discharge into natural environments require that wastewater treatment methods evolve from treating pollution linked to suspended organic matter, to include:

- treatment of industrial pollution
- nitrogen and phosphorus removal
- treatment of bacterial pollution (UV, ozone, others)
- the use of coupled membrane treatments (MF, UF, NF, RO)

For domestic effluent of average composition, traditional processes are sufficient for eliminating suspended matter through simple gravitational settling, and degrading organic pollution through conventional activated sludge. The usual efficiencies of processes and residual concentrations in major pollutants are shown in Table 4.9.

Table 4.9 (Opposite) Efficiency of the main wastewater treatment processes (Valiron, 1994).

	Treatment process	SS	BOD5	COD	NK	NGL	Total P	Total Germs	Fiability	Recei-ving water
0	Raw water	350	300	600	60	15	10^8/100 ml			
1	Primary settling	50-65% 120/175	25-35% 195-225	25-35% 390-450	weak 55-60°	weak 55-60	Very weak 12-15	week < 1 log	unfair	
2	Physico-chemical Treatment	60-90% 35-140	50-70% 90-150	50-70% 180-300	weak 55-60	Weak 55-60	80-95% 1-3	weak < 1 log	Very good	Non sensible
3	Activated sludge high load (3 kg DBO5 m3/j)	70-92% 30-100	65-87% 40-100	60-83% 100-240	< 10 % ≥ 55	< 10 % ≥ 55	< 20 % ≥ 12	1 log.	Bad	Non sensible
4	Activated sludge medium load	88-95 % 15-30	90-93 % 10-20	85-88 % 70-90	35-60 % 25-40	30-40 % 35-40	< 20 % ≥ 12	1 - 12 log.	Good	Normal
5	Activated sludge low load	90-95 % 10-30	93-97 % 10-20	88-92 % 50-70	40-90 % 6-35	30-85 % 10-40	< 30 % ≥ 12	2 log.	Good	Sensible for nitrogen
6	Lagooning	70-90 % 30-100	65-90 % 30-100	60-85 % 90-240	Partial reduction	Partial reductio n	> 20 % ≥ 12	2 - 4 log.	Good	Non sensible to normal
7	1+Extende d aeration	90-95 % 10-30	> 90 % 5-20	85-90 % 50-70	> 90 % > 5	> 80 % ≥ 10	30 %	2 log	Very good	Sensible for Nitrogen
8	4+Phopho rus precipitati on	90-95 % 15-30	90-95 % 10-20	85-88 % 70-90	35-60 % 25-40	30-40 % 35-40	80 % < 3	1 - 2 log	good	Sensible for Phosphorus
9	7+biologic al Phophorus removal	90-95 % 10-30	90-95 % 15-20	85-90 % 50-70	> 90 % < 5	> 80 % ≤ 10	≥ 80 < 2	2 log	Very good	Sensible for Nitrogen and Phosphorus
10	(2,4 or 5)+tertiary filtration/s and or expanded clay like Biolite	> 95 % < 5	> 95 % < 10	92-95 % ≤ 50	60-90 % 5-35	40-90 % 10-35	40 à 90% 2-9	1 – 2 log	Good	
11	9+Tertiary Phosphoru s removal	> 95 % < 5	> 95 % < 10	> 95 % < 50	> 90 % < 5	> 90 % < 10	99 % < 1	2 log	Very good	Sensible for Nitrogen and Phosphorus
12	2+Biofiltre	90 % 20-30	90-95 % 5-20	85-90 % 50-70	30-50 % 20-40	< 10 % 55	80-95 % < 3	1 – 2 log	Very good	Normal sensible for P
13	2+Biofiltre C, N, DN	> 90 % 10-20	> 90 % 5-20	> 90 % 50-70	> 90 % ≤ 5 ≤ 10		80-95 %	1 - 2 log	Very good	Sensible for Nitrogen and phosphorus

Recent studies (Ginestet, 1999) show a need for better knowledge of the product to be treated (the residual water) in order to optimise wastewater treatment systems (sizing, modelling, risk evaluation, treatment choice and performance). It is necessary to go beyond the traditional parameters (BOD, COD, SS, NTK, P, TAC) and fraction the product according to the behaviour of the pollution during both water treatment and sludge treatment processes. Typing residual water in this way makes it possible to determine:

physico–chemical type: settleable, non-settleable–coagulable, and non-settleable–non-coagulable fractions.

biological type: easily biodegradable, rapidly hydrolysed, slowly biodegradable, and/or non-biodegradable organic matter, all of which have different degradation, de-nitrification, and phosphorus removal potentials and speeds.

The overall objective is to better match the characteristics of the residual water not only to the local situation, but also to the operating risks of the facilities concerned, which are designed in light of the results of raw water characterization tests.

When necessary, conventional biological treatments may be used in conjunction with physico-chemical coagulation flocculation treatments to assist in primary settling. This would apply to wastewater with especially high levels of suspended matter, or to the removal of "hard" organic pollution that does not easily biodegrade. Used in conjunction with filtration to further refine the clarification of treated effluents, these physico-chemical treatments may also be used to remove phosphorus through precipitation of insoluble phosphates. To meet discharge standards in bathing or shellfish breeding zones (Table 4.9), it may be necessary to occasionally or continuously disinfect the treated water. Various technologies such as ozonation or ultra-violet light may be implemented in these cases, but chlorination must be avoided due to the disinfection by-products (organo-chlorinated cyto-toxic addition compounds).

More recently, membrane coupling techniques from micro filtration to low-pressure reverse osmosis, including ultra filtration and even nano filtration, have produced water qualities acceptable for recharging aquifers threatened with salt intrusion, for irrigating landscaped areas, and for various industrial uses. New changes in membrane filtration techniques have made it possible to implement technically and economically acceptable solutions on a large scale. The example of the West Basin (described in 4.6.1) provides a glimpse into how these technologies might be applied in regions without abundant water resources.

4.6.4 Compact, low-impact equipment for urban environments

The choice of whether to use extensive processes (lagooning, extended aeration) or intensive processes (physico-chemical, biological filtration) will depend on the special circumstances of the treatment plant site. Design engineers must consider various parameters in seeking solutions to minimise cost, plant nuisance factors (noise, odour), and the effects of parameters associated with urban areas (land cost, topography, etc.).

Among the selection criteria, it is especially advisable to consider overall and operational) and solid waste disposal costs (sand, grease, screenings, excess sludge).

In the most challenging areas where land cost is very high or where mountains create special restrictions, or in coastal areas with dense urban populations, added costs may be as high as 30–50% of the generally accepted base costs.

Moreover, in some cases, residences may be located near wastewater treatment plant areas despite prior zoning plans. In these cases, plant structures must be covered and ventilated or, better yet, the treatment volumes must be deodorised. Compact structures such as accelerated clarifiers with lamellar structures, or biofilters with high purification yields and appropriate volume loads are best suited to these types of situations.

A research project having last several years was aimed at determining which plants have the lowest nuisance factors (noise, odour, waste).

Recently built and commissioned, the Seine Centre Treatment plant for the city of Paris (France), provides a good example of such an integration (Chèze and Gousailles, 1998). The "Seine Centre" treatment plant located in the heart of the city of Colombes, France, is a model of urban integration. Here, biofiltration techniques are used in association with primary treatment by lamellar settling with the addition of reagents. This choice was dictated by three considerations: dynamic equipment control, treatment selection based on the requirements of the receiving environment, and strict environmental restrictions (in particular, the olfactory nuisance factor).

Commissioned in 1998, the plant has an inflow capacity of 2.8 m^3/s to 12 m^3/s. During dry weather (low flow), the plant removes carbon, nitrogen, and phosphorus; in wet weather, treatment efforts focus on organic pollution.

The system uses pre-treatments followed by lamellar settlers (9x140 m^2) for treating phosphorus and particulate matter; soluble pollution is treated during a biological phase with three biofilter stages: 24×104 m^2 for carbon (stage 1), 29×112 m^2 for nitrification (stage 2) and 12×104 m^2 for the final post-de-nitrification, with the addition of a carbonaceous substrate (stage 3). Normal operation during dry weather has maximum performance values (80% for phosphorus, 70% for NGL, 95% for NH_4). The number of filters may be adjusted according to the incoming flows. However, lower doses of the injected reagents may be used at certain times to reduce operations costs; slightly lower performances will be achieved while still meeting the quality requirements of the specifications manual (before the 1994 rulings).

Under **dry weather conditions** (Figure 4.37) three complementary options are available:

- treatment with maximum or partial de-nitrification: the de-nitrified water (end of stage 2) is partially recycled at the beginning of stage 1, making it possible to treat the carbon and recirculated nitrates; this is in the pre de-nitrification phase. The methanol normally added in stage 3 (post de-nitrification) is not used. However, stricter quality requirements may be satisfied if pre and post de-nitrification take place simultaneously to achieve maximum nitrate removal.
- treatment without de-nitrification: the three biofilter stages operate in parallel and fulfill the same functions but at much lower pass speeds.
- treatment without de-nitrification or phosphorus removal: It is no longer necessary to inject reagents (ferric chloride and methanol); stages 1 and 3 operate in parallel to eliminate organic pollution; the intermediary stage 2 continues to provide nitrification.

Under **wet weather conditions**, (evaluated at 80 days per year), two treatment methods are operational, and de-nitrification is no longer supplied (Figure 4.38):

- wet weather incoming flow rate of 8.5 m^3/s for 8 hours: this configuration is followed by a dry weather period. Stages 1 and 3 operate in parallel (carbon treatment) and stage 2 continues to provide nitrification.
- wet weather incoming flow rate of 12 m^3/s: the three biofilter stages operate in parallel, and nitrification is reduced (35%).

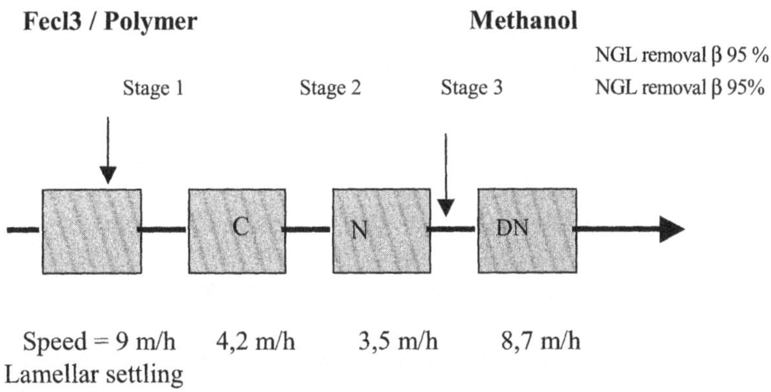

Figure 4.37 Operation of the Seine-Centre plant under dry weather configuration

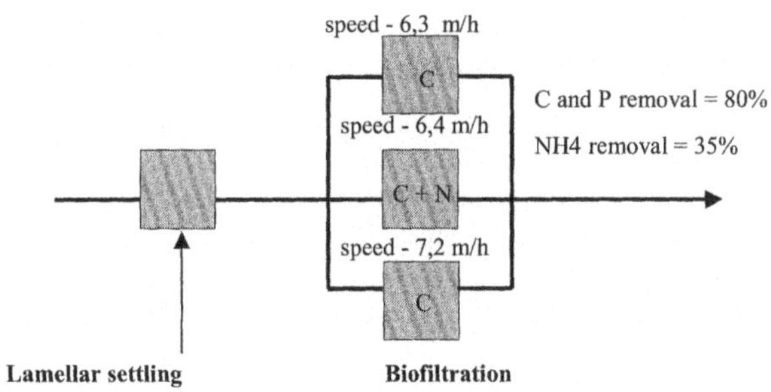

Figure 4.38 Operation of Seine-Centre plant under wet weather configuration (12 m^3/s)

The plant has been operating for more than a year, and nearly all configurations have been tested; results are in line with expectations. This is a clear example of adapting technology to quality requirements that vary over time. In this type of situation, the optimum technical–economic solution is sought using the most flexible processes available.

4.6.5 Process integration (sludge, waste, energy)

Processes for collecting and treating wastewater and stormwater generate various types of waste that are discharged via systems which, to ensure maximum environmental protection, must be designed in accordance with the characteristics of the waste and local conditions. All of this waste is high in pollution and treatments must be specifically designed to achieve its final removal.

It is now generally agreed that waste should be removed and treated at the source, in the sanitation networks, using simple collection methods (sand traps, grease interceptor tanks, screening) to avoid needlessly overloading process operations at the treatment plant site.

In some cases, processes suitable for treating wastewater can be also used to treat wastes generated by urban communities, including solid household waste (household refuse) general characteristics of which resemble those of certain sanitation wastes (sludge and grease). Common solutions must be sought to achieve economies of scale and make use of common facilities, thereby reducing construction and operating costs. To find these solutions, it will be necessary to take the various availabilities into consideration during the first feasibility studies.

It is impossible to collect wastewater without also collecting road sand. The solids resulting from the regular cleaning of the collection systems and any road sand collected in various traps upstream of the distribution systems must be treated in specific ways. It may be possible to recycle these materials for various uses such as pipe installation bedding: the wash water from this process must then be sent to the wastewater treatment plant. Special treatment units may be installed for these purposes, and the industrial plant based at Evry (France) is providing a good example of what can be achieved.

Sewerage systems (wastewater and stormwater) are cleaned regularly in order to safeguard their hydraulic capacities. The solid residues resulting from this cleaning process are sometimes referred to purification by-products. Due to the upcoming closure of landfills scheduled for 2002 in European countries, these products must now be treated.

Treatment consists of separating solids according to type (sand, plant debris, and oversize waste) from the liquid phase, laden with organic matter.

The Evry purification by-product treatment plant uses purified water from the wastewater treatment plant to wash the sand so that it can be recycled as trench backfill material with an organic matter content of less than 5%.

The treatment plant has a nominal capacity of 11.5 tons per hour (17,500 to 23,000 tons per year) and can handle the purification by-products of the entire department of Essonne, France.

The facility includes:

- a truck weighing station

- a deodorised building housing three storage pits and the sand washing treatment unit
- a physical–chemical treatment unit for the wash water
- a specific sludge dewatering unit

The automated loading station, equipped with a skip that has 500 litre buckets, supplies a rotating screen that is used to separate the solid elements according to size, sort them, and store them in skips (gravel and oversize waste between 10 and 30 mm in size, and large oversize waste above 30 mm).

The water containing sand and plant debris first enters a hydrocyclone before entering a densimetric separator with a rotating strainer to remove the plant debris. A second hydrocyclone followed by a vibrating dandy provides counter-current washing and collects the washed sand into a silo (120 m^3).

The wash water undergoes physical–chemical pre-treatment in order to precipitate the suspended matter rich in heavy metals generated by the purification by-products. The sludge obtained from this process is dewatered and stored separately from the treatment plant's biological sludge.

The wash water is purified in the plant's biological treatment facilities.

Screening refuses containing various materials which are generally not fermentable are assimilated into solid household waste and placed in landfills after draining or, better yet, pressing, to avoid leachates.

Grease collected either from the distribution system in special traps located at certain service connections, or from the wastewater treatment plant must be treated appropriately. Commonly used methods are incineration, for the largest plants, and biological treatments specially designed for this product's low biodegradability. These processes generate sludge and liquid wastes that are then mixed in with those of the wastewater treatment plant.

Excess sludge resulting from the degradation of organic pollution constitutes the greatest elimination difficulty because of its composition and associated olfactory pollution. The various wastewater treatment processes generate sludge with different characteristics (concentration, composition, fermentability), which can require additional treatments to achieve their final elimination.

Three solutions are available today. They must be evaluated in light of local conditions and sludge quality:

- recycling for agriculture
- incineration
- landfill placement

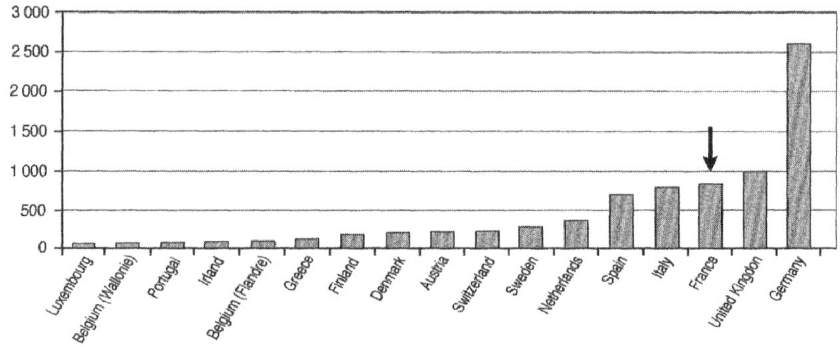

Figure 4.39 Sludge production (thousands tonnes of dry solids per year)

A study (ADEME, 1999) of sludge production rates (Figure 4.39) shows that the 15 countries of the European Union produce 7.5 million tons of dry solids, about the same as the North American countries (over 8 million tons), for respective populations of 380 million and 300 million. Can industrial sludge explain the disparity?

The amount of sewage sludge recycled through agriculture varies greatly from country to country (Figure 4.40). Land application accounts for an average of 30–50%, with extreme values of 4% for The Netherlands and 90% for French-speaking Belgium (Wallonie); but national averages mask great regional or even local disparities resulting from differing agricultural practices, climates, and regulations.

The European Directive of 1986 (86/278/EEC, amended in 1988) relating to residual sludge is being revised to modify a great many sludge quality parameters and the conditions for agricultural recycling. The acceptable heavy metal values will be considerably reduced and new organic pollutants will be added (oestrogens, polyaromatic hydrocarbons, PCBs, etc.). In addition to mandating sludge traceability, new requirements on conditions for agricultural recycling include the disinfection of all sludge used in agriculture.

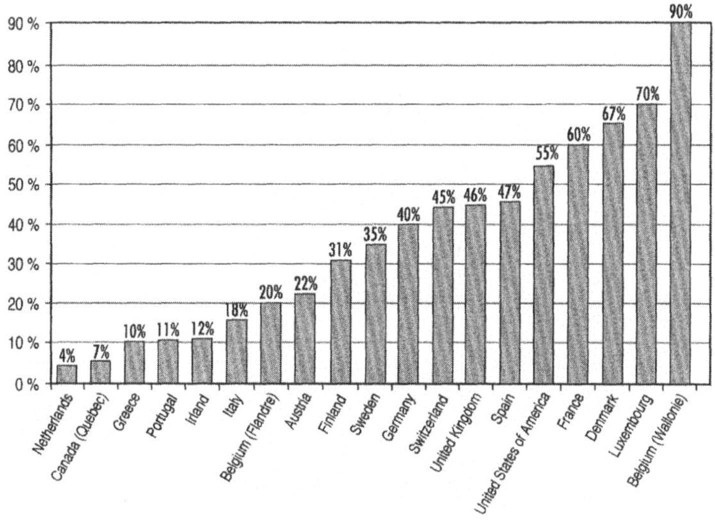

Figure 4.40 Ranking of European countries by sludge to land disposal

These provisions are designed to reassure farmers and consumers concerning the quality of sludge used in agriculture. However, they may unfortunately have the opposite effect, as was the case in France following new regulations on land application of residual domestic sludge. A study on the social acceptability of sludge analysing the behaviours of the different players underscores the fact that passage of national regulations does little to promote the local consensus necessary to successfully confront the problem.

In this climate it is now quite difficult and probably risky to advise local communities on a sludge removal technique. The "multiple re-use" concept (Figure 4.41) may be a way of making the requisite investments pay: sludge needs to be converted into dry pellets or granules before it can be used as an energy source (co-incineration, gasification) or fertiliser, or for placement in a landfill should a lapse in quality occur due to pollution in the sanitation network.

In a few cases, it has been possible to combine sludge removal efforts with efforts to remove other urban wastes. It seems possible to achieve important synergies along the lines of those in Bordeaux where sludge has been co-incinerated along with household refuse after the existing capacities of the incineration plant were increased due to better sorting and recycling of household garbage.

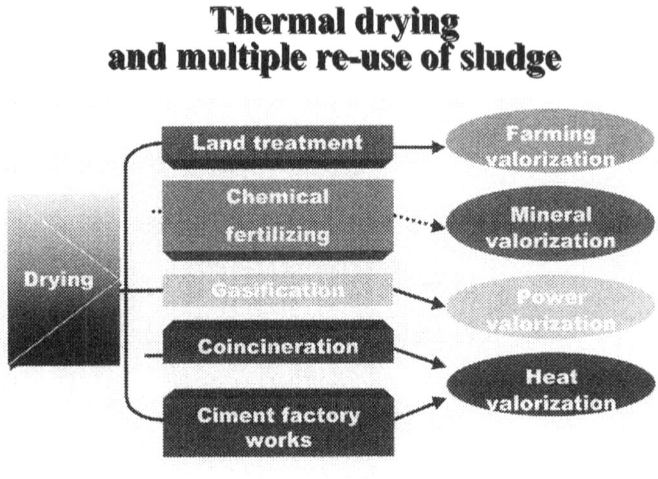

Figure 4.41 Sludge multiple re-use concept

The ten treatment plants of Greater Bordeaux have a total capacity of 800,000 PE. They produce 23,000 tons of sludge per year. This sludge was re-used in agriculture or sent to landfills until 1997, when the exceeding of micropollution standards led to suspension of the land application programme. Because of the complexity of Bordeaux's combined system, the many different industries connected to it, and the advanced age of some of the distribution systems, it was impossible to conduct diagnoses and take action to bring the micropollutants under control. Therefore, a new solution was sought, and Greater Bordeaux considered the possibility of incinerating urban residual sludge within the framework of its household waste incineration programme.

In 1997, Greater Bordeaux built two incineration plants:

Cenon: two furnaces, with an annual incineration capacity of 120,000 tons of household waste per year;

Astria: three furnaces, with an annual capacity of 250,000 tons of household waste per year.

In 1999, 85% of the sludge from Greater Bordeaux's wastewater treatment plants was incinerated. Landfill placement remains an alternative solution.

The five incineration furnaces use a Degremont process to inject the sludge into the furnaces. Cenon is expected to process 18,000 tons of sludge per year; Astria will handle 17,000 tons. The sludge arrives daily by truck, at a rate of 140 tons per day. Sludge is stored in a closed location and deodorised to reduce olfactory pollution as much as possible. The ratio of injected sludge is approximately 10% for an 18–28% dry solids content.

Such a full-scale example clearly indicates that coordinated solid waste management policies are very necessary to achieve such solutions, as they make it possible to take availabilities and flow rates into consideration during the treatment facility design phase.

4.7 CONCLUSION

In concluding this chapter, close liaison between the different departments in cities, including exchanges with those responsible in areas outside cities should be encouraged. And of course with users themselves. Table 4.10 on page 226 synthesizes the strong points of it.

Table 4.10 Dialogue required for overall management as a guarantee of effectiveness

Improvement purpose	Action required	Dialogue to be established
Water supply		
Catchment protection	- waste water and stormwater discharge - agricultural work (fertiliser, pesticides)	- water and drainage - town and outside area
Limitation of groundwater over-abstraction	- transfer of groundwater needs to surface water - recycling industrial water - reuse of waste water - reduction of wastage	- town and outside area - water and drainage - water and industry - water and user
Protection of sheets of water	- control of all waste: waste water, stormwater, industrial and agricultural water	- town and outside area - water and drainage - water and industry - water and farming - water and user
System leaks Wastage on user's premises	- geology - soil settlement (groundwater oversupply) - internal system leaks	- water and town planning - water and user
Distributed water quality improvement service safety	- interconnection	- design and maintenance between systems
Low revenue city service	- maintenance - organisation with local officials	- new works and maintenance - water and user
Sanitation		
Unit separator system	- connecting control	- water and roads department
Environmental protection	- control of all waste	- town and outside area - water and industry - water and maintenance - water and waste - water and user
Connection	- connection control	- water and roads department - water and town planning - water and user
Reuse of waste water	-waste water treatment	- water and town planning - water and industry - water and farming
Individual drainage	- disposal and final waste control	- drainage and groundwater protection - water and town planning - water and user
Flooding and overtopping	- flood forecasting - rain control - protection of inhabited areas	- town and outside area - water and town planning - water and drainage - water and user
Technical management		
Measures and controls	- groundwater - sheets of water - systems	- water and drainage - town and outside area
Plant output	- all structures including pipes	- structural design and maintenance
Maintenance	- all structures and systems	- design and maintenance

4.8 REFERENCES

ADEME (1999) *Situation du recyclage agricole des boues d'épuration urbaines en Europe, et dans d'autres pays du monde*, ADEME, Paris.

Anselme, C., C. Cabassud & M.R. Chevalier (1990) Journées Informations Eaux, pp 1–13, Poitiers.

Anselme, C., I. Baudin, & M.R. Chevallier, (1993), *Aqua* **42** (5), 295–300

Azzout, Y., S. Barraud, F.N. Cres, & E. Alfakih (1994) Techniques alternatives en assainissement pluvial, Agences de l'Eau et STU, Lavoisier Tec et Doc, Paris.

Baudin, I. & C. Anselme, (1995) Membrane Technology, 65, pp 6–8

Bos, M. & Jarrige, A., (1990) Specialised Conference IWSA, Amsterdam, pp 143–147

Bourrelier, P.L *et al.* (1999) Natural risk prevention policy, rapport de l'instance d'évaluation des politiques publiques de prévention des risques naturels, documentation française, Paris.

Caulet, P., B. Bujon, & J.M. Andic, (2000) Optimisation of Carlon and Nitrogen removal infull scole Wastewater treatment plant (IWA 2000 – Paris – special report)

Centre d'information sur l'eau (1998) Les usages de l'eau en chiffres, Paris.

Chèze, J.M. & M. Gousailles, (1998) La station d'épuration de Seine Centre, SIAAP, Paris.

Commission Mondiale sur l'eau pour le XXIème siècle (2000) Rapport au Forum de La Haye, Conseil Mondial de l'Eau, Marseille.

Cosgrove, W.J. & F.R. Rijsberman (2000) World Water Vision. Making water everybody's business. Conseil Mondial de l'Eau, EARTHSCAN, New-York.

Dégardin, F. & P.A. Gaide (1999) Valoriser les zones inondables dans l'aménagement urbain, repères pour une nouvelle démarche, CERTU, Lyon.

Ginestet, P. (1999) Typologie des eaux résiduaires urbaines, CIRSEE, research report, Paris.

GRAIE (1998) Innovative technologies in urban storm drainage, 3rd. International Conference proceedings, Novatech, GRAIE, Lyon.

Haeringer P. (1998) La mégapolisation, un autre monde, un nouvel apprentissage, in *De la ville à la mégapole: essor ou déclin des villes au XXIème siècle, Techniques, Territoires et Sociétés*, n°35, Ministère de l'Equipement, des Transport et du Logement, Paris.

IAURIF (1997) Eau, ville et urbanisme, cahiers de l'IAURIF, n° 116, Paris.

Instituto per studie ricerche sui servizi idrici (1998) Le Tariffe Idriche, Pro Aqua.

IWSA (1999) International Report No. 5 and the 18 Contributions and National Reports, Water Demand Management and Conservation including Water Loss Control, ISWA general assembly, Buenos-Aires.

Levine, B. and V. Lazarova, (1999) The West Basin Water Quality Research Program, CIRSEE research report, PEP Process 00.01, CIRSEE, Paris.

OECD (1999), The Price of Water – Trends in OECD Countries.

Roche, P.A. & P.L. Bourrelier (1998) Prévention des risques naturels, in PCM-Le Pont, n°10, Paris.

Roche , P.A. (2000) L'eau au XXI ème siècle: enjeux, conflits, marchés, in Ramsès 2001, IFRI, Dunod, Paris.

STU - Agences de l'eau (1994) Guide technique des bassins de retenue d'eau pluviale, Lavoisier Tec Et Doc, Paris.

Valiron, F. & J.P. Tabuchi (1992) Maîtrise de la pollution urbaine par temps de pluie, Tec. Doc. Lavoisier.
Valiron, F. (1994) Mémento du Gestionnaire de l'Alimentation en eau et de l'assainissement, Lavoisier, Paris.
Valiron, F. & M. Affholder (1996) Guide de conception et de gestion des réseaux d'assainissement unitaires, Agences de l'Eau et STU, Lavoisier Tec et Doc, Paris.
Valiron, F. in IAURIF (1997).
Valiron, F. (1999) Gestion des eaux, Coût et prix de l'alimentation en eau et de l'assainissement, Presses de l'ENPC .
UNCHS (1999) Managing water for African Cities, jointly with UNEP, Nairobi.
UNESCO & Académie de l'eau (1997) Proceedings of the symposium on water in cities and urban planning, Paris.
UNESCO & Suez-Lyonnaise des Eaux, (1999) Solutions alternatives à l'approvisionnement en eau et à l'assainissement conventionnels dans les secteurs à faibles revenus, Paris.
UNEP (2000) Global Environmental Outlook 2000, Earthscan, New York.
Wade, N. M. (1998) Technical and economic evaluation of distillation and reverse-osmosis desalination processes, *Desalination*, **93** (1993) 343–363
Wanggnick, K. International Desalination Association, report n° 15
WHO (1999) World health report, Genève.
World Bank (1999) *World development report 1999/2000: Entering the 21st Century*, Oxford University Press, New York.

5

Emerging paradigms in water supply and sanitation

Saburo Matsui, Mogens Henze, Goen Ho and Ralf Otterpohl

5.1 NEED FOR PARADIGM SHIFT OF NEW TECHNOLOGY IN URBAN WATER INFRASTRUCTURE

5.1.1 Water in the 21st Century

The statement of the World Water Vision 'Making Water Everybody's Business' was released and discussed at the Second World Water Forum, the Hague, 17–22 March 2000 (Cosgrove & Rijsberman, 2000). It sets the goal of a radical improvement of water conditions for human-beings with better health,

 Frontiers in Urban Water Management: Deadlock or Hope
Edited by Čedo Maksimović and José Alberto Tejada-Guibert. ISBN: 1 900222 76 0

better nutrition for the entire population by a wiser water resource management, more powerful community commitment and higher farm yields by 2025.

The statement also stresses the importance of the water allocation for ecosystem sustainability of our planet. The statement envisions a scenario of renewable water use by 2025, where municipal water consumption is expected to rapidly increase in the developing countries while stabilisation or decrease occurs in developed countries (Table 5.1). The statement emphasises the necessity of a drastic change in water resources allocation, where the 'business as usual' scenario that simply would extend the current trend of water consumption is difficult to achieve and sustain. It will be very difficult to support in a sustainable way water demand of the expected population increase. The agriculture sector is the largest water consumer, and must limit irrigated agriculture activity, while more water should be allocated into municipal uses.

This analysis requires as prerequisites a drastic improvement in urban water use efficiency for both quantity and quality. Although the analysis clarifies the necessity of a drastic change in the future water allocation, it does not fully describe how improved water use efficiency for both quantity and quality will be achieved or managed. It expects a drastic technological breakthrough to be introduced by the water utility sector. Thus urban water management in the future must be critically improved in the context of global environment issues.

5.1.2 Urban sanitation is the first step for control of water pollution

Safe drinking water comes first for urban water management. If we cannot protect drinking water source from either ground water or surface water from contamination and over use, we cannot properly manage the urban water issue. Developed countries may be able to overcome contamination of drinking water and limit water use by introducing advanced technology and new financial arrangement such as the introduction of some form of privatisation. Developing countries face difficulty in all aspects of water management. The problems facing developing countries are elaborated further below.

Developed countries introduced sewage works to overcome problems (chaos) occurring in the middle of 19th century. It took one century for Western countries to shape them up. The sewage works installed a single pipe drainage with water flush toilet for sanitation practices. Most developed countries even accept storm water to the single pipe drainage system that transports all urban water into a sewage treatment plant, if properly installed. Is this only way to regulate the chaos? The performance of the sewage treatment plant may not be

Table 5.1 Renewable water use in the World Water Vision (Cosgrove and Rijsberman, 2000)

In the Vision the water for irrigated agriculture is drastically limited, with 40% more food produced (partly from rain-fed agriculture) consuming only 9% more water for irrigation. Industrial use goes down in developed countries, but the decline is more than offset by increases in the developing world. Municipal use goes up sharply in developing countries, to provide a minimum amount for all, and down in the developed world. Recycling and increased productivity lower the ratio of water withdrawn to water consumed for all uses.

User	Cubic Kilometres 1995[a]	Cubic Kilometres 2025[b]	Percentage increase 1995–2025	Notes
Agriculture				
Withdrawal	2,500	2,650	6	Food production increases 40%, but much higher water productivity limits increase in harvested irrigated area to 20% and increase in net irrigated area to 5–10%.
Consumption	1,750	1,900	9	
Industry				
Withdrawal	750	800[c]	7	Major increase in developing countries is partly offset by major reduction in developed countries.
Consumption	80	100	25	
Municipalities				
Withdrawal	350	500[d]	43	Major increase and universal access in developing countries; stabilization and decrease in developed countries.
Consumption	50	100	100	
Reservoirs (evaporation)	200	220	10	
Total				
Withdrawal	3,800	4,200	10	
Consumption	2,100	2,300	10	
Groundwater over-consumption	200	0		Increase recharge of aquifers makes groundwater use sustainable.

Note: Totals are rounded.

(a) The 1995 uses are provided for reference.

(b) World Water Vision staff estimates.

(c) For industry it is recognised that developing countries need a major expansion in industrial water use. For the roughly 2 billion people in cities in developing countries that need livelihoods (both the current poor plus the increase in population) an average of 200 litres per person per day is used. This means a 400 km^3 increase in diversions for industry in developing countries. At the same time, diversions for industry in developed countries can be drastically reduced. Better management and reduced losses the ratio of water withdrawn to water consumed.

(d) Residential water use of people in developing countries needs to be drastically increased. Residential use in developed countries stabilised and is reduced.

enough to protect the conditions of the receiving water environment. An example is nitrogen and phosphorus removal to control eutrophication of lakes, reservoirs, bays and enclosed seas. There are other problems that need solutions such as heavy metals, persistent organic pollutants, endocrine-disrupting chemicals, pathogens and cyanobacteria toxins. We also have to look at related questions such as water economy and energy saving for advanced treatment of water supply and wastewater, in order to solve those problems. Under these new situations of urban water management in the complex global context, it is a good opportunity to think of new systems and technology for rehabilitation of aged water infrastructure for developed countries as well as installing new infrastructure for developing countries.

A sustainable society requires a strong coordination between urban and rural components for nutrient and biomass recycling. The most profound problem is food production and consumption in a sustainable way. After consumption, the question is to how to deal with urine and faeces and put them into the recycling of nutrients and biomass. The current Western approach is not sustainable in terms of recycle of nutrients. This is the point of reflection and the start of opening into a new institution and technology for 21st century society.

Developing countries must solve sanitation problems first, which will require technology that saves water and incorporate water conservation for both for water supply and sanitation. The solutions can be shared by both developed and developing countries.

5.1.3 Diversion of urine from sewage pipes for eutrophication control

One solution that can have a profound impact is diversion of urine and faeces or urine only out from the conventional sewage pipe drainage outlet. It is even better if urine can be diverted at the source, especially for the urban situation. It can be brought via a different pipe to a storage tank in the vicinity, where vacuum trucks can transfer it to a nutrient processing plant for recovery of nitrogen and phosphorus into chemical fertiliser. Doing so, we can drastically save water for flushing and improve biological nitrogen and phosphorus removal treatment process efficiency. Figure 5.1 depicts solutions applied by (a) WM-Ekologen ab, (b) BB-Innovation – Sweden, and (c) Roediger - Germany. In these cases, urine and flush water are collected by the separate urine pipe; faeces and flush water go to the sewage pipe.

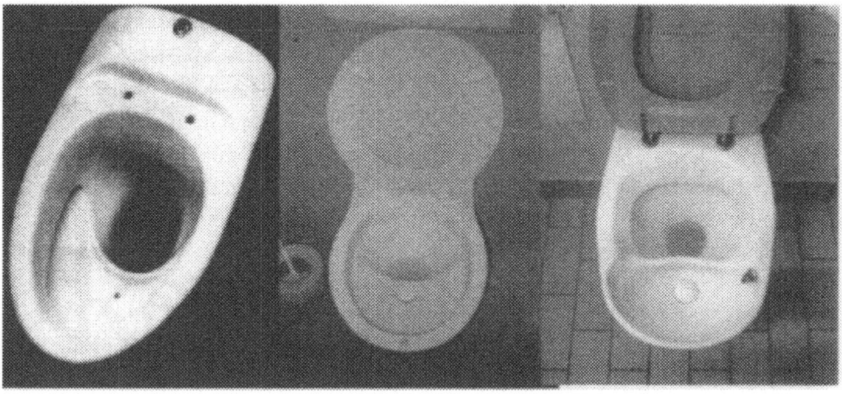

Figure 5.1 Several types of urine separation toilet. The separate urine pipe collects urine and flush water separately from faeces and flush water, which go to the sewage pipe. (Courtesy: WM-Ekologen ab-Sweden, BB-Innovation - Sweden and Roediger –Germany.)

Table 5.2 Nutrient distribution between urine and faeces (after SEPA, 1995)

Parameter	Urine		Faeces		Total Discharge	
	g/person/day	%	g/person/day	%	g/person/day	%
Wet weight	900–1200	90	70–140	10	1000–1400	100
Dry weight	60*	63	35	37	95	100
Nitrogen	11.0	88	1.5	12	12.5	100
Phosphorus	1.0	67	0.5	33	1.5	100
Potassium	2.5	71	1.0	29	3.5	100

* A large proportion of this dry substance is rapidly biodegradable. Much of it already degrades in a sewage pipe.

Table 5.3 Pollution loads of wastewater generated by a person (modified after Sewage Works Guideline, Ministry of Construction, Japan 1996)

Item	Average g/person/day	Std. Deviation	No. of Data	Urine+Faeces %	Grey water %
BOD_5	58	18	99	32	68
CODmn	26	9	96	36	64
SS	44	16	99	47	53
Nitrogen	11	3	9	75	25
Phosphorus	1.0	0.2	8	75	25

Note: In Table 5.3 SS means suspended substances; CODmn means COD measured with potassium permanganate.

Urine generally contains more nutrients than faeces, containing 88% of the nitrogen, 67% of the phosphorus and 71% of the potassium (Table 5.2), which is different from what most people perceive. Urine is generally free from pathogens that can be found in faeces. It is also important to appreciate the pollution load of wastewater daily generated by a person. Grey water contributes more BOD and COD load than urine and faeces, but its content of nitrogen and phosphorus is the opposite. When we combine the data from Tables 5.2 and 5.3, we can deduce that urine contains 66% of the nitrogen and 50.3% of the phosphorus discharged by a person in the form of sewage. Urine diversion at source from a sewage pipe can contribute significant reduction of nutrients to municipal wastewater treatment plants. We can calculate from Table 5.3 that the average sewage pollution load consists of BOD_5 and Nitrogen and Phosphorus in the proportion of 100:19:2. This proportion indicates excess nitrogen and phosphorus to BOD when a conventional biological system treats sewage, in which the ideal proportion is 100:5:1. Wastewater engineers have developed various types of biological denitrification processes that are modifications of the conventional biological system. It can remove about 80% of nitrogen when the proportion of sewage is such typical one. However, it is not enough to control eutrophication under many receiving water circumstances. If it is necessary to remove more, extra BOD load, such as food processing and other high BOD industrial wastewaters, must be introduced to the wastewater.

Another solution is the diversion of urine from the wastewater, which can change the proportion of BOD:N:P to 100:8.2:1, improve loading conditions and can enhance denitrification and phosphorus removal. Wastewater engineers with microbiologists have developed a biological phosphorus process that is also modification of the conventional biological system by introducing strong biological reduced and oxidised conditions in the process that allow generation of phosphorus rich cells in the waste sludge. Phosphorus is removed from water in the form of phosphorus rich sludge. This system also requires a high BOD to phosphorus ratio by introducing anaerobic reduced conditions in the process. When urban wastewater is treated with high BOD strength trade wastewaters, it can achieve such high BOD to phosphorus ratios. However, this kind of conditions does not generally occur.

5.1.4 How urine recovery is advantageous for nutrient recycle

Chemical fertiliser use is inevitable to sustain modern agriculture. The so- called Green Revolution took place in the agriculture sector during 1960s and 1970s, which enabled us to halt widespread starvation but it created new problems in the water environment such as eutrophication and agrochemical contamination. Nutrient recycle from urban areas to agriculture area can give us advantages of nutrient saving specially phosphorus, because of its limited availability on the earth. Figures 5.2, 5.3 and 5.4 give us interesting figures on nitrogen, phosphorus and potassium consumption among various countries and potential supply rates of urine nutrients to the consumption (FAO Country Tables, 1995). Nutrient recovery from urine is in general not enough for the agriculture consumption in any country except for Japan for nitrogen and Egypt for potassium, which indicates that urine recovery can supplement nutrient requirement for the agriculture sector. The current practice of overuse of chemical fertilizer must be controlled to stop eutrophication and groundwater contamination by nitrate. The World Water Vision (Cosgrove & Rijsberman, 2000) predicts an increase in water consumption by the agriculture sector from 1,750 km^3/year in 1995 to 1,900 km^3/year in 2025, only a 9% increase, while human population would increase from 5.5 billion to 7.5 billion, a 36% increase. This means agricultural water use will be intensified giving possibly discharge of more contaminated run-offs. Eutrophication of surface water and nitrate contamination of groundwater is likely to intensify in the future, if the current 'business as usual' management continues. Thus these analyses show the necessity and great potential benefit of recovery of nutrients from diverted urine from sewage works.

5.1.5 An emerging issue: endocrine-disrupting chemicals in the ecosystem and human hormone discharge

'Environmental hormones' is a new term to describe endocrine-disrupting chemicals that upset the homeostasis of animals. The distinctive disruption is related to the development of sex in animals regardless of whether they are vertebrate or invertebrate. Endocrine disrupters are harmful to the development of reproductive organs as well as the reproductive function. Among endocrine disrupters oestrogen mimics and androgen mimics are of much concern for humans as well for fish and other wild animals. Conventional toxicity tests are not adequate for evaluating endocrine disruptions with different animals.

It is now understood that many xenobiotics may fit into the category of oestrogen mimics / androgen mimics. Among oestrogen mimics DDTs, DDEs, chlorinated pesticides, and herbicides are included. Surfactants such as octyl-phenol, nonyl-phenol are potential oestrogen mimics. Bis-phenolA, di-2-ethylhexyl phthalate that are basic building blocks of plastics and plasticizers are also potential oestrogen or other endocrine mimics. They are released through sewage effluents, industrial wastewater treatment plants and agriculture wastewaters. The contamination levels of those chemicals are very low but well

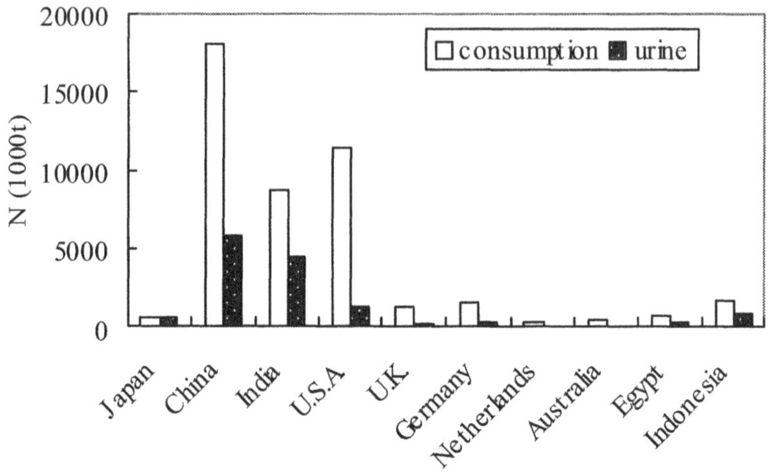

Fig. 5.2a Comparison of nitrogen consumption by fertilizer and potential nitrogen supply by urine among various countries (1993)

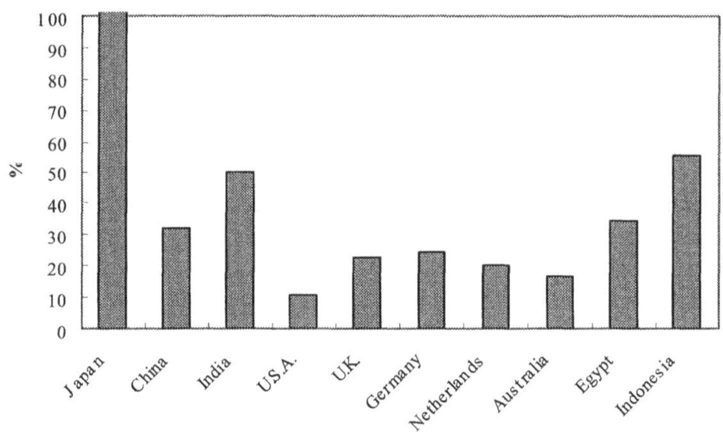

Fig. 5.2b Potential supply rate of urine nitrogen to the consumption of nitrogen by various countries (1993)

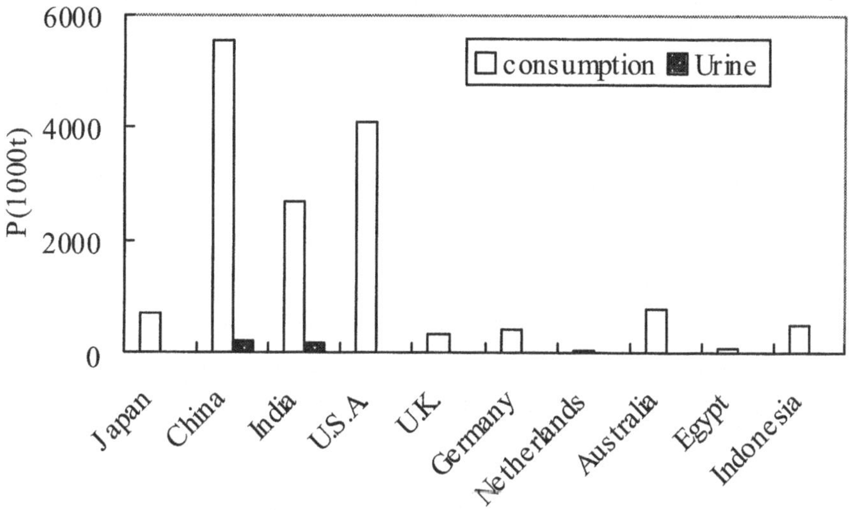

Fig. 5.3a Comparison of phosphorus consumption by fertilizer and potential phosphorus supply by urine among various countries (1993)

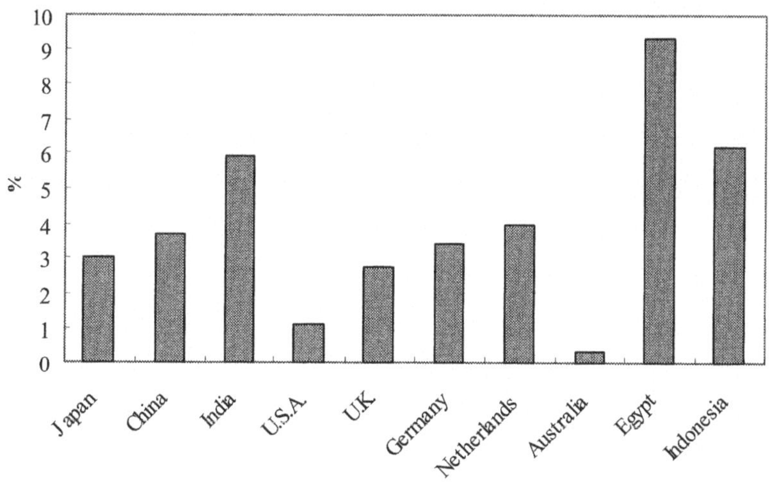

Fig. 5.3b Potential supply rate of urine phosphorus to the consumption of phosphorus by various countries (1993)

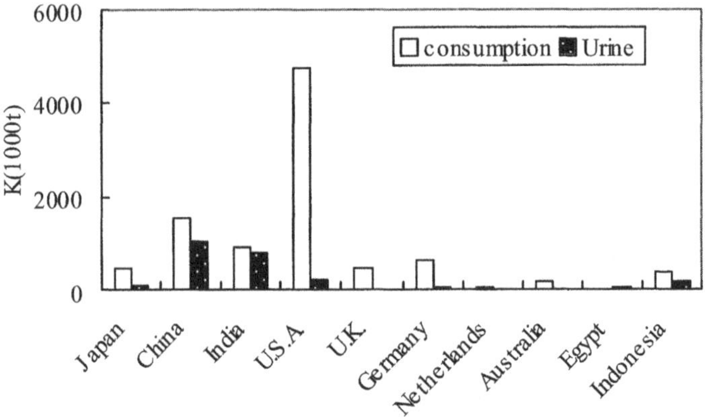

Fig. 5.4a Comparison of potassium consumption by fertilizer and potential potassium supply by urine among various countries (1993)

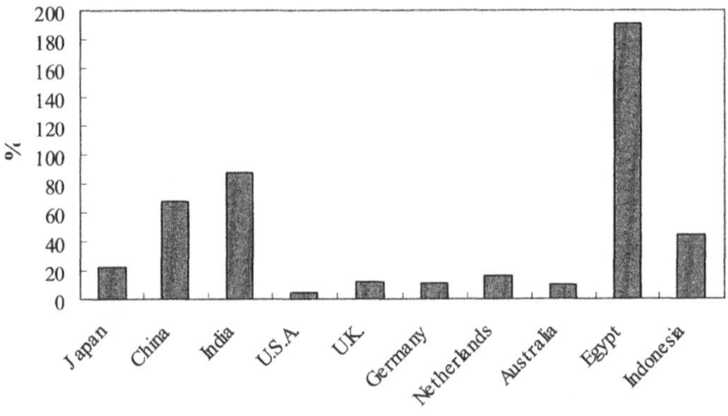

Fig. 5.4b Potential supply rate of urine potassium to the consumption of potassium by various countries (1993)

detected in rivers and lakes by advanced analytical and genetically modified bioassays. However, it is necessary to widen our view of endocrine disruption with fishes. There are significant amounts of discharge of 17ß-oestradiol, and oestrone that are true natural oestrogens, from sewage treatment plant effluents. Both men and women discharge these natural oestrogens in urine. They are, of course, not harmful to humans, but are harmful to fish in their reproductive activity under high contamination.

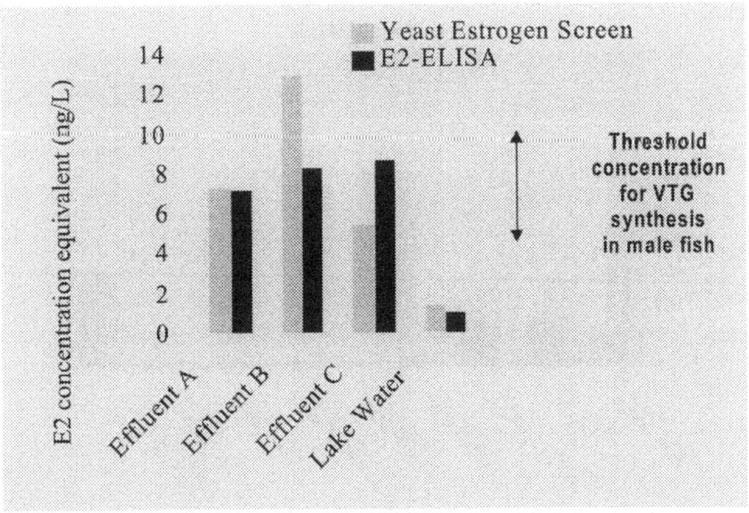

Figure 5.5 Human oestrogen is causative in the oestrogenic activity remaining in sewage effluent

Matsui (1999) applied the yeast oestrogen screen method and the 17ß-oestradiol (E2) ELISA method to evaluate the contamination levels of sewage effluents and lake water for oestrogen and its mimics (Figure 5.5). The yeast oestrogen screen method used genetically modified yeasts that can possess human oestrogen receptors so that any oestrogen mimic as well as true oestrogen could be detected by the receptors. The 17ß-oestradiol ELISA assay is an immunological assay method that detects 17ß-oestradiol and very small amounts of other oestrogens. Figure 5.6 shows E2 concentration equivalent (ng/L) in sewage treatment process waters evaluated by the Yeast screen and ELISA method. The oestrogenic activity in the biologically treated effluents could be mainly attributed to 17ß-oestradiol of human urine origin. Male fish are affected by 17ß-oestradiol, oestrone and other oestrogen mimics such as octyl-phenol and nonyl-phenol in the effluents so that their testes contain eggs among the sperm. Vitellogenin (VTG) is a yolk protein for egg formation that can be found in the blood of affected male fish. The threshold concentration of oestrogen and its mimics in water for stimulating the formation of VTG in blood is just about that of the treated sewage levels shown in Figure 5.6. The spread of the use of oestrogen pills significantly increases contamination of sewage with oestrogen released with urine.

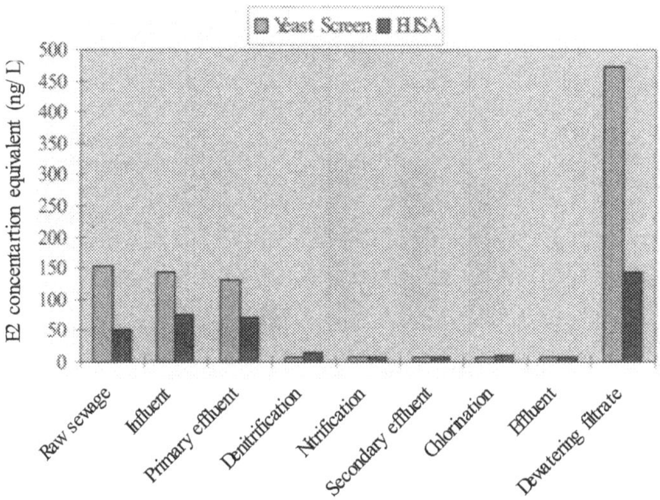

Figure 5.6 E2 concentration equivalent (ng/l) in sewage treatment. Process waters evaluated by the Yeast screen and ELISA methods.

There are also phyto-oestrogens in natural food, which are potentially harmful, if over consumed, to animals as well as humans. They are also discharged in urine. It is also important to look at pharmaceutical drugs including anticancer drugs and antibiotics that are passed on through urine, because they are detoxified in the liver and discharged through kidneys in the form of conjugates. Many of them may revert from their conjugate forms to the original forms by biological treatment processes and discharged into receiving waters. It is obvious that new science and technology is necessary to examine urine contents in addition to our current concern for nutrients. Large-scale urine recovery will need new approaches for the development of suitable treatment technology.

5.1.6 Another important option for urban sanitation: Ecological Sanitation

Urbanisation among major cities in developing countries has been predicted to occur at such a fast pace that it may be difficult to provide adequate water supply and sanitation for all citizens. We have to prepare another important option for such a condition. Ecological Sanitation – a dry toilet with diversion of urine and faeces at source is such an option (Winblad *et al.*, 1998). This concept is now gradually accepted among developing countries and pilot projects are taking place. If urine and faeces are collected and used for agriculture in the practice of Ecological Sanitation, sewage characteristics may change into so-called greywater, which is easier to treat by conventional wastewater treatment processes. The collection of greywater may not be necessary with a pipe system. A pipe system is necessary for collecting wastewater from multi-storey buildings, but not for collection along the street where open channels are still acceptable for minimum conditions. The collected greywater can be treated locally and discharged to surface waters in the vicinity. This approach is not satisfactory from the standard of developed countries but may greatly improve the current situations of urban water and help reduce financial burdens on municipalities that have to install water infrastructure. The Ecological Sanitation approach is an important option that sanitary engineers can use to solve the complex problems of urban water.

This paradigm shift can provide a new framework for the management of human wastewater from the traditional one to various options such as new sewage treatment technology, recovery of nutrients from urine, treatment of drug and hormone conjugates in urine, and better treatment of pathogens in faeces including *Escherichia coli* type 157 and *Cryptosporidium*.

5.2 ENVIRONMENTAL ISSUES RELATED TO REHABILITATION OF AGED WATER INFRASTRUCTURE

5.2.1 Introduction

Aged water infrastructures are experiencing a devious affliction. Seldom do the systems break down, but their function is slowly deteriorating. At a certain point in time decisions have to be taken and this presents several questions. Some are technical and economic, some sociological and others directly environmental. All the questions end up as environmental issues, which are addressed below.

Considerations

- person load development
- type of future wastewater
- effect of water savings
- wastewater design
- public health
- chemicals risks in the wastewater
- pollution of receiving waters
- soil pollution
- air pollution (noise and odour problems)
- local situation
- technical alternatives
- resources (energy, land, water and materials)
- economy
- social–cultural aspects

5.2.2 Development of person load (PL)

The actual contribution from a person living in the sewer catchment, the Person Load (PL) can vary considerably (see Table 5.3). The reasons for the variation can include: working place outside the catchment, ecological lifestyle, household installations like garbage grinders, etc. In the context of new development of the water infrastructure, the development of the person load in the future is important.

Table 5.3 Variations in Person Load (Henze *et al.*, 2000)

Component	Level
BOD g/(person·d)	15–80
COD g/(person·d)	25–200
Nitrogen g/(person·d)	2–15
Phosphorus g/(person·d)	1–3
Wastewater m^3/(person·d)	0.05–0.40

5.2.3 Type of wastewater in the future

It is considered that most wastewater to be handled in the future is the classic mixed wastewater. However, by separation, the wastewater fractions defined in Table 5.4 can be created.

Table 5.4 The wastewater palette (definition of wastewater fractions from households) (Henze and Ledin, 2000)

Type	Content
Classic	toilet, bath, kitchen, wash
Black	toilet
Grey	bath, kitchen, wash
Light grey	bath, wash
Yellow	urine
Brown	faeces

5.5 Water savings

Table 5.5 Fractionated water consumption. Values from Northern Europe, l/cap·day (Henze and Ledin, 2000)

Water consumption	Today	With savings (new installations, sewer rehabilitation)	Primary water consumption with savings and use of secondary water (for toilets and laundry)
Toilet	50	25/0	0
Bath	50	25	25
Kitchen	50	25	25
Laundry	10	5	1
Infiltration	80	25	—
Total	240	105	51

Through water savings and sewer rehabilitation, the figures shown in the middle column of Table 5.5 can be obtained. Only a part of the future water supply need to be drinking water (primary water), as showed in the right column. The remaining part of the water consumption could be covered by secondary water (rainwater or treated grey water).

The water consumption for the various fractions can be reduced considerably. Without much effort the water consumption in the households can be reduced by 50%, and even more by rehabilitation of the sewers. It is a realistic strategy to install water saving devices when old devices need replacement. This means that it will take some 20–40 years in developed countries before most installations have been replaced.

Water savings have two important implications on society. One is the reduced amount of energy used for the water supply and wastewater treatment, and the second is the reduced exploitation of freshwater resources. Water for toilets and washing of clothes, can realistically be replaced by secondary water, for example rain water. It is thus obvious that small water consumption is more sustainable than a high one seen from an energy consumption point of view.

The water savings will increase the concentrations of pollutants in the wastewater (the quality of pollutants remains constant), but the treatment costs will decrease due to less water to be treated without an increase of the quantity of the pollutants that should be removed.

5.2.5 Wastewater design for households

The use of one or more of the earlier mentioned waste handling technologies in households in combination with water savings, makes it possible to design a wastewater with a specified composition, that will be optimal for its further handling. If the goal is to reduce the pollutant load to the wastewater, Table 5.6 illustrates the possible reductions in waste loads to the wastewater by the different actions suggested.

Table 5.6 Households. Reduced waste load to wastewater by different technical actions, g/cap·day (Henze and Ledin 2000)

Technology →	Present	Toilet separation[+]	Cleantech cooking[#]
COD	130	55	32
BOD	60	35	20
Nitrogen	13	2	1.5
Phosphorus	2.5	0.5	0.4

[+] water closet → compost toilet

[#] part of cooking waste from sink → solid waste

The coupling of water savings and load reductions gives one more degree of freedom for the design of a given wastewater composition. The results are shown in Table 5.7

Table 5.7 Concentration of pollutants in raw wastewater by separation of toilets and cleantech cooking (assuming: non-P detergents) (Henze, 1997).

Wastewater production		250 l/cap·day	160 l/cap·day	80 l/cap·day
COD	g/m^3	130	200	400
BOD	g/m^3	80	125	250
Nitrogen	g/m^3	6	9	19
Phosphorus	g/m^3	1.6	2.5	5

5.2.6 Public health

The handling of waste, they be waterborne or solids, has big implications on public health. Our present day technology has been developed to give maximum protection against transmission of diseases via the waste handling. Wastewater is heavily polluted with all kind of microorganisms (see Table 5.8). Table 5.8 gives an idea of the concentration of microorganisms in raw domestic wastewater.

Table 5.8 Microorganisms in wastewater (number/100 ml) (Henze and Ledin, 2000)

Type	High	Low
E.Coli	$5\cdot10^8$	10^6
Coliforms	10^{13}	10^{11}
Cl.perfringens	$5\cdot10^4$	10^3
Faecal streptococcae	10^8	10^6
Salmonella	300	50
Campylobacter	10^5	$5\cdot10^3$
Listeria	10^4	$5\cdot10^2$
Staphyllococus aureus	10^5	$5\cdot10^3$
Coliphages	$5\cdot10^5$	10^4
Giardia	10^3	10^2
Roundworms	20	5
Enterovirus	10^4	10^3
Rotavirus	100	20

5.2.7 Pollution of water, air and soil

All wastewater handling creates pollution of air, soil and water. Depending upon the technology involved, the amount of pollution caused varies. Centralised wastewater handling creates point-source pollution, while decentralised handling creates diffuse pollution. Domestic wastewater contains metals, hormone-like substances and endocrine disrupters and much more. It is not easy to decide whether it is better to pollute the soil and groundwater/water course near every house to a small degree or create heavier pollution from a big treatment plant. Many local waste handling systems will ultimately have to deposit its waste either to centralised treatment plants for wastewater or solid waste, or if possible to agricultural land.

5.2.8 Local factors

The optimal water and waste handling system and technology will vary with the location. Population density, climate, peri-urban agricultural areas, distance for waste transport, culture and comfort all plays a role when the decision on the optimal system has to be made. In order to reach the optimal solution a detailed assessment needs to be made. Such a framework for assessing the sustainability of wastewater solutions is outlined in Figure 5.7 (Eilersen *et al.*, 1999). The important elements are an analysis of the local context (on-site analysis), which includes an environmental analysis and a stakeholder analysis, a listing of alternative technical solutions, and a multiple-criteria evaluation and prioritisation of the alternatives followed by the final (political) decision. The on-site analysis and the evaluation will depend on local and national environmental regulations, but the idea is that the assessment framework should give some feedback to the policy level, eventually leading to changes of regulations where necessary.

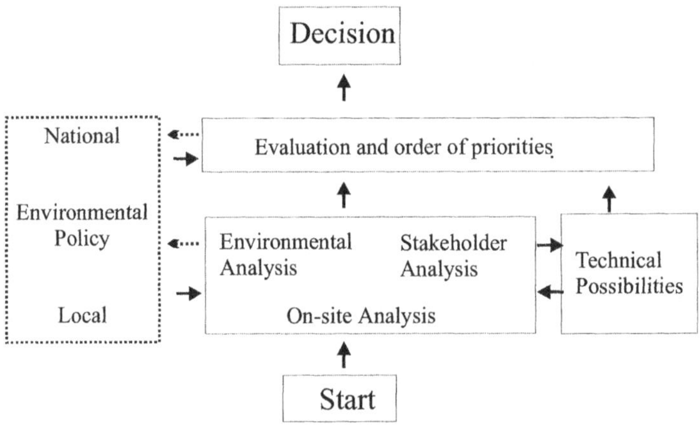

Figure 5.7 Principal outline of the assessment method (Eilersen *et al.*, 1999).

The multiple-criteria evaluation and prioritisation should be based on assignment of scores and weighting of criteria in a quantitative manner. It is simple to make a long list of criteria important to the evaluation of the sustainability of a given system. However, it is far more complicated to organise them into independent groups of criteria, avoiding double counting and overestimation of some and thereby underestimation of others. There is a basic set of criteria that are relevant for all assessments; these are mentioned in section 5.2.1. In addition to what has been discussed above, the socio-cultural aspects need also to be dealt with. This includes:

- Transparency and demonstration effect
- Involvement of users in the operation of the systems
- Visibility of local material cycles.

The set of criteria will constitute an explicit core around which a locally dependent assessment can be made. The assignment of scores for each criterion is critical, as is the selection of an appropriate weight for it, relative to the weighting of the other criteria. Since the criteria cover a wide range of impacts of concern, they will result in comparisons of dissimilar elements, and weighting of the more qualitative criteria will be judgmental. The scoring system must accept both hard and soft data in a similar format, and the weighting system should be constructed partly by default values, partly by user modifiable values. The relevant stakeholders can weigh the locally dependent assessment criteria and they may find it necessary to add extra criteria to ensure that all the significant issues are covered. The performance of the different alternatives for wastewater handling will be compared to that of a reference condition, a benchmark performance. In summary, the assessment framework should be transparent, locally based, have a holistic view on technological choices and finally be robust and flexible.

5.2.9 Strategies

There are several strategies for handling aged water structures:

- renovation
- stepwise development towards new technology
- throw away the present technology and implement a new one.

There is no standard solution to be chosen in every case. The above-mentioned factors must be evaluated.

When implementing new technologies, there is always the risk that they are not maturely developed. It is important to apply high safety factors/reliance in the solutions used, in order not to compromise new technologies through failures.

At the same time the technologies must be flexible. The lifetime expectancy of many waste-handling systems is 30–40 years, and many boundary conditions might change before the investments have paid off.

In most cases, the immediate discarding of an existing system will not be sustainable. Possibilities for stepwise development/construction of new systems must be looked for and if new technologies will be a success, they must be brought into use stepwise.

5.3 NEW BIOTECHNOLOGY AND PHYSICO-CHEMICAL METHODS FOR THE PARADIGM SHIFT

5.3.1 New biotechnology for end-of-pipe treatment

Treatment methods in municipal wastewater treatment plants are still not very different from those in use more than 100 years ago. One major change has been the introduction of nutrient removal for phosphorus and nitrogen. Quite remarkable is the overwhelming success of source control by nationwide introduction of phosphate-free household chemicals in many countries. This has been possible with massive public intervention against strongly visible water pollution. The spread-out of nitrification and denitrification treatment of wastewater in many countries has been due to dramatic algal blooms, e.g. in the North Sea, and international political agreements. However, most smaller plants do not have to perform nutrient removal yet. The discussion is now on micropollutants (Matsui *et al.*, 1999; Daughton and Ternes, 1999) and very recently on the possible spread-out of resistance against antibiotics through wastewater treatment plants (WWTPs) (Feuerpfeil *et al.*, 2000) that can counteract efforts for better medical care. Biological reactors are a perfect environment for exchange of DNA information by microorganisms; even blocks of such resistances can be transferred easily among bacteria in the form of plasmid exchange.

With respect to treatment methods, the simple activated sludge system is still surprisingly competitive. The adoption of biological aerated filters (BAFs) of many types is rarely economically competitive, except when land is very expensive – other alternatives require more of it, and possibly higher operating costs of the alternatives are taken into account. Activated sludge reactors are built deeper thus with better oxygen uptake and increasing cost effectiveness further. The latest step is the combination of activated sludge reactors with membrane technology replacing secondary clarifier and including further treatment. The first large-scale applications are successful. Economics comes down to the question of the value of the additional cleaning performance. It does include hygienisation and can decrease COD and P-concentrations significantly. If legislation specifies technical development there may be a bright future for this type of system. A strong advantage is also that even the smallest on-site treatment plant can use this technology and large scale is not a limitation either. Today significant savings are found with upgrading of plants when additional tanks can be avoided. Another market will be for wastewater reuse, also for greywater treatment (see below). Concerning micropollutants it becomes obvious that source control in industry and industrial household chemicals does make more sense than an expensive further treatment stage.

A major change of technology is emerging for countries with warmer climates with the Upflow Anaerobic Sludge Blanket (UASB) technology (Zeemann and Lettinga, 1998). Warmer wastewater temperatures allow anaerobic treatment of solids, and adsorbed matters that are held back in a filter either build up or are formed with the biological process. Solids get long retention times while the liquid phase remains in the reactor for only a few hours. This process is in use in many industrial plants but application in municipal wastewater is also increasing. There are some major problems with odour from sulphur in biogas but this is relatively simple to solve by dosing small amounts of air into the digestor or into the biogas. Another, more severe, disadvantage is the need for post treatment in many cases. If the effluent is used as a combination for irrigation with fertilizer this is not a big problem. However there are limitations for usage as hygienisation is not included in the process.

Besides the major changes outlined above there is remarkable development in sludge treatment technology. Further big steps for the end-of-the-pipe treatment are not emerging yet. Introduction of special microorganisms to the municipal activated sludge has minor effects, if any. This type of work seems to be limited to special industrial wastewaters where good progress can be achieved. Anaereobic systems promise significant improvements due to small growth rates of many organisms under these conditions.

The potential for further improvement of municipal WWTPs at low cost seems to be limited. Breakthrough can be expected from source control systems especially in newly constructed settlements. For changes of existing infrastructure the time scales are long.

5.3.2 Source control sanitation and treatment options

Efficient reuse of nutrients and wastewater will require new designs of sanitation in order to allow for appropriate systems. Engineering today has produced a wide variety of basically one system: Flush sewers, mechanical, chemical and aerobic biology of different types. Exception is the UASB-technology that allows anaerobic treatment of diluted municipal wastewater in warmer climates, however it is not well designed for reuse. Conventional wastewater systems are often capable of handling greywater and municipal wastewater without urine in a reuse-efficient way. The weakness of the systems is their incapability to deal with soluble matter like ammonia, phosphorus, potassium, calcium, magnesium and so on with respect to full reuse. The sludge fraction contains only a small percentage of the landborne inorganics but accumulates hazardous matter.

Source control systems have a strong potential for new high- and low-tech treatment methods (Henze *et al.*, 1997, Otterpohl *et al.*, 1997, 1999 and 2000, Winblad *et al.*, 1998). High concentrations from low-diluted blackwater or urine call for treatment aiming at products that are ready for safe reuse, and preferably for sale. There will be production units that produce different types of fertilisers, soil improvers, energy and maybe more. It may be considered as an industry that gets its raw material not only for free but can even obtain a fee. Some economically feasible solutions are there and others will soon follow. New biotechnology and physico-chemical methods are foreseeable to a certain extent but distinction of the best ways is not yet possible. There are many pilot projects existing or on the way, there are chances to research and learn from real life applications.

With emerging source control sanitation there is no longer just 'Wastewater'. An established distinction is blackwater (toilet wastewater) and greywater (household wastewater without blackwater). Both are raw products for reuse, but for blackwater dilution should be limited in order to produce fertilizer and energy. Greywater is suitable for reuse as it has not been used in the toilet. Due to low concentrations of nutrients (supposing phosphate free detergents have been used) it can be treated with simple biological methods preferably including filtration. As soon as blackwater is further separated with no-mix-toilets there are the two fractions that we suggest designating 'brown-water' and 'yellow-water'.

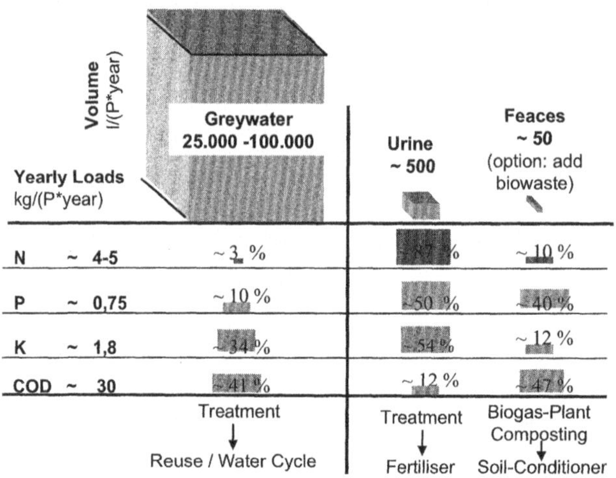

Figure 5.8 The reasons for source control: loads in fractions of household wastewater

These fractions can be produced with the respective sanitation concept. Design will have to be adapted to the geographical and sociological situation, economic optimisation will answer the question of how central or decentralised the sewers and treatment units should be. Treatment methods can be economic on small scale when sustainable savings can be achieved with fewer pipes.

5.3.3 Treatment options for fractions of household wastewater

Treatment methods for highly concentrated flows will be more like those typically used in industrial wastewater treatment. Only greywater is close to conventional municipal wastewater and relatively easy to treat because of the low nutrient concentrations. Brown- and blackwater have a good potential to improve soil fertility; degrading soil quality is a major problem in many countries around the world (Pimentel, 1997). The following list gives an overview on possible treatment systems:

- Treatment of blackwater

Separate collection of blackwater will mostly only make sense if it can be collected with little dilution. In this case anaerobic digestion can be a good choice especially if mixture with kitchen waste is feasible. For colder climates digesters are a rather high-tech solution where economics require a certain size. Tropical climates make digesters much more simple and energy generation more rewarding. Dewatering after the biogas plant has the disadvantage that all the soluble contents that are useful in fertilizer and harmful in receiving waters will escape. Therefore in almost all cases collection and transport with little water are the main problem to be solved. Conventional flush-toilets are prohibitive for their high dilution. For high-tech sanitation vacuum-toilets with vacuum blackwater transport is feasible (Otterpohl *et al.*, 2000). Separating or no-mix-toilets would be a good choice: if both pipes are connected they can produce blackwater in small dilution, too. The most economic solution is toilets that are installed over or at the side of a biogas plant where little flushing is required.

New biotechnology may prove to be helpful in the construction of digesters for blackwater that have high efficiency. The obstacle for anaerobic blackwater treatment is the high concentrations of ammonia/ammonium that can well be inhibiting thermophilic treatment. However, there are biowaste plants that could be operated with such concentrations. Careful startup and proper operation are required.

Physical or chemical treatment does not seem to have advantages for blackwater (see Figure 5.8 brown/yellow water)—drying is far too energy consuming, membranes would be hard to operate with faeces. New types of material may lead to solutions, but still there is a need to hold back nitrogen, phosphorus, potassium and others. This would require multiple step filtration down to reverse osmosis. Obviously there would be more simple ways with separate collection of brown and yellow water.

- Treatment of Brownwater

Low-diluted faeces without urine can be treated in a straightforward way. Treatment options are anaerobic digestion, composting and for warm climates desiccation. Collection and transport are in the most simple way free fall to a treatment facility as a composter or a desiccation toilet. Desiccation is a great option if there is enough sunshine all year round. This type of treatment combines an extreme reduction of volume combined with excellent hygienisation. Two chambers are used in yearly exchange to allow good results. Problems can occur in countries with wet anal cleaning because of the entry of water into the chamber and urine should then preferably be diverted. (Winblad, 1998).

For transport over some distance there is the no-mix or separating toilet. Even though the faecal bowl is flushed with 4–6 litres this flush is only for faeces. As most visits to the toilet are for urinating the overall dilution of the brownwater will be acceptable. This mixture can still be transported in pipes with appropriate gradient. Separation of excess water is possible as most of the soluble nutrients have been collected beforehand. Biological treatment would preferably be by anaerobic or composting in two chamber 'Rottebehälter' (see Figure 5.9). There is a need for intense research and pilot projects for both technologies as sanitation concepts of this type seem to be ecologically and economically very advantageous.

A compact system for very densely populated areas could be based on separating vacuum toilets or other toilets with very little water consumption. It would be possible to combine waterless urine collection with highly concentrated brownwater. Digestors could be thermophilic around 55 °C or with identification of consortia of extremophilic microorganisms even at higher temperatures. Hygienisation would be included and there could be economic biogas units with mass production of such technology.

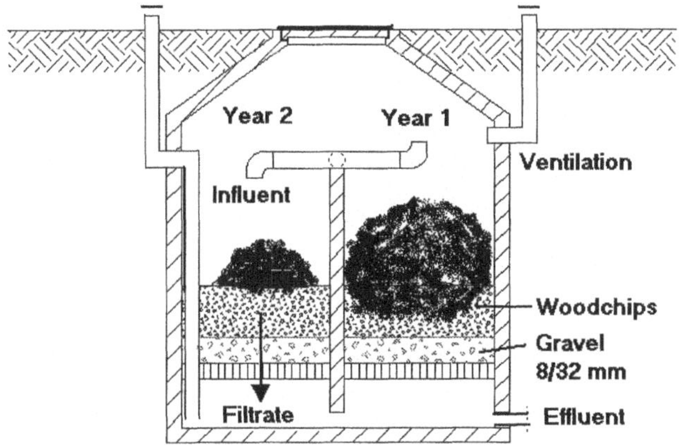

Figure 5.9 Combined dewatering and composting unit

- Treatment of Yellow-water

The big advantage of urine or yellow-water is that the basic 'treatment' can be just storage for 6 months. Within this time span the initially small amount of pathogens can be diminished and degradation of medical residues can take place. In its natural composition yellow water is a well-balanced liquid fertiliser returning many useful substances to the soil that would be lost to the receiving waters otherwise. Application to brown land and mixing into top soil is the most advantageous way to minimise nitrogen losses. If used on plants directly it should be diluted before usage to avoid damage to the plants (Hellström, 1998).

Direct reuse of yellow water may be an economic method in many cases. However, there may be feasible ways to further concentrate it to make it easy to store, transport and use (Hellström, 1998). Traditional ways are to direct it to flow down porous walls where it is sucked in. Water evaporates and the salts will crystallise on the surface where it can be harvested after some years (Winblad, 1998). Keeping the pH values low will be necessary and knowledge is still limited. Producing higher concentrations by heat will lead to major losses of nitrogen. Lower temperatures combined with vacuum can be a possibility. From the farmers point of view there should either be undiluted urine or it should be converted to a solid form of a nutrient rich fertiliser, because the concentration of other salts would make determination of application dosage difficult.

- Treatment of Greywater

One of the advantages of separate collection and treatment of greywater is the possibility of safe and simple reuse. Typical greywater does not have an excess of nutrients; therefore simple biological treatment can be applied. However, COD-concentrations can be as high as in mixed municipal wastewater. Pre-treatment is needed for separation of solids or grease from kitchen wastewater. Activated sludge systems that rely on settling of sludge can get problems with a lack of nutrients – sludge can lose the ability of flock forming. Therefore biofilm methods such as trickling filters, bio-filtration, rotating disk or constructed wetlands (bio-sandfilter) are advantageous. On-site treatment can be combined with watering of plants in the garden, experiments with local soil and plant species are recommended. Even though greywater does not contain toilet wastewater it may contain faecal pathogens to a certain extend from showers and washing machines. This will make treatment methods based on sand filtration and membranes preferable. Greywater contains a high amount of household chemicals (Ledin *et al.*, 2000). High-quality effluents and reuse without restrictions will be more simple with a choice of household chemicals that can easily be mineralised, not only degraded from the initial product.

5.4 NEW APPROACHES FOR SOLVING THE NEED FOR DEVELOPING COUNTRIES

5.4.1 Introduction

The need for paradigm shift described above has been derived largely from experiences in the developed world. It is clear that the same paradigm shift is needed in the developing world. The need for source control and separation at source, and the need to recycle resources are universally applicable. It has been argued that the implementation of the new paradigm could be undertaken first in developing countries, because infrastructure in the developed world still has many years before it needs replacement, whereas the service infrastructure in the developing world is in a much more incipient stage. It should be borne in mind, however, that the developing world is often burdened by the problem of 'poverty, population and pollution'. Attempts to solve just one of these (and in our context it is pollution) are bound to meet head-on with the other equally large problems of poverty and population. The problems facing developing countries are therefore first summarized. These have been discussed widely and in the context of environmentally sound technologies have been documented in several United Nations Environment Programme – International Environmental Technology Centre publications (Ho, G.E. 1997, 1999) and IETC (2000).

There is further the paradox that professionals and others in developing countries consider that technology currently practiced in the developed world is 'best-practice' and hence should be adopted in their countries, perhaps not realizing there is an imminent paradigm shift in the developed world. We should promote the concept vigorously in both the developed and developing worlds, and the developed world should make it a point to illustrate the application of the new approaches. There is a glimmer of hopes, because even in developing countries there is beginning to be a realization of the need for the paradigm shift, especially because of the other two problems of poverty and population, which would logically support the new paradigm shift. Part of this shift is the employment of small-scale systems, which by their nature lend support to the implementation of the paradigm shift. Examples of these new technologies are mentioned including examples of approaches developed and applied in South America (see 5.4.3), together with research needs that need to be supported.

5.4.2 Problems facing developing countries

5.4.2.1 Water shortage

- Shortage of water due to large and growing population, particularly in urban and peri-urban areas.
- Deterioration of quality of existing water resources from pollution from domestic and industrial wastes.
- Tension: There are forces of modernisation, which result in the introduction of appliances that utilise more water per person, and hence produce more wastewater (e.g. flush toilet).

5.4.2.2 Lack of funding

- Low gross national product and low income of the population, but generally improving and there is a desire to modernise (hence the problem in (5.4.2.1) above).
- Tension: higher priorities are given to modern conveniences and improving dwellings, lower priorities are given to sanitation services (solid waste, storm water and wastewater). An example of this is a mansion with a satellite dish, but with only a septic tank for wastewater, poor collection of municipal solid waste, and inadequate drainage for storm water (usually carries greywater as well), frequent flooding.

The lack of funding is not only for capital investment, but also for operation and maintenance. For instance, sewers from the colonial period in many countries that gained their independence from the 1950s are deteriorating and treatment plants may not be operating to specification or not operating at all (by-passed). There are also problems related to lack of expertise (see below) and there may be a culture of 'dependence' on external aid.

5.4.2.3 Lack of expertise

- There are not enough trained people with the skills to choose the correct technology, to operate and maintain water, storm water and wastewater collection, treatment and disposal systems. There is also a lack of availability of information on alternative technologies.
- Tension: There is a desire to adopt modern technology, and this is usually equated to the conventional sewerage system and activated sludge process (because of its common use in developed countries), and yet the costs are high, and the expertise required is highly specialised.

5.4.2.4 Lack of institutional support

- Institutions are required to plan for water supply and sanitation services, to implement them and to operate, maintain and monitor them. In developing countries these institutions are generally not well developed. Peri-urban development occurs without proper planning, resulting in 'illegal' or informal settlements. Hygiene promotion is generally inadequate. Communities seem to be powerless to implement community-based sanitation, or do not give high enough priority to it compared to housing, education and health (illness).
- Tension: The model used in developed countries tends to be followed, with its highly structured government departments, requiring highly educated and trained managers and personnel.

5.4.3 Current solutions in developing countries

Current solutions adopted in developing countries will be described highlighting the tensions in the adopted solutions. These solutions include:

a. Conventional sewerage and activated sludge process

The problems associated with this solution are well known (see 5.3.1 above). This system is likely to be financially sustained only in capital cities in most developing countries. Its extension to informal peri-urban areas is out of the question from the financial point of view. The sustainability of the conventional system is questionable even for developed countries (see 5.2 and 5.3 above).

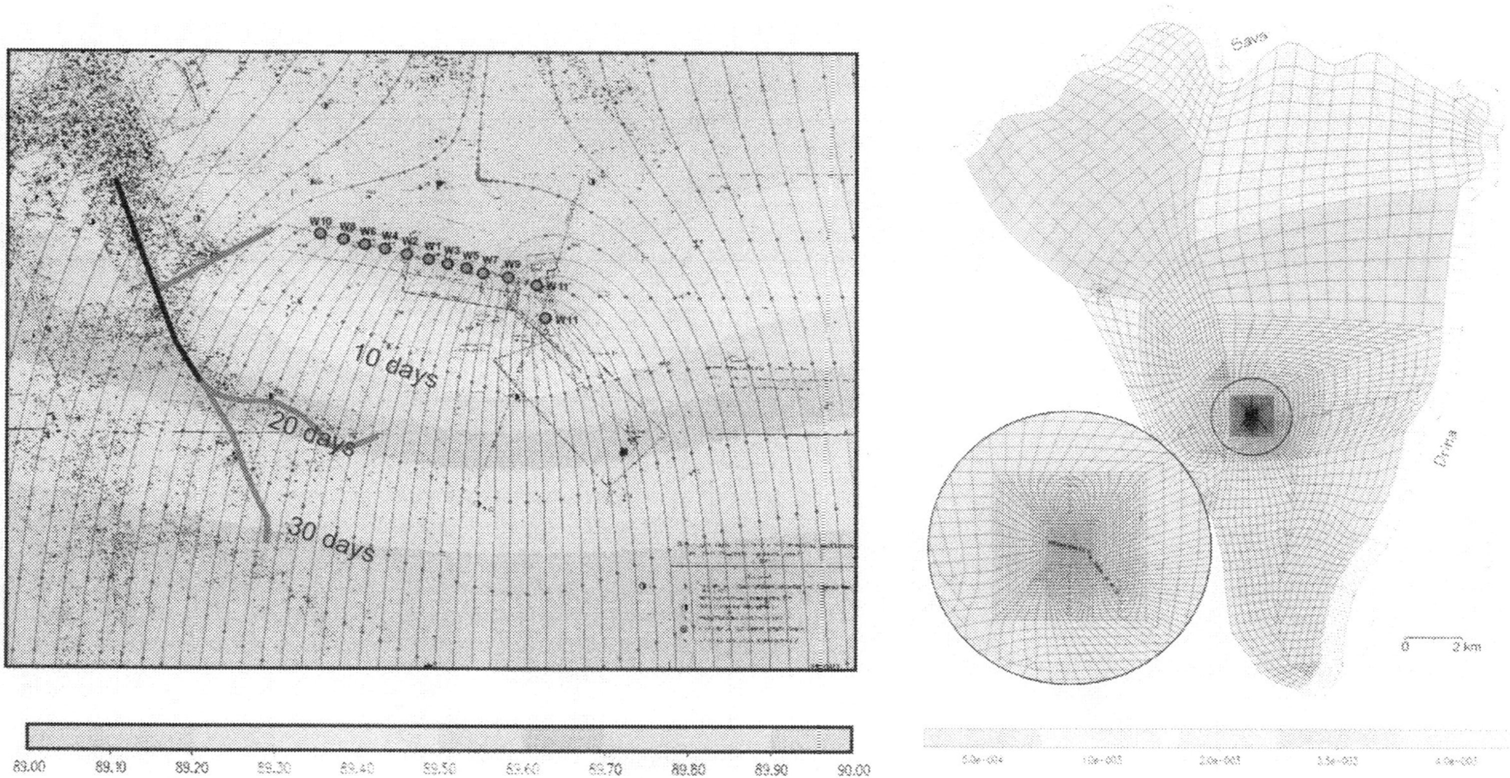

Plate 3.1 Pathlines and pollution travel times (left) used as a tool for analysis of the pollution remediation of the Grmić aquifer (right) Well field area zoomed on the grid map colour codes; Piezometric heads (left) and conductivity (right) (after Pokrajac – 1999)

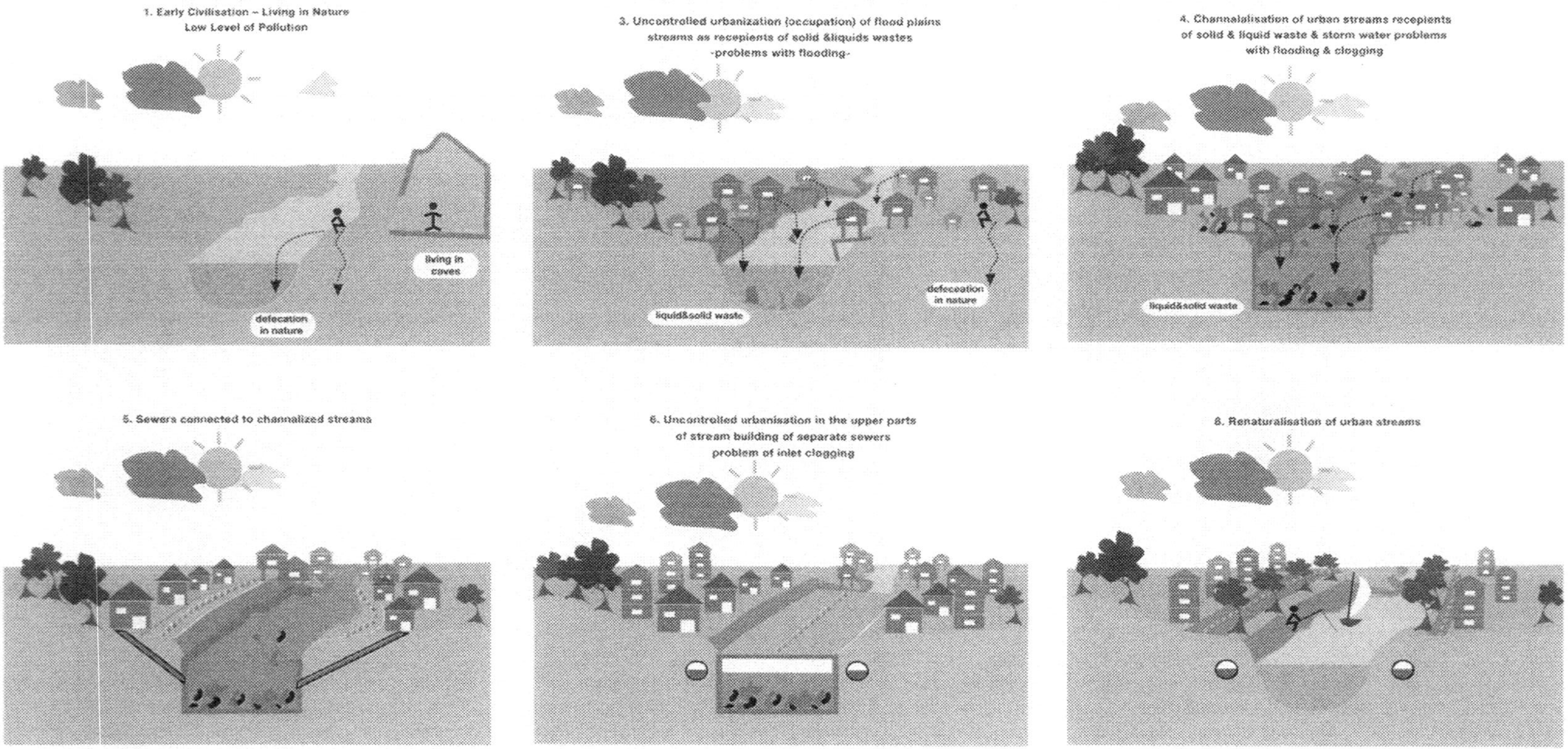

Plate 3.2 History of urban streams – from the early settlement's pollution to daylighting (After Maksimovic & Ho– UNEP - 2001)

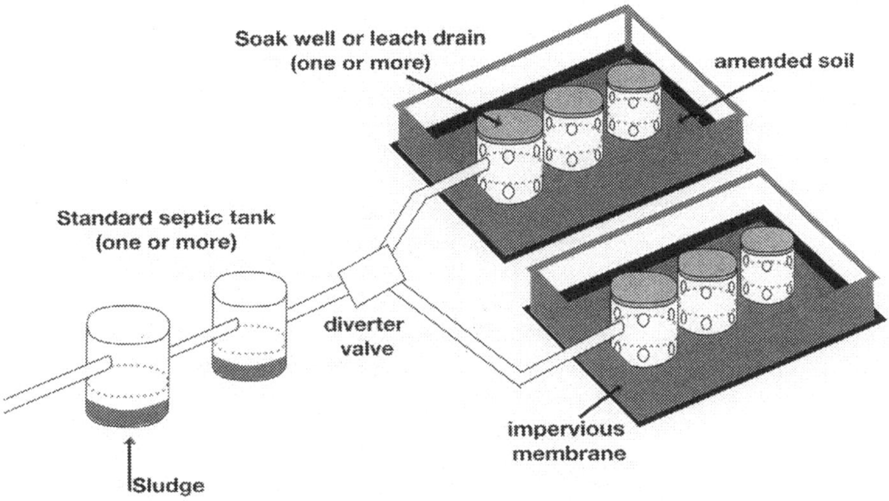

Plate 5.10a Econmax on-site wastewater treatment unit; modified leach drain to reduce groundwater pollution – general view

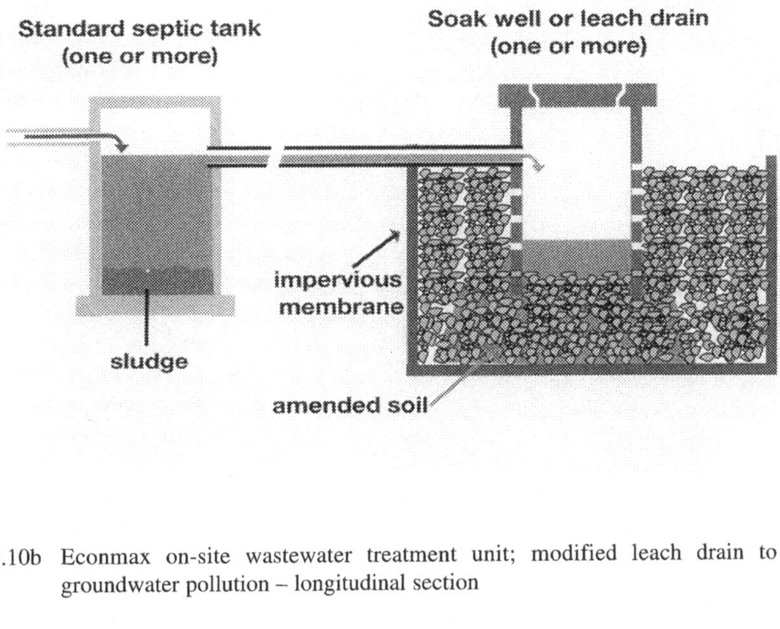

Plate 5.10b Econmax on-site wastewater treatment unit; modified leach drain to reduce groundwater pollution – longitudinal section

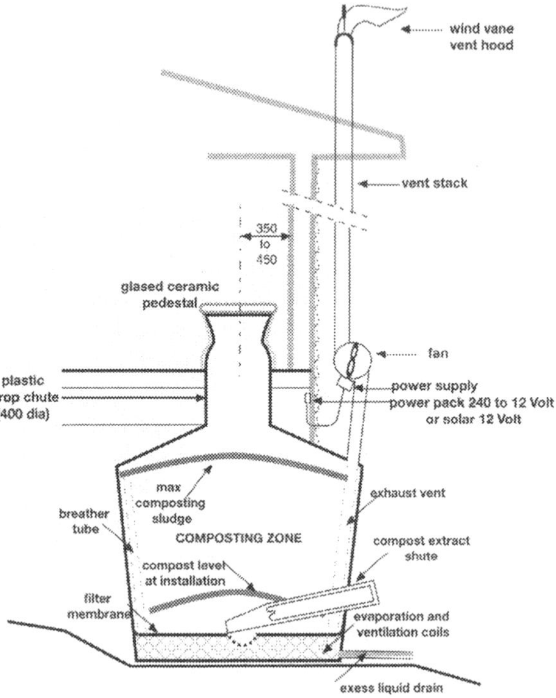

Plate 5.11 Downmus composting toilet

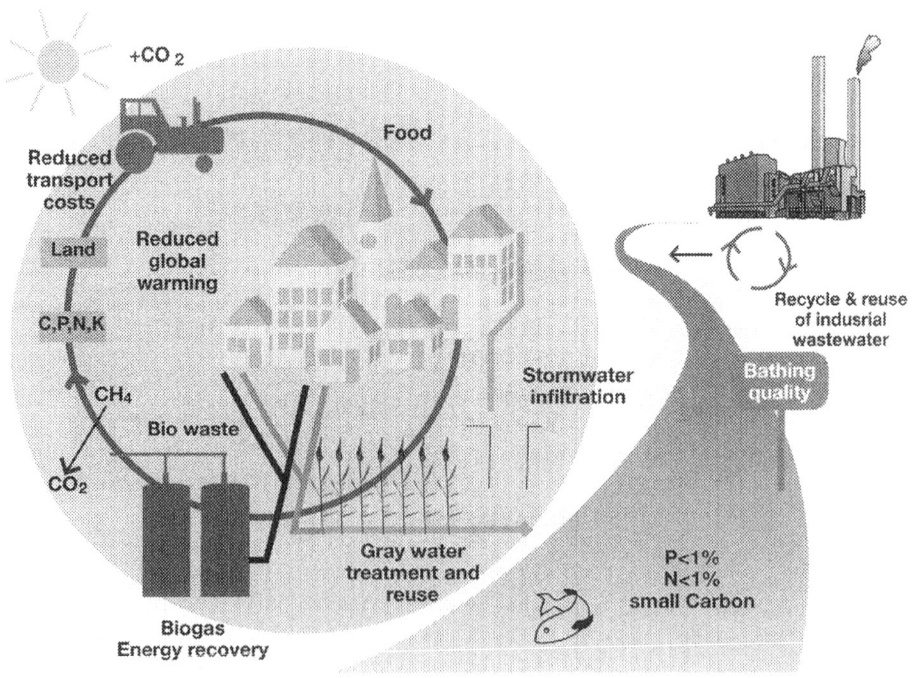

Plate 5.12 Separate blackwater and greywater to achieve sustainable development

b. Low cost sewerage

- This includes shallow sewerage and settled sewerage. Shallow sewerage carries wastewater as in the conventional sewerage, but is laid shallower and pipes have smaller diameters. Condominial arrangement has also been used. South America has many examples. Settled sewerage carries only wastewater that has been sedimented in a septic tank (single or communal) and can therefore run full of water. Examples are in South America. Implementation required a high degree of community participation.
- Tension: Social workers and engineers need work as a team on a partnership (50:50) basis.

c. Low cost sanitation (John Pickford - WEDC (1995))

- Systems generally employ variations of pit latrines. The ventilated improved pit (VIP) latrine or its modification is the most widely used in sub-Saharan Africa.
- Tension: Groundwater pollution occurs with low cost sanitation systems. Pit latrines cannot be used when groundwater is near the surface or where the ground is rocky. Alternatives like evapotranspiration beds are available. Hygiene promotion and training in maintenance and handling of full pits are required.

d. Ecological Sanitation (Winblad, 1998)

In situ separation of urine and faeces. The hygiene practice (soft-ware) that goes with this system needs to be more fully developed. Trials are on-going in several regions.

e. Household-centred sanitation

- This is an attempt to introduce sanitation at the grass roots level (bottom-up) rather than the sanitation dictated by governments or government department/agency/experts.
- Tension: This model relies on the western type model of democracy and its application in many developing countries needs to be tested.

f. Decentralised wastewater systems

Under this heading are concepts/ideas that have been proposed to overcome the centralised systems associated with the conventional sewerage and activated sludge process. The basic premise is that if we avoid the problems associated with the conventional system, which is now recognised to be an inappropriate solution for developing countries, there will be an improvement.

Solutions outlined in (b) to (f) above are solutions that are largely developed in developed countries for developing countries. It appears that what we need to do is to obtain solutions that are applicable to both developed and developing countries, so that we avoid the dichotomy of approach, which is largely artificial, and fortunately can be overcome by considering the concept of sustainability.

5.4.4 Glimmer of hope

Sustainability in wastewater management can be achieved by reuse of the water and nutrients in the wastewater for agriculture or aquaculture. In this way pollution from the organic carbon and nutrients (nitrogen and phosphorus) in the wastewater is prevented, and cycling of carbon, nitrogen and phosphorus in the closed (complete cycle). To achieve this, industrial wastes must be separated treated (and reused) and not discharged with domestic wastewater. Storm water is allowed to flow along natural streams and water courses. Domestic waste is separately collected and treated/reused.

Examples of separate systems are:
On-site system:

- Ecomax (amended soil filter after septic tank), (Presented in colour as Plates 5.10a and 5.10b);
- Dowmus (enclosed pit vermi-composting process for blackwater), (Plate 5.11).

Off-site system:
Composting for blackwater, wetland for greywater (Otterpohl, 2000); lagoons followed by aquaculture.

- In Plate 5.12 (presented in colour), separate blackwater and greywater treatment to achieve sustainable development
- Figure 5.13, wastewater reuse through aquaculture.

Issues that have to be addressed are pathogens in the wastes, and the software, hygiene practices to be observed, and equipment for maintenance (equivalent to the very well developed practices, cleaning equipment, cleaning agents for the flush toilet and downstream system).

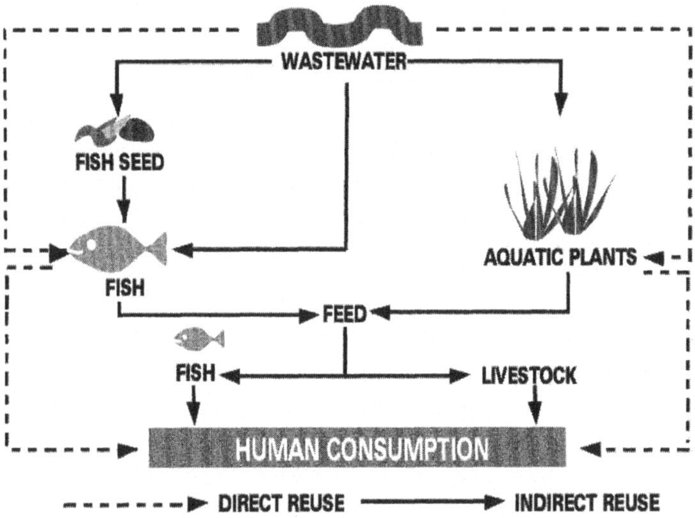

Figure 5.13 Wastewater reuse through aquaculture

5.4.5 Future scenario

It is likely that numerous ideas will be tried and systems developed to achieve the objective of recycling/reusing water and nutrients contained in the water thus closing the cycle and achieving environmental sustainability. This can be considered to be a healthy development. Successful technologies will be those that will enable environmental sustainability, but also accompanied by the necessary software to ensure acceptance (convenience, availability of cleaning equipment and agents, ease of maintenance). Trials can occur in both developed and developing countries. The concept of sustainability applied equally to both. Low cost systems can be developed using local materials and labour for application in developing countries.

5.5 RESEARCH AND DEVELOPMENT NEEDS

5.5.1 Research and development needs for source control and recycle of nutrients

With respect to source control in urban and rural sanitation there is a lack of research. Today we find about ten principal ways to deal with wastewater, but research is still almost exclusively directed to the more than 100-year-old conventional systems. With regard to dramatic situations now and even worse in the near future for around 70% of the world population there has to be a fast change in focus. Too many research institutions work in parallel on a limited set of ideas simply to optimise for small gains in systems that are very advanced but not made for reuse. Going for source control allows a quantum leap in overall efficiency including the full range of high-tech to very economic solutions. The wastewater research community should take its responsibility and work towards water saving zero-emissions total reuse systems. Compared to other branches of technology we are just starting—the number of new solutions has increased with treatment and reuse technology that has been built for industry but is applicable to low-diluted brown- and yellow water.

New physico-chemical technology is necessary to give clean and safe water supply that should preferably originate from greywater. Unrestricted usage as tap water should exclude previous mixing with human excreta. As we know there are many substances in urine and faeces that should not be taken up by humans again. If ten-fold reuse will be achieved with economic technology this will enable a life of ease with 10 litres of fresh water per person per day. With such low numbers even limited resources will be suitable for growing populations. The technologies are available – research will have to make them work under a wide range of socio-economic conditions.

Pilot projects are necessary to give real-life results and many of these will have to be started in industrialised countries. Developing countries tend to copy the 'West'; therefore changes will have to be promoted there as well. Many ways are open for responsible private-public partnership if appropriate solutions are available rather than copying yesterday's models... However, research has to take a lead to open these ways to their full potential with latest technological knowledge.

5.5.2 Research for problems of developing countries

Based on the emerging and likely solutions the research needs can be classified into research needs in technology, public health, social and cultural factors.

5.5.2.1 Technology

The conventional sewerage and activated sludge process has evolved over several decades, and we appear to have forgotten the research and development efforts that went into toilet design (now with low volume or dual flush), cistern design, cleaning equipment, cleaning agents, sewer system, its construction and maintenance, the microbiology of activated sludge, aeration systems. The scientific understanding has resulted in advances and improvements in technology. A similar effort is required in systems that will support the solutions described in above.

Two examples are given for illustration purposes:

- Composting toilet technology needs further research to optimise its performance (effect of temperature, aeration, hydration, maturation; with vermin-composting the optimum conditions for the worms).
- Soil filter vegetation system needs further research in the soil filter media and removal of BOD, N and P. Effect of climate and suitable local vegetation.

5.5.2.2 Public health

Much research is required on the fate of human pathogens (bacteria, viruses, parasites and helminths) in blackwater and greywater in the proposed technologies. Because reuse of water and biosolids are proposed to achieve a closed loop, the handling of the water and biosolids requires good scientific data on the presence and viability of the pathogens in the recycled materials. The safety of pharmaceutical drugs as well as natural hormones in urine must be investigated when urine are collectively treated and converted for new type of fertiliser.

5.5.2.3 Social and cultural factors

Introduction of any new technology requires acceptance of the technology by the community. Factors that facilitate acceptance have been hinted above and include ease of use and maintenance (convenience factors), performance as good or better than existing technology, cost competitive or preferably cheaper.

5.5.3 The contribution of information and automation technology to new biotechnology

The contribution of information technology to new biotechnology can be strong in the following three areas:

(1) Improvement of availability of information about new technology and exchange of experience,
(2) Better performance of treatment facilities by better monitoring and control devices,
(3) Improvement in operation by remote supervision and pricing for services.

It is important to explore the potential and possibility of information and automation technology in the field of urban water management. Water and wastewater engineers do typically utilise the classic type of automation technology for the operation and management of water and sewage networks. With growing modern information technology a step up in performance at lower costs is possible. This is especially true for smaller decentralised units, where lack of professional supervision can be overcome by Internet systems with advanced monitoring technology and analysing software. The key for upgrading is interdisciplinary work where possibilities and demand can be discussed on a state-of-the-art-level of both the wastewater and the information technology. There are numerous technological advances in other fields that can fairly easily be adopted to wastewater management, such as biosensor techniques which could be applied for monitoring water and wastewater quality.

5.6 REFERENCES

Cosgrove, W.C. and R. Rijsberman (2000) World Water Vision-Making Water Everybody's Business, *the Second World Water Forum, the Hague*, 17-22 March.

Daughton, Ch.G. and Th. A. Ternes, (1999): Pharmaceutical and Personal Care Products in the Environment: Agents of Subtle Change? *Environmental Health Perspectives,* **107**, Supplement 6, Dec., pp. 907.

Eilersen, A.M., S.B. Nielsen, S. Gabriel, B. Hoffmann, C.R. Moshøj, M. Henze, M. Elle, & P.S. Mikkelsen, (1999): Assessing the sustainability of wastewater handling in non-sewered settlements. In: Kløve, B., Etnier, C., Jenssen, P. & Mæhlum, T. (eds.): *Proceedings of the 4th International Conference Managing the Wastewater Resource. Ecological Engineering for Wastewater Treatment, Ås,* June 7-11, Jordforsk, Department of Agricultural Engineering, The Agricultural University of Norway & IEES, Ås.

FAO Country Tables – basic data on the agricultural sector,Reve, Roma,1995.

Feuerpfeil, I., J. López-Pila, R. Schmidt, E. Schneider, R. Szewzyk, (1999): Antibiotikaresistente Bakterien und Antibiotika in der Umwelt (Bacteria with resistances against antibiotics in the environment) *Bundesgesundheitsblatt - Gesundheitsforschung – Gesundheitsschutz ISSN: 1436-9990 (printed version) ISSN: 1437-1588 (electronic version)* Springer, Volume 42 Issue 1 pp 37-50 (in German)

Guideline for the investigation of a basin-wide sewage works, *Ministry of Construction, Japan* (1996).

Hellström, D. (1998): Nutrient Management in Sewerage Systems: Investigation of Components and Exergy Analysis. *Department of Environmental Engineering, Division of Sanitary Engineering.* Lulea, Lulea University of Technology.

Henze, M. and A. Ledin (2000) Types, characteristics and quantities of classic, combined domestic wastewaters. *Chapter 4 in: Decentralised Sanitation and Reuse (Lens P. and G. Lettinga , Eds).* IWA Publications, London 2000.

Henze, M. (1997): Waste design for households with respect to water, organics and nutrients. *Water Science and Technology,* **35** (9), 113–120.

Henze, M., L. Somolyódy, W. Schilling, and J. Tyson, (1997): Sustainable Sanitation. Selected Papers on the Concept of Sustainability in Sanitation and Wastewater Management, *Water Science & Technology,* **35** (9).

Henze, M., P. Harremoës, J. la Cour Jansen, and E. Arvin, (2000): *Wastewater Treatment - Biological and Chemical Processes. 3.Edition,* Springer Verlag, Berlin 2000.

Ho, G.E. (Ed.) (1999): *Proceedings of the International Regional Conference on Environmental Technologies for Wastewater Management* held at Murdoch

University, Perth, Western Australia, 4-5 December 1997. UNEP International Environmental Technology Centre, Osaka (ISBN 0-86905-708-1).

Ho, G.E. (Ed.) (1998): *Workbook for Training in Adopting, Applying and Operating Environmentally Sound Technologies.* Implemented 8–13 December 1997 in collaboration with Murdoch University. UNEP International Environmental Technology Centre, Osaka (ISBN 92-807-1478-0).

Ho, G.E. (Ed.) (1998): *Proceedings of the Workshop on Adopting, Applying and Operating Environmentally Sound Technologies for Domestic and Industrial Wastewater Treatment for the Wider Caribbean Region.* Implemented 16-20 November 1998 in Montego Bay, Jamaica. UNEP International Environmental Technology Centre, Osaka (ISBN 0-86905-679-4).

IETC (2000) *International Source Book for Environmentally Sound Technologies for Wastewater and Stormwater Management.* To be published by UNEP International Environmental Technologies, Osaka.

Ledin, A., E. Eriksson, and M. Henze, (2000) Aspects of groundwater recharge using grey wastewater. *Chapter 18 in: Decentralised Sanitation and Reuse* (Lens, P. and G. Lettinga, Eds). IWA Publications, London 2000.

Matsui,S., H. Takigami, T. Matsuda, J. Taniguchi, H. Adachi, H. Kawami, and Y. Shimizu, (1999) Oestrogen and oestrogen mimicas contamination in water and the role of sewage treatment, *Proceedings of the International Symposium on Development of Innovative Water and Wastewater Treatment Technologies for the 21st Century*, Hong Kong,8-10 October.

Otterpohl, R., M. Grottker, and J. Lange, (1997): Sustainable Water and Waste Management in Urban Areas, *Water Science and Technology,* **35** (9), 121–133 (Part 1).

Otterpohl, R., A. Albold, and M. Oldenburg (1999) Source Control in Urban Sanitation and Waste Management: 10 Options with Resource Management for different social and geographical conditions, *Water, Science & Technology,* **39** (5), 153-160.

Otterpohl, R. (2001) Design of highly efficient Source Control Sanitation and practical Experiences in: *Decentralised Sanitation and Reuse* (Lens, P. and G. Lettinga, Eds). IWA Publishing, London, UK.

Pickford, J (1995) Low-cost sanitation*: A survey of practical experience. Intermediate Technology Publications.* London.

Pimentel, D. (1997) Soil Erosion and Agricultural Productivity: The Global Population/Food Problem, *Gaia* **6** (3).

SEPA (1995) report 4425 *What is the content of sewage from households?* Swedish Environment Protection Agency.

Winblad, U., Eds. (1998) Ecological Sanitation, *SIDA,* Stockholm, ISBN 91 586 76 12 0

Zeemann, G., G. Lettinga (1998): The role of anaerobic digestion of domestic sewage in closing the water and nutrient cycle at community level, *Water Science & Technology,* **39** (5), 187-194.

6
Problems of developing countries

David Stephenson

This chapter succinctly describes common problems encountered in developing countries concerning urban water management, particularly water supply and sanitation concerns in LCD's (Less Developed Countries), and points out some possible approaches to face them).

6.1 THE CONSTRAINTS

6.1.1 Population explosion

The end of the 20th century is one of relatively high population growth. Of the world's population of 6 billion, 75% (4.5 billion) can be classified as poor; they earn less than 20% of the world's income. Of these, 1.5 billion live in abject poverty (insufficient means to maintain nutrition and health). The majority are in Asia now but within 25 years, the majority (1000 million) will be in sub-Saharan Africa.

Also 1.5 billion people do not yet have access to safe drinking water and 3 billion do not have adequate sanitation facilities. Over 4 million people die each

 Frontiers in Urban Water Management: Deadlock or Hope
Edited by Čedo Maksimović and José Alberto Tejada-Guibert. ISBN: 1 900222 76 0

year from water related diseases. UNCED (1993) indicated that about $20 billion should be spent a year on water supply and sanitation in developing countries, which is only $5 per capita, so is probably too low to achieve a significant improvement in standards. The 1980s was seen as a decade of hope. 1.6 billion more people received water. The number of urban people with access to water increased to 80% and with sanitation 50%. But still the number of urban people without sanitation increased by 70 million.

Bhatia and Falkenmark (1993) suggest $11–14 billion is required for water supply and sanitation infrastructuring urban areas in developing countries. Forty-five percent of the world's population lived in cities in 1990 and over 50% in the year 2000, thus compounding the population growth rate. The majority of the urban growth will in the future be in developing countries. There are now 320 cities with over 1 million inhabitants, and this could double to 600 by the year 2025.

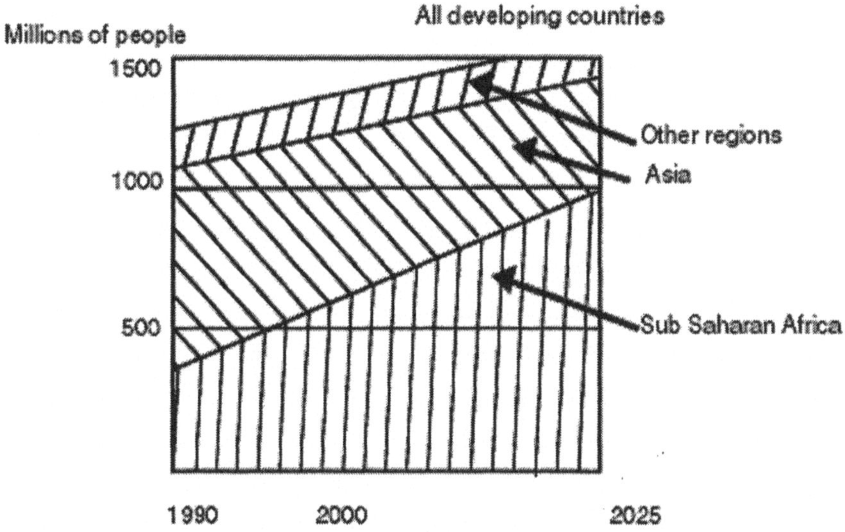

Figure 6.1 People in poverty in developing countries

Poorer countries have doubtlessly often been affected adversely by their historical relations with developed nations. Some of the reasons for this maybe, to a lesser or greater degree, dependent on the country:

- Colonisation, raising expectations and leading to breaks in social customs.
- Foreign aid packages.
- First world dealers placing their goods by unfair practices.
- Inability to create institutions capable of handling aid efficiently.

- Modern medicine resulting in longer lives and thereby higher populations (even though now the raging AIDS epidemic has brought in a new dimension, especially in Africa).

The African population growth rate is the largest in the world. It can, however, be controlled as was demonstrated in Kenya. There the total fertility rate (TFR) was 7.9 children per woman in 1977 and population growth rate 4.0% p.a. By 1989, the TFR was 6.7. Also the average desired family (ADF) declined from 5.8 to 4.4 in the same period. The same cannot be said for the rest of Africa, with TFR averages of 6.3 and ADF of 6 to 9. High TFRs of 7.5 in Ethiopia and Malawi and 8.3 in Rwanda are noted, compared with 1.9 in the USA and UK (Petersen, 1993).

The African urban population is expected to increase from 140 million in 1990 to 5000 million in 2020. They will need an estimated $47 billion to service (Marshall, 1996). Political uncertainty, corruption and finances present large hurdles.

Table 6.1 Access to safe drinking water in developing countries, by region, 1980, 1990 and 1994 (millions) (WHO, 1996, Gleick, 1998)

Region and country	1980 Population	Percent with access	1990 Population	Percent with access	1994 Population	Percent with access
Africa						
Urban	120	83	201	67	239	64
Rural	333	33	432	35	468	37
Total:	453	46	633	45	707	46
Latin America & the Caribbean						
Urban	237	82	314	90	348	88
Rural	125	47	126	51	125	56
Total:	362	70	440	79	473	80
Asia & the Pacific						
Urban	549	73	829	83	955	84
Rural	1823	28	2097	53	2167	78
Total:	2373	38	2926	61	3122	80
Western Asia						
Urban	28	95	45	87	52	98
Rural	22	51	27	63	29	69
Total:	49	75	72	78	81	88
TOTAL						
Urban	**933**	**77**	**1389**	**82**	**1594**	**82**
Rural	**2303**	**30**	**2682**	**50**	**2789**	**70**
Total:	**3236**	**44**	**4071**	**61**	**4383**	**74**

The World Bank estimated that in 1985, 65% of the rural population in the developing world was without access to safe, convenient water. 75% had no satisfactory sewage disposal facilities. This is despite the fact that over US$10 billion had been invested in rural water supply and sanitation in the preceding decade.

As fast as investment now is, the populations, particularly of Africa, are increasing faster. The population in Africa is increasing at a rate of over 2% per annum, adding 12 billion people per annum, requiring over US$1 billion per annum to be invested in water supply.

Table 6.1 shows world coverage of water supply increased from 30% to 70% in rural areas from 1980 to 1994, but in urban areas coverage has remained almost static percentage-wise at 80%, although numbers-wise it increased from 718 million to 1315 million. Sanitation has fared even worse, with urban percentage decreasing from 67% in 1990 to 63% in 1994.

6.1.2 Financial limitations

The debt of developing countries stood at about US$1.7 trillion in year 1995 and was increasing at 5% p.a. with little sign of improvement in payback ability. In fact conditions are deteriorating at an increasing rate. Until the 1980s, aid by the developed countries was largely in the form of capital grants to the poorer nations. The British Commonwealth and French colonies and earlier the Spanish and Portuguese settled territories received preference by their respective mother countries during the colonial period. Money was used to build infrastructure for use largely by the expatriates or to benefit the colonial power. When colonization stopped, so did much of the capital flow. Further aid was largely token aid. War risks in certain areas have made economic investment even less attractive. Figure 6.2 shows the relative distribution of world income.

Now the pattern has changed even more drastically. The new emerging countries of Eastern Europe will absorb a lot of the capital from Western Europe and the US. The strategic importance of Africa as a destination of massive investment has been lost. The Far East is showing signs of economic development and is also attracting capital. The economic incentive to invest in Africa, the Middle East, the rest of Asia and South America has, in general, been deceased. The poorer countries are now left to service debts far greater than their gross domestic production, in many cases. The World Bank indicates a cash flow of $157 billion to developing countries in 1992, of which 60% is now from private organisations as opposed to public aid money.

The total input is still falling behind the debt, which increased at about 4% p.a. from $137 in 1986 to $179 billion in 1992 from the poorest countries. The debt service charge for Sudan and Somalia is 2000 times the value of their exports; it looks like they can never catch up on repayments. In addition, aid is

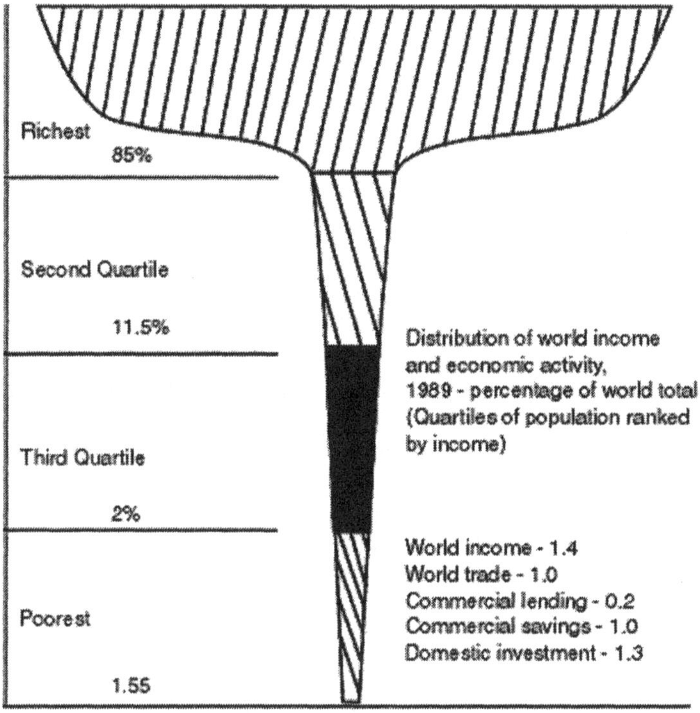

Figure 6.2 Global income and economic disparities (UNDP 1992)

decreasing to Africa at about $2 billion per year. Of Africa's 51 countries, half have debts in excess of their GNP. Mozambique's is 400 times the GNP. However, the economies of these countries are very low. The debts in Latin America are far higher in absolute terms, e.g. $616 billion in 1992.

The new policy of structural adjustments to obtain IMF funds is causing confusion in developing countries and favouring developed countries who still subsidize crops.

The changing focus of aid from capital to institutional projects is also a source of greater corruption and less money may get through to the workers in the short term. It may be a recipe for degeneration of morality in much of Africa, for corruption spins off down the line. The elite in these countries remain relatively untouched by removal of subsidies on transport, food and welfare, as they hardly use them.

The World Bank (1988) found that the rate of failure of its projects in sub-Saharan Africa was twice as large (20%) as the average. The average rate of return was half that of Asian projects. Institutional projects were even less successful (28% success).

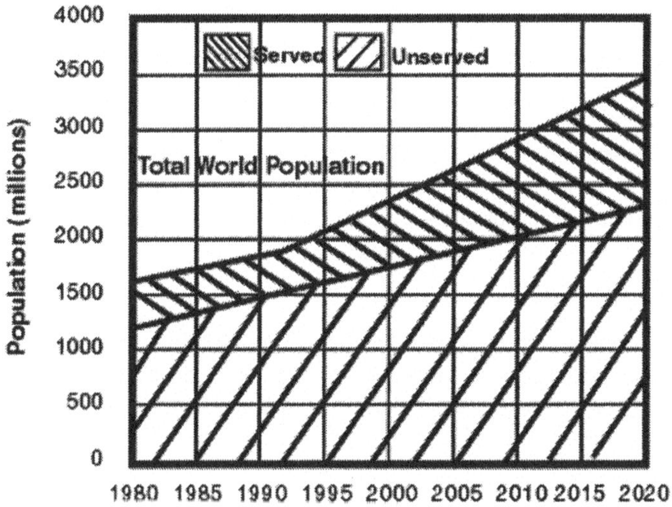

Figure 6.3 Urban population and water supply coverage in developing countries (Bhatia and Falkenmark, 1993)

Evaluations are generally made at the end of the project duration. But after five to ten years, all African projects dropped in productivity (from 10% to 3% average annual return). Livestock and plantation projects showed no success at all or negative returns.

In the early 20th century, water-funding agencies concentrated on irrigation and hydro power. Towards the 1970s, emphasis swung to water supply. This was spurred by the UN Water Decade in the 1980s, the goal being to supply everyone with at least 20 *l*/d of water. A mass balance was not carried out on finances available and finances required, with the result that installation fell woefully short of expectations.

6.1.3 Institutional problems

Problems due to poor institutional involvement can be:

- Reluctance to accede to community wants, as it appears to diminish officer's standing.
- Perception that alternative technologies are better, notwithstanding public preferences.

- Under-recovering of costs reduces financial sustainability.
- Subsidised costs annoy the subsidisers.
- Cross-subsidies result in decreased usage by the high payer, and over-usage by the low payer.
- In the limit, the utility may provide free services to avoid costs of collection.
- Low contributions of poor users means suppliers do not feel obliged to extend or maintain services.
- The priorities of public servants will put richer users and public salaries before poorer users.
- Cuts in services to non-payers wastes the capital installation cost and would result in users finding alternatives, possibly not as hygienic.

6.1.4 Water supply needs

Where water consumption records are not available, consumption per capita can be estimated by talking to residents. However, by upgrading the supply system, consumption may change and the following figures may be used to estimate typical consumption (Tables 6.2 and 6.3).

Table 6.2 Typical rural domestic water usage (IRC, 1981)

Type of water supply	Typical consumption (litres/capita/day)	Range
Communal water point		
Well or standpipe at considerable distances (> 1000m)	7	5–10
Well or standpipe at medium distance (250 - 1000m)	12	10– 15
Well nearby (< 250m)	20	10–25
Standpipe nearby (< 250m)	30	20–50
Yard connection	40	20–80
House connection		
Single tap	50	30–60
Multiple taps	150	70–250

An improved estimate could be obtained by studying existing water supply systems in the same area.

WHO (Tejada-Guibert, 1998) give the minimum quantities per person per day as ranging from 15 litres to over 50, with a mode of 20, but if an amount required for safe hygiene and food preparation is added this goes up to nearly 50 litres per capita per day.

Water requirements for non-domestic purposes are difficult to estimate and, where possible, field measurements should be taken.

Table 6.3 Distance people are prepared to go for water (Tejada-Guibert, 1998, from WHO, 1996)

Distance (Metres)	No. of countries	
	Urban	Rural
50	20	10
100	6	1
250	3	6
500	8	17
1000	1	4
2000		4

Alternative levels of participation were tested by Narayan (1994), i.e. by contribution, influence, sharing power, control, resources, benefits, knowledge and skills. Thus the input may be real, or just a desire for transparency. Participation becomes less effective where agencies retained control. Sarah *et al.* (1998) went further to define a Demand Responsive Approach as one where:

- Water should be increasingly managed as an economic as well as social good
- Management should be at the lowest appropriate level
- A holistic approach to se water resources should be employed
- Women should play a key role in the management of water.

Much attention has been paid to the provision of potable water to communities in the last two decades. In particular, the years 1981-1990 were declared the International Drinking Water Supply and Sanitation Decade by the United Nations. A lot was learnt from the attempts to supply all those without potable water with a reasonable supply. The objective was to supply all the world's people with potable water and sanitation. In fact, the program barely reached 30% of those without formal schemes, and of those schemes which were investigated, the majority are no longer operative.

The program started with the idea of providing scaled down or mini water supply schemes to as many villages as possible. The engineers and planners were largely from developed countries and they had no doubt undergone a tremendous learning experience, and in particular, realized that there is much more to be considered than purely installation of equipment. The major factors not sufficiently considered included reliability, the ease of repairs and more particularly, the social consequences and the creation of institutional capacity to manage the schemes.

6.2 WATER QUALITY AND HEALTH

More people die of poor water quality than due to lack of water. Yet people will sacrifice more to get water than to purify it or otherwise ensure its safety.

A minimum of 2–3 litres of water is required per person per day to live. Additional water would be required for washing people and more for utensils. As the community becomes more sophisticated, it will require more cleansing and even luxury water, for example for gardens, washing motor cars and topping up swimming pools. The demand of sophisticated communities can exceed 300 litres per person per day.

The quality of water is particularly important for poorer communities. They may not have fuel to boil water and in any case may be unaware of the health hazards of polluted water. Up to 80% of all diseases in the world are associated with impure water. In developing countries, in particular the tropical areas, these disease such as schistosomiasis (bilharzia), malaria and blind fly are endemic. The diseases may be transmitted by insects, such as mosquitoes or snails or flies. They can even be obtained from fish or other foods obtained from the water. However, many of the diseases are contracted directly from the water which transmits cholera, typhoid, dysentery, hepatitis and other illnesses from person to person. The interlinking of sanitation and water source is therefore of great importance. Other diseases may be due to the lack of water, such as skin problems, lice, stomach problems and fevers, as well as worms and parasites.

The alternatives perhaps need more careful consideration before embarking on piped potable water supplies to poor communities, however. The cost of a piped purified water supply is considerable, especially when considered by the recipients. It can cost thousands of dollars per person for purification and pumping and storage plus sanitation facilities. The cost per capita increases the smaller the community because of the effect of scale being non-linear. Economically therefore it is impossible to supply more than half the world with such facilities until urbanization absorbs a greater proportion of populations. This funding could not be supplied by the recipients and free aid is certainly not going to be forthcoming on this scale.

Alternate methods of transporting water in greater quantities, such as cart or even light delivery vehicles are the cheapest solution after hand transport, and in fact hold potential for economic development. For example, the purchase of a delivery vehicle for $10,000 would provide transport for water for hundreds of people, the alternative of which was to supply piped potable water and sanitation which would cost hundreds of thousands of dollars. In addition, the delivery vehicle will provide transport to schools, work or of agricultural produce to markets, etc.

In some situations, the standards of supply for example, and the rate of supply per capita, may be relaxed compared with western standards. The main consideration should perhaps be water quality.

Conventional water treatment is often beyond the economic ability or even comprehension of poor and transitory people. Alternatives need consideration to reduce the problems of pollution viz health education, water resources management, alternative sources. Even disinfection is unlikely and if uncontrolled could pose problems in itself. Fuel is expensive for boiling water and bottled water is expensive. So polluted water is a particular problem of poor communities.

Mwabu (1999) identified ten health determinants in developing communities:

- Individual level variables, e.g. education, time endowment, gender
- Household variables, e.g. safe water, social infrastructure, health services
- Purchased medicines, housing
- Region specific factors including diseases, climate and natural disasters
- Demographic variables such as population size, growth rate
- Structure and institutional including urbanisation, trade
- Armed conflict
- External economic pressures
- Population migration
- Social security, including food, income and health services.

Cross sectional analysis for Africa revealed that water supply reduced infant mortality more than any other factor. A 10% improvement in access to safe water reduced IRM by 5%. The factor was followed by secondary education and then income.

Pritchett and Summers (1996) support the wealth factor, and Sen (1995) supports development, but the water factor is implicit and its importance revealed above.

6.2.1 Effects of polluted water

Pollution is inevitable in overcrowded, underprivileged and poorly-positioned communities. Pollutants in water can be due to:

- Suspended solids - floating debris, settleable sediment or buoyant algae
- A quality unrelated necessarily to solids, e.g. temperature, acidity, color, odor or taste. The latter are often associated with matter in the water, however
- Dissolved chemicals: many cations, anions, with different effects, e.g. on health, equipment or toxicity
- Bacteria, pathogens, viruses, organic micropollutants.

There are further subclassifications, e.g. nutrients which include phosphates and nitrates. Sometimes, organic matter is measured by TOC, or related to the oxygen requirements of the water, measured in terms of BOD or COD.

6.2.2 Protection of public health

The probability of contracting a water-related disease is statistically related to the concentration of pathogens or viruses. Improved laboratory techniques have resulted in the identification of harmful organic inorganic matter. These may occur in water sources naturally but also from industry and agriculture. Of concern to health also are heavy metals such as mercury, cadmium and lead as well as pesticides such as DDT and carcinogenic compounds.

Diseases may also be transmitted by water-based parasites, e.g. snails, and vectors, e.g. mosquitoes, especially in the tropics.

Pathogenic bacterial-related diseases include cholera, typhoid and dysentery, whereas viruses, which are smaller, require a host and include hepatitis. It is difficult to detect bacteria in water, so evidence of excreta is sought and indices such as *E. coli* are used.

6.2.3 Water-related diseases

Water-related diseases may be grouped as follows:

a) Waterborne diseases:

Waterborne diseases are those which are mainly spread through contaminated drinking water. The main infecting organisms are bacterial (*Vibrio cholerae, Salmonella typhi, Shigella*), viruses (hepatitis A, orgiviruses, rotaviruses and enteroviruses) and protozoa (*Giardia lamblia, Histolytica*). Contamination of the water occurs through faecal matter entering the water – that is, through poor hygiene and sanitation.

b) Water-washed diseases:

These diseases are mainly infections that can be significantly reduced by an improvement in domestic and personal hygiene. They depend on the quantity of water that is available, rather than on the quality. All diseases with faecal-oral transmission fall into this category – such as typhoid and cholera. Others are skin and eye diseases, such as skin sepsis and trachoma, and infections carried by parasites on the skin surface, such as lice. Most of the intestinal worms also belong in this group, including roundworm, threadworm, whipworm and pinworm.

c) Water-based diseases:

In the case of water-based diseases, the pathogen has to spend part of its life cycle in the water. The best known of these is *Schistosomiasis* (Bilharzia). It is a water-contact disease that has infected many millions of people in the tropics. It is spread through schistosome eggs in human excreta, which hatch on reaching water. The resultant larvae invade suitable snail hosts and multiply.

d) Water-related insect vectors:

Water provides the environment necessary for the development of many insects that transmit diseases. Malaria is a water-habitat vector-borne disease, certain mosquitoes being the host. Other such diseases are filariasis and elephantiasis (also transmitted by mosquito), and onchocerciasis (transmitted by the black fly).

The transmission of disease is a complex process and hence no direct relationship between water supply and sanitation improvement and the occurrence of disease can be found. However, the following steps in combination with education may cause a marked reduction in the occurrence of water-based diseases:

- disinfection of domestic water supplies
- provision of well-designed and constructed latrines
- increased quantity of water for domestic use
- provision of laundry facilities, thus reducing contact with open water bodies
- provision of adequate drainage and disposal of waste water
- management of open water surfaces, e.g. level variation, or spraying.

Water quality refers to substances or living organisms dissolved or suspended in water. Water used for domestic purposes need not be completely pure but could have limited dissolved salts for taste and to minimize the corrosive potential of the water. Furthermore, it has been estimated that only one out of every 20,000 strains of bacteria is pathogenic, and the mere presence of bacteria in drinking water is not necessarily cause for concern.

The approach to water quality control in water supply projects should therefore include the following steps:

- protection of all components (including the source, storage units and pipelines) against possible contamination by pathogenic organisms
- removal of basic pollutants, e.g. suspended solids
- improvement of the existing water quality to ensure aesthetic acceptability (turbidity, odour and taste)
- neutralise possible irritants, e.g. pH, temperature, hardness
- educate consumers as to basic precautions for the collection, storage and use of water.

Apart from diseases, many other factors in water can render it unsafe. High dissolved salts can lead to high blood pressure. Alkaline water can affect the arteries and cause skin disorders. Heavy metals and some organic pollutants concentrate in the life chain and can reach lethal proportions. The factors causing cancer are only beginning to be understood.

Then there are physically dangerous factors: flooding, drowning, land slides due to groundwater pressures, pipe bursts, dam break, wave action, temperature and drought.

6.2.4 Water quality standards

Water quality alone is not the only determinant in the good health of a community. Evidence is that the availability of water, sanitation and hygiene are also significant factors.

Table 6.4 Importance of water related activities to health

Infections	Water quality	Water availability	Dispose excreta	Excreta treatment	Hygiene cleanliness
Agents:					
Viral	2	3	2	2	3
Bacterial	3	3	2	2	3
Protozoal	1	3	2	2	3
Poliomyelitis and Hepatitis	1	3	2	2	3
Worms:					
Ascaris	1	1	3	3	1
Hookworms	1	1	3	3	1
Enterobius	1	3	2	2	3
Tapeworms	0	1	3	3	1
Worms aquatic:					
Schistomiasis	1	1	3	2	1
Guinea	3	0	0	0	0
With 2 stages	0	0	2	2	0
Skin, eye and hair from lice	0	3	0	0	3
TOTAL	**14**	**22**	**24**	**23**	**22**

These factors in a developed society assume a lesser important because of the institutional infrastructure Table 6.4 (Briscoe *et al.*, 1986) demonstrates their relative importance.

6.3 URBAN AND PERI-URBAN STRUCTURES

6.3.1 Reasons for patterns

The reasons for the establishment of a community vary, but the reasons for recent rapid growth around cities in developing countries appear to be:

Economic	-	Hopes of earning money
	-	Access to markets
Destitution	-	No other hope of survival
Outcast	-	From city
Immigration	-	Because of anonymity in the case of illegals
	-	Hope in the case of nationals
Transit	-	Temporary housing for people moving to the city, hiding, displaced

The problems associated with these communities include:

Lack of housing
Lack of services - water and sanitation in particular
Lack of employment opportunities
Rapid growth rate
Outbreak of disease
Crime
Exposure to risk - health, natural hazards (floods, pollution, erosion, etc.)
Lack of authority or organization
Little money

In this context, peri-urban does not refer to relative high income suburban sprawl, but to rapidly growing unstructured settlements springing up around in developing areas.

Typical peri-urban towns include the favelas of Rio de Janiero/Sao Paolo, the villas miserias of Buenos Aires, Kampung in Jakarta, bidonvilles of Port-au-Prince, pueblos jovenes of Lima, zonas invalidas of Limeira, the squatter towns around Johannesburg, urban sprawl in India and Asia are refugee camps. The allocation of space is somewhat random. Title deeds to land are rare and landlords may profit illegally, with little regard for safety or facilities.

There are attempted improvements, e.g. site and service areas and even township layouts in many cases, but in others, immigration rate is greater than the rate of extension of services.

Table 6.5 Phases of peri-urban development (Habitat & UNEP, 1996)

Characteristic	Evolution			
Description	Informal	Developing Intermediate	Quasi-formal	Formal
Description	Scattered dwellings, poorly constructed	Numerous dwellings, better constructed	High density with scattered formal housing	Formal housing
Infrastructure	Non-existent	Alleys etc. develop, community organisation created	Street systems develop, strong community organisations and informal governments	Infrastructure exists, municipalities formed
Need for investment	High	High	Moderate	Low
Source of investment	External	External: Cash Internal: Labour	Becoming internal	Internal
Land tenure	None	Very few land owners	Some land ownership	High level of ownership

Under these conditions, it becomes difficult to initiate and install services. Advisors may not have adequate knowledge of the communities' abilities, even if they can conceive the necessities. The desire of people to pay for services is hindered by not having tenure, in fact being in danger of being relocated in many cases. They may be aware of unjust or illegal enrichment. Services may be inadequate and even unhealthy. Breakdowns and maintenance are nobody's responsibility it may appear. People, uneducated in economics, are often under the impression that water is a free right.

Not only is there an attitudinal problem to project conception, there are also physical obstructions to construction. Access may be difficult, there are no servitudes and poorly located buildings in the way. In many developing areas, the physical problems of location may be extreme, e.g. mountain slopes, swamps, rock or mud. In fact they are often located on waste dumps of the cities.

Climate may also be severe, e.g. precipitation and temperature. Diseases prevalent to the tropics will attack imported workers more than the locals. Food, housing and any equipment may be difficult to come by and communications (transport, telephone) are generally poor. There may be theft of equipment or difficulties with labour organisations.

These factors, coupled with the additional administration and social interaction, push up the costs of services considerably.

UNDP and World Bank (1998) brought together participants in water development from a number of countries, to find that there is no single blueprint for success. The conference focussed on demand responsive approaches and empowerment, and a number of case studies were discussed.

In Bolivia, the low cost sanitation program, PROSABAR, and in Brazil the PROSANEAR programme, combined low cost technologies, innovative approaches and health education. The project to service Yacupaj was divided into two; an investment arm and a technical assistance component. It emerged that enforcement of rules, dissemination of information and financial policies required attention. For example, the rule of requiring a 5% cash contribution plus some labor and materials was not always enforced. Communities expressed a desire to select the monitoring agencies and control the contractors. Nevertheless, institutional arrangements proved tedious. It was suggested pilot projects be constructed and guidelines for further projects be issued. Water systems operators, on the other hand, appeared to be doing a good job.

China implemented three community participation projects within the UNDP framework. The proportion of urbanised people (16%, of 900 million) is relatively low in China, and the problems of water supply, sanitation and hygiene education are concentrated on rural populations. The policy of government proposals, department co-ordination, society support and individual participation still takes a lot of responsibility away from the individual and the community responsive approach may take a while to mesh with current World Bank ideas. One result of top down decision-making was an over-emphasis on cast, with a resulting dissatisfaction with the type of service provided. Complicated procurement procedures also bogged down projects to an extent.

The success of participating projects was pronounced in Africa (Ghana, South Africa). This was perhaps because these countries are in the throes of reforming government. Mvula Trust in South Africa issued the following challenges: Communities had to demonstrate some organizational ability when applying for assistance. They even had to appoint an engineer to conceive a feasibility study. They also had to select a level of service. There are a number of risks associated with this approach, viz. consultants may not be paid if funding is not forthcoming. Local payment in kind, e.g. by labour, was not successful as laborers demanded some payment. Another problem was the attitude of politicians that clean water and sanitation is a right of the people.

In India, the policy of cost recovery was initially met with resistance. However, the creation of partnerships between government, NGOs and water and sanitation committees has started to break down initial resistance. The Swajal project under the World Bank Rural Water Supply and Environmental Sanitation Project, enabled the central government to retract from its responsibility of service provider. The traditional reticence of women to participate was also eroded. Again the lugubrious nature of the process of implementation proved a frustration.

Similar problems were encountered in Indonesia but central control is perhaps even greater there with good and bad consequences. Training will be speeded up, but responsibility may be lacking.

An assessment of problems in Africa (Mwanza, 1997) revealed the following endemic problems:

- Multiplicity of organisations without clear responsibilities
- Inadequate financial resources due to low income, lack of other sources
- Shortage of qualified and experienced manpower
- Poor operation, maintenance and management
- Low morale of consumers
- Poor billing and collection (less than 50% on average).

These factors are repeatedly blamed, but the root cause, lack of responsibility, may be due to earlier colonisation, and low incomes, which only has a long-term solution.

6.3.2 Indian experience

A fresh approach to solving problems of slums and drainage together was described by Parikh (1999). The Indore Habitat Project proved a successful pilot project and is being followed by Baroda and Ahmedabad.

All cities have strong natural drainage paths. Without these, villages and towns would drown in their own waste long before they ever grow into cities. The paths are nature's own means of disposal and, if properly exploited, also become ideal routes for the manmade urban infrastructure systems of sewerage, storm drainage, water supply and roads. The environmental skeleton of city greens and water bodies also lies on the same paths. Studies of several cities in India and in other parts of the world showed that slums are consistently located along these natural paths. Once this connection between slums, urban infrastructure and environment is clearly understood, it is easy to see how slums can be used to transform cities.

The slum fabric is seen in the context of the whole city and interventions proposed are mutually beneficial to the slums as well as the rest of the city. The objective is not to find solutions unique to the slums but, instead, explore the commonality between the slums and the better parts of the city to integrate the two. As slums are not the causes of urban degradation but the consequences of distorted development, the solutions likewise must treat the slums as mere symptoms and use them to work back into the city fabric to the origins of the problems.

There is a close correlation between the slum locations and the natural drainage paths of a city. This again helps to build up low cost service trunks, particularly for gravity based systems of sewerage and storm drainage, together with environmental improvements such as creation of fresh water bodies, cleaning up of polluted rivers, development of green pedestrian spines and restoration of waterfront structures. The slums naturally benefit from the improved city level support. For the city too, the slums offer opportunities of change through this symbiotic process.

Unconventional concepts such as topography management, earth regradation and constructive landscaping are introduced. These coupled with the locational attributes of the slums with respect to the water courses and the marginal lands have certain ramifications,. The natural water courses and low lying areas tend to form nuclei around which slums cluster. By sensitive treatment of these lands several advantages are possible.

Firstly, areas prone to flooding and waterlogging can be lastingly improved by earth regrading at marginal costs.

Secondly, the natural drainage paths are the most efficient routes for the gravity based city drainage with the added advantage that the problems of land acquisition and demolition normally encountered in built up areas during installation are avoided here. In Indore, just by providing the missing links between the slums, it was possible to build up city level sewerage at costs less than half those for the conventional system as proposed by Public Health Engineering Department. This in turn intercepted the sewage from polluting the water bodies and paved the way for creation of water bodies and gardens around them.

The integration of both the scales and the activities intrinsic to Networking opens up exciting possibilities missing in other development strategies. Many solutions at first thought unviable at micro level become quite economic. A comparative study for Indore showed that the cost of house to house piped sewerage by Networking was about Rs.1500/- per family for the lines and Rs.1000/- for the off-site collection and treatment. Against this, the cost of a shared UNDP twin pit latrine, often considered to be "appropriate" for developing countries, worked out at about Rs.2500/- per family. Whereas the sewers also take care of the foul waters from kitchens and bathrooms, UNDP latrines do not. The additional advantages of the networked sewers were, firstly, that all the families had individual facilities, and, secondly, that the families other than in slums could also be connected to the same system without incurring the off-site costs – i.e. the cost per family decreased as the contributing families increased. In an extension of the concept to the micro level, coordinating the roads, storm drainage and sewerage top natural gradients in each settlement results in better function and economy. As far as possible, all roads are placed in cuts and have positive downward slopes from high points to the water courses. The surface cleanliness of the margins is achieved with grading and soft landscaping instead of expensive paving. The service infrastructure is simplified and modified so that individual services (instead of shared facilities) can be offered to slum families at low costs. Thus, the communities can participate in the execution of the works as they have the best knowledge and sensitivity of their surrounding environment. At the same time, the maintenance burden is reduced and can be shifted from the local government to the individual householders.

The population of Indore city in 1991 was 1.25 million, out of which slum dwellers accounted for 0.35 million. The slums in Indore were characterised by overcrowding, dilapidated housing, unhygienic conditions, grossly inadequate basic amenities, unplanned layouts and poor accessibility. These areas housed economically weaker sections of the community often engaged in casual service occupations. Over two thirds of the slum families in Indore lived below the poverty line. On average 40% of slum dwellers were illiterate, with the female illiteracy rate as high as 53%. A large proportion of persons reported being sick within the fortnight before the survey. In addition to the working days lost, about 8% of the monthly income was reportedly spent on medical expenses.

Table 6.6 Upgrading of the slums of Indore in the 1990s

Total length of new roads	360	km
Total length of new sewer lines	300	km
Total length of new storm drains	50	km
Total length of new water lines	240	km
New trees to be planted	120,000	nos
Total area of grassing/shrubbing	500,000	m^2
New community halls	158	nos

In a project executed by Indore Development Authority, and financed by Overseas Development Administration U.K., the Slum Networking concept has been demonstrated successfully in the city of Indore. Over a period of six years, the slum matrix of the city covering 450,000 persons was upgraded with quality environmental and sanitation improvement, together with extensive community development programmes related to health, education and income generation. The quantum of physical work in each slum pocket may be small but, as seen in the preceding table, the aggregate impact on the city as a whole is high.

Most development alternatives designed for the urban poor rarely transcend beyond the slum boundaries. In contrast, as a by-product of Slum Networking, Indore now has 90 km of piped sewer mains installed in the non-slum areas in a city which, until recently, had no underground sewerage to speak of. As a result, the polluted rivers of the city are being converted to fresh water lakes in stages and associated with that, the historical riverside structures are restored and new pedestrian greens formed. A recent study has shown that the quality of water in the wells around these areas has improved.

Out of the 360 km of roads provided in slums, about 80 km on the slum peripheries were linked up at the city level to reduce the traffic congestion on the existing trunk roads. Similarly, the storm drainage runs in the slums were placed in such a manner that large areas of the remaining city were also relieved of flooding.

6.3.3 African experience

Lagos in Nigeria (1997) exhibits conditions typical of an African city. The population increased from 1.1 million in 1963 at independence to 10.1 million in 2000 and is expected to be 28.9 million in 2025. The target is to provide 120 *l*/c/d. This may be ambitious considering only 35% of the population received public water in 2000. Of these, 10% obtain water from public standpipes, 14% from yard standpipes and 6% from household connections. The remainder obtain water from vendors (37%) or other sources.

Sanitation practices in the city of Lagos are also poor compared with the western world. 30% of the inhabitants use pit latrines, 53% use pan flush toilets, 12% have water closets and 1% use public toilets (which cost N10 and require an average travel time of 6 minutes). Other sanitation methods include bushes and plastic bags.

Luanda in Angola typifies a city ruined by war and neglect. Many water works plants were destroyed during the war. Only 30% of the meters work. At $100 per inhabitant, it will need $550 million to rehabilitate water services and $212 million for sanitation services. A longer-term strategy has been spelled out. This includes establishing an institutional framework, public awareness, financial sources and rehabilitation and maintenance of physical assets.

6.3.4 Environmental degradation

It is generally recognized that low cost urban areas are of poor quality and unaesthetic, unhealthy, dangerous and uncontrolled. There is little incentive to upgrade. The underlying causes of urban environmental degradation are:

- No land ownership or pride in property
- Not an integrated approach to health (disciplines and departments "stovepiped" approach
- Inadequate total spending by providers
- Because of low payments, providers do not feel obliged to provide an adequate service
- Skewed spending, in favour of those who have
- Skewed spending, in favour of water, without as much attention to sanitation (and without paying much attention to what happens to grey water)
- Spending on domestic services, and not available for e.g. treatment
- Inappropriate level of service and technology

- Over-utilisation (unsustainable) use of resources due to living from hand to mouth
- Aesthetics and the natural environment are not high on the priorities of poor populations.

6.4 WATER SUPPLY OPTIONS

6.4.1 Sources of water

The source and method of delivery of water have a wide variation from one community to another. They depend on, among other things:

Distance from source to consumer
Quality of water at the source
Cost of raw water and delivery methods
Scale of supply
Alternatives available
Method of return drainage
Method of sanitation
Social customers, preferences
Ownership of water and land
Risk of shortfall, outage, pollution

Sources include:
Springs
Rivers
Lakes
Ground aquifers
Rainfall
Snowmelt
Reclaimed wastewater
Desalinated seawater

6.4.1.1 Surface bodies

Rivers are the easiest source of water to tap, but they pose many problems.

- They can dry up in droughts or seasonally;
- They can be polluted by upstream waste discharges or human or animal sewage;
- Surface water bodies are breeding areas for many water related diseases;
- The water level can vary with flood or drought.

The construction of dams solves some problems but is too costly for many small communities. Foundations and spillways cannot be scaled down for small dams.

6.4.1.2 Groundwater

Almost 13% of water for consumption is abstracted from groundwater. If hydrological conditions are favourable, quantities of water to support a small town are even possible from secondary aquifers.

Since most precipitation infiltrates the ground, it is a big source and relatively stable. The ability to tap the source may however be limited. In high or rocky areas, it is generally not worthwhile. Boreholes are more yielding in porous aquifers, e.g. alluvial soils, or where dykes catch underground flow. Artesian conditions could exist in secondary aquifers, enabling the water to be brought to the surface naturally, but in other cases the water table could be deep. The assistance of a hydrogeologist is advisable to avoid costly boreholes yielding insufficient water. The sustainability of the yield must also be estimated.

Subsurface pollution also needs thought. The proximity to pit latrines, or waste dumps or naturally saline water needs to be considered.

6.4.2 Delivery

The sustainability of the sources may feature less in the minds of poor recipients that it should in the minds of planners and providers. Therefore locally identified sources are more prone to risk (of drying up, causing environmental damage, pollution). Methods of delivery include:

Hand transport in containers
Vehicular transport; containers, tankers
Pipelines – gravity, pumped
Channels
Use at source, e.g. washing
Pipelines to: troughs
reservoirs
standpipes
communal ablutions
individual households
businesses
public places
irrigation of fields

The location of source, delivery, houses, latrines, liquid and solid waste disposal sites, and obstructions, need consideration in planning supplies.

Some storage is advisable at the user end, to balance rate of use with rate of delivery, and for emergencies.

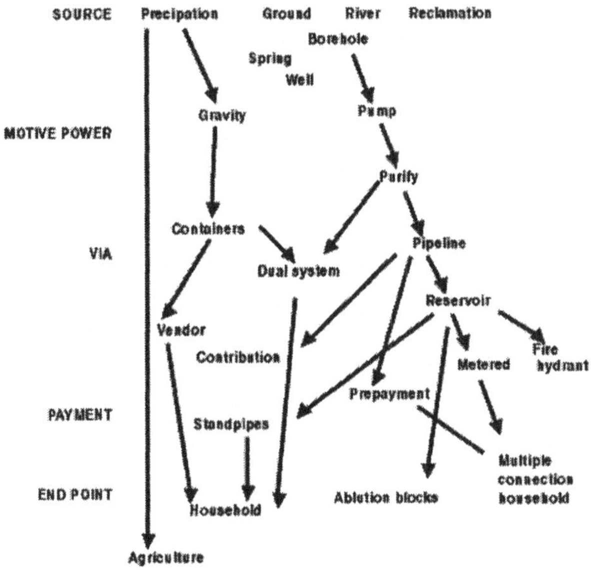

Figure 6.4 Distribution of water

Piping is the way most people prefer. Pipes are secure, unobtrusive if buried, long life and easily controllable. They are relatively economic but require a large capital outlay which is often impossible by individuals. It is for this reason municipalities or governments, or even aid organizations usually provide the basic system.

Although there has been considerable emphasis on sustainability and user participation in recent years, the first cost is the conveyance system, and one of the last tasks which could be taken on by the community is design and fabrication of pipes.

The geographic or demographic pattern of development will influence the capacity of pipes to be installed. That is, whether to cater for a fixed area and population and standard, or to allow for extensions, densification of housing and increase in standard of living. The latter could affect rates of consumption as well as type of receiving system. A communal standpipe system may over time be expected to supply individual households so the capacity may have to be built in. The alternative of relaying new pipes may be expensive and cumbersome, owing to dense housing and pathways covering the area.

6.4.3 Level of service

Depending on the investment and requirements of the community, different levels of supply could be planned. Figure 6.5 shows a progressive range of options. They could be one of the following:

(1) *Simple source and disinfection.* Protection of sources, particularly springs or flowing streams, against contamination is perhaps the most basic form of protection of the water system. Transport of the water would still be containers.

(2) *Pump and supply line to a reservoir or standpipe.* This would shorten the distance to travel to collect water but still provides a communal collection point. The number of standpipes should be sufficient to ensure there is not a long queue or else an open reservoir for dipping containers in, in which case disinfection is important.

(3) *Individual standpipes or connections.* A standpipe for each house or else each pair of houses back to back further improves the availability. Going a step further, it could be provided in a basic lean-to laundry-cum-kitchen structure. At this stage it may also be possible to consider water-borne sewage in the way of a communal toilet.

(4) *Individual household supplies.* At this stage, the householder may have multiple connections to kitchens, ablution facilities and toilets. He may also consider a roof tank for storage if the supply rate or pressure are inadequate for meeting peaks. Generally at this stage there is also communal storage for balancing the demand against the supply rate which would be at a uniform rate in order to optimally use the pipes.

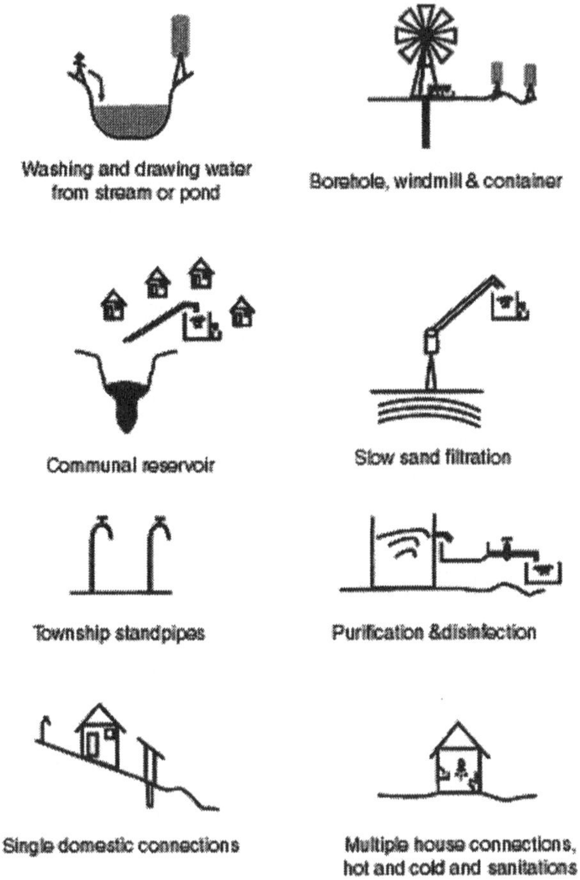

Figure 6.5 Evolution in water supply standards

The minimum water requirement is 5–10 litres per day if access is difficult. With piped water supplies, the consumption can increase to between 10–20 litres per day in the case of standpipes, and 30–50 litres per day in the case of individual household connections with minimal ablution facilities. The per capita consumption will only increase above this with communities earning above the breadline salaries. Provision should also be made for community centers and commercial development. If the water is economic, it may be made available for limited agricultural purposes, but generally piped and purified water cannot be economically used for irrigating.

Losses can also be considerable in sub standard systems in particular. Leakage and illegal connections can account for up to 50% of water use, where a realistic target is less than 10%.

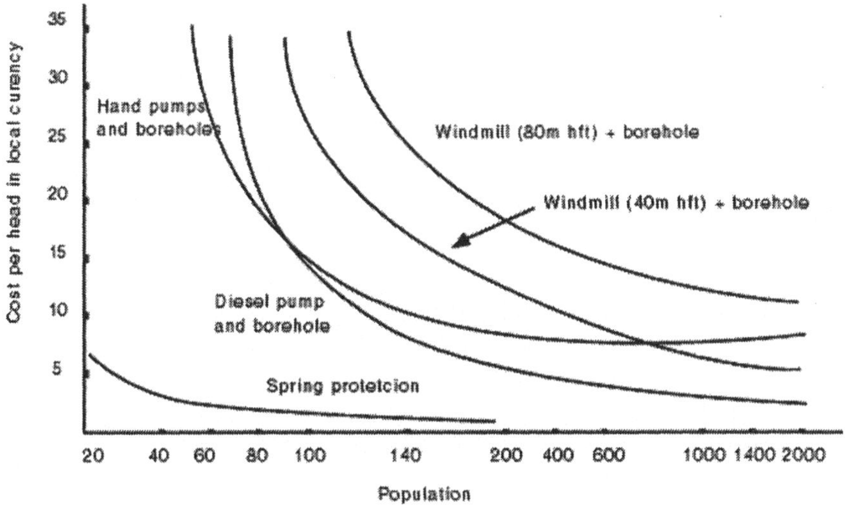

Figure 6.6. Cost graph for comparison of various types of water supply against population (Cairncross *et al.*, 1980)

The City of Durban (Macleod, 1997) embarked on a water supply program to supply some 200 000 families within 20 years. The problem of installing standpipes was that residents refused to pay if they had to walk 200m to collect water.

They adopted a 3-option policy:

- Water tank system: Small bore pipes feed 200 litre tanks next to houses. Water is paid for when the tank is filled
- Semi-pressure system: Water is fed through a restriction to roof tanks in each house so that peaks are reduced
- Full pressure system to cope with multiple connections under normal supply conditions.

The installation program is accompanied by extensive consultation and an information program.

Non-payment of water accounts was initially a problem, but defaulters' water was then restricted to 200 *l*/day with a restrictive washer, resulting in a lowering of bad debts to 0.1%.

6.4.4 Alternative systems

To reduce costs and make water supply systems more flexible, the following are possible:

Use pipes for grade B water (partially treated wastewater) for some washing operations and gardening. Pipes should not go into houses to reduce the possibility of drinking the water. Potable water could be provided by tanker, container or bottle, or a separate supply pipe network (dual system). But considering the amount of pure water actually drunk without boiling, this requirement is most likely to be met with containers. The problem of cost thus arises. Bottled water is notoriously expensive in the western world. This is largely due to bottling, transport and marketing. Subsidisation of containerisation of water will in many cases be less than purifying all water to potable standards and reticulating it.

Small bore water supply piping can be a means of supply control, and ultimately even population density control. Demand control by financial means such as tariffs (Stephenson, 1999) is not feasible in sub-economic communities, as they do not pay, or are subsidized anyway. Small bore systems (sometimes with distributes storage) can be extended eventually from standpipe distributions to household distribution if roof tanks are used. Then the pipes supply over 24 hours at an average rate and peaks are met from the tanks. This is one way of minimising costs and phasing development. It also reduces pipe pressure classes as hydraulic gradients are high. Distribution by container is one way of inducing payment and therefore responsibility into communities.

Conjunctive use of alternative sources can reduce costs. For example, when a local spring dries up, there may be a further, reservoir source. Or rainwater could be collected for peak supplies, if the piped supply was inadequate.

6.4.5 Stages of development

Old suburbs taken over by squatters may have been reticulated to a reasonable standard originally, so densification of the population may not overload the system. Lower per capita consumption and peak reductions may be offset against increased numbers.

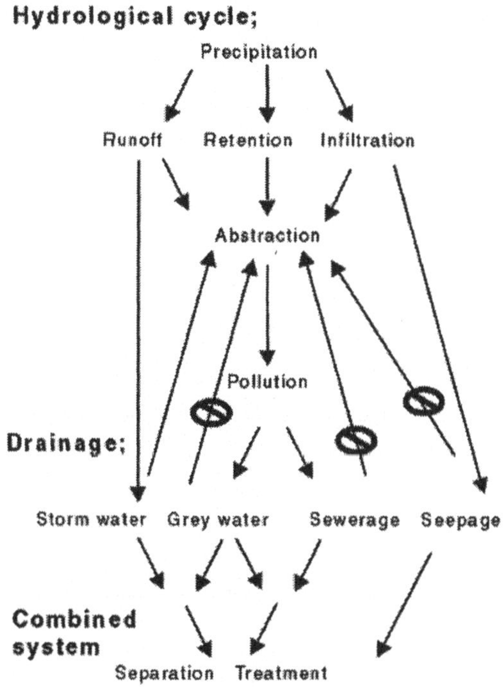

Figure 6.7 Flow paths

The problem of loss control, i.e. leakage, illegal connections and intentional damage, becomes greater as settlement expands.

In the case of newly serviced townships, the piping can be correctly sized, but the question of what ultimate density and standard to cater for still arises. If the township is temporary (by refugees or transitory in seeking to upgrade elsewhere) the level of service may be established. Ideally, the new reticulations will be planned for access, security and upgradeability.

The rate of installation of water supplies to poor communities will be limited by the availability of finance, from sponsors, or the community themselves. The installation of water mains to many existing cities took centuries. Collective sanitation occurred even later, and took place in stages (Lyonnaise des Eaux, 1998), viz.

- Collection of stormwater and grey water, with separate individual sewage systems, e.g. septic tanks, cess pits
- Wastewater collection networks once the volumes became too great
- Wastewater treatment at point of discharge, to protect the environment and downstream water resources.

Lyonnaise des Eaux classified the factors influencing the choice of water supply under three headings:

(1) *Institutional*, including land tenure, administrative structure and finance.
(2) *Population*, including income, dwellings, social, behaviour and existing supply.
(3) *Technical*, viz. scale, demography, urban layout, resources, topography, alternatives.

The matter of land tenure affects attitudes, ease of finance, type of supply and the ability to charge. Table 6.7 indicates possible scenarios and the precautions necessary.

Table 6.7 Forms of land tenure and low-income groups (Lyonnaise des Eaux, 1998)

Examples of possible courses of action according to the attitude of the authorities		**Public land**			**Private land**	
		Inalienable	**Alienable**	**"Available"**	**Registered**	**Un-registered**
Attitude towards an area	"Officially accepted occupation"	Rare case ... Caution	Possible case ... Possible intervention	Highly possible... Possible intervention	Rare case ... Clarify or steer clear	Possible case ... Caution
	"Implicitly" accepted or rejected occupation	Analyse, evaluate with care, do not make any commitments, even verbally, without an official decision being made by the authorities [1]				
	"Officially" rejected	Steer clear, if the communities concerned are large, analyse and evaluate so as to be in a position to react in the event of a change of status				
Attitude towards a specific status	Land with recognised title deeds	Extreme caution	Caution	Possible intervention	Avoid	Clarify or steer clear
	Land with unlawful title deeds	Avoid	Clarify or steer clear	Clarify or steer clear	Avoid	Avoid
	Occupation with no title deeds	Only intervene by area (see above)				

6.4.6 Problems in supply (Lyonnaise des Eaux, 1998)

In view of the complexity and diversity of the situations encountered, we are obliged to simplify matters considerably and only take two groups into account.

(1) *Poor neighbourhoods which are to be equipped*

These areas may be in cities or in marginal urban areas. They are equipped with a minimal street system and sometimes networks. The residents' situation is either in line with the law or is due to be regularised by the authorities. This category includes the "barrios humildes" of Argentina, most of El Alto (Bolivia), etc.

(2) *Areas of spontaneous improvised and temporary housing*

These areas may also be found in cities or in marginal urban areas; they are characterised by illegal occupation of land and lack of public services.

They are not definitely established from a strictly legal point of view, even if they have existed for a number of years, until the situation is regularised.

The latter group is extremely heterogeneous and can be broken down into many subdivisions, taking account of the social organisation of the community especially. It may seen absurd, for example, to group together the highly-structured "zonas invadidas", where streets systems are planned, homes are "solid" but land occupation is (for the moment) illegal, and the shanty towns of Port-au-Prince, where buildings are impermanent and services non-existent.

Private concessionaires cannot take over from the authorities to "officialise" areas of spontaneous housing, nor can he replace the town planning service. They can, however, "support" the town planning services, making it easier to integrate these communities into an official framework by providing them with adequate services.

Generally speaking, two main difficulties can be highlighted for poor neighbourhoods.

(1) *Connection costs are too high*;

For example, in Buenos Aires expansion areas, users have to pay infrastructure charges which are set by the regulator and are intended to cover expansion costs amounting to around \$500 for a water connection and \$1 000 for a sewer connection. In certain areas, the average monthly income stands at \$240.

(2) *Customer management costs are too high*;

There are various reasons for this, and they are not always specific to poor areas. Related problems are:

- A high percentage of unpaid bills
- A high rate of unbilled or fraudulent consumption

- Low consumption resulting in relatively highly collection costs
- High network maintenance costs.

(3) *Consumers' water procurement strategies*
Each city should be considered as a specific case in a given socio-economic and cultural context. However, Table 6.8 illustrates a situation which is frequently observed in large urban centers in emerging countries. As regards water supply, populations may be:

- supplied by the network
- supplied by other method
- standpipes, wells, with or without intermediaries
- water deliverers
- not supplied, resorting to uncontrolled sources (ponds, rivers) or wells.

Table 6.8. Water procurement strategies (Lyonnaise des Eaux, 1998)

Supply method	Conditions for producing water	Use	Consumption (l/c/d)	Water quality	Payment method	Cost of water per m^3
Individual service line	Private tap in the home or close by (yard)	All use	40 to 250	Generally good	Monthly billing	P
Public fountain	Carrying, waiting at water supply point	Food	10 to 50	Possible contamination at water supply point	Payment in cash	10*P
Delivery	Delivery to the home	Food	50 to 50	Variable	Payment in cash	30*P
Private well	Access close to dwelling	Toilet, washing etc.	50	Variable (user is not always aware of risks)	Individual maintenance	"Free" (except electricity)
Uncontrolled sources (pond, river)	Access close to dwelling or carrying	Toilet, washing etc.	50	Bad (contamination)		"Free"

6.4.7 Alternative approaches

For those who do not have access to the networks, the problems of water availability and service costs lead them to search constantly for the right consumption level: the users thus aim to reconcile their vital water requirements with their financial capacity while reducing the drudgery involved as far as possible. This strategy often leads people to adopt appropriate additional supply methods according to the quality requirements of different water uses. They therefore choose, by preference:

- unlimited use of "free" resources (private wells, ponds, rivers, etc.) for uses other than human consumption, if these resources are close to the home
- to buy the quantity of potable water required each day (home delivery, tanker truck, standpipe, neighborhood resale, etc.), using it parsimoniously to limit costs.

Additional supply methods are also used by populations who are linked to the networks, when the service is intermittent. A recent study in Delhi revealed the strategies adopted by users connected to the networks, to deal with an unreliable water supply. There are many such strategies and, on average, households use more than two. Frequently the best situation to supply informal housing is a juxtaposition of alternative methods.

6.4.8 Payment methods

The procedure of post use billing is not really acceptable in communities living from hand to month. Prepayment on the other hand is frequently acceptable. In South Africa, prepaid cards and filling of tanks have been successfully introduced. Cards have been applied institutionally and private vendors have applied the tank or container system. Unfortunately, the unit cost of water is increased by these methods but the significantly lower consumption means a lower monthly bill than for richer consumers. Alternative financial solutions are listed in Table 6.9.

Table 6.9 Alternative financial solutions (Lyonnaise des Eaux, 1998)

	Connection meters	**Infrastructure (secondary networks)**	**Water tariff**	**Work on private property**
Cross subsidies			*	
Welfare price			*	
Payment method	*	*		
Labour participation	*	*		
Grants	*	*		*
Welfare connections	*			
Work funds		*		
Micro-loans	*	*		*
Pre-payment			*	
Tontines, mutual aid funds, etc.		*		*

6.4.9 How much do the poor in urban areas pay for water?

The problem of lack of water services hits the poor in the slum areas of the large cities in developing countries. Often, the only choice for low-income households than can not afford a house connection is to buy water of substandard quality from private vendors at a relatively high price, sometimes 100 times more than that provided by public authorities. Examples of price distortion are shown in Table 6.10.

Table 6.10 Ratio between prices charged by vendors and by public utilities (Bhatia and Falkenmark, 1993)

Country	**City**	**Ratio**	**Source**
Bangladesh	Dacca	12–25	1
Columbia	Cali	10	2
Ecuador	Guayaquil	20	2
Haiti	Port-au-Prince	17–100	1
Honduras	Tegucigalpa	16–34	1
Indonesia	DKI Jakarta	4–60	3
	Surabaya	20–60	1
Ivory Coast	Abidjan	5	1
Kenya	Nairobi	7–11	1
Mauritania	Nouakchott	100	2
Nigeria	Lagos	4–10	1
	Onitsha	6–38	4
Pakistan	Karachi	28–83	1
Peru	Lima	17	1
Togo	Lome	7–10	1
Turkey	Istanbul	10	1
Uganda	Kampala	4–9	1

Sources: World Bank, 1989; Whittington *et al.*, 1989.

A compromise in charging of public water supply may be necessary. Free water will be abused, but full costs would make it too expensive and therefore lead to chaos and health hazards, thus a life-line rate is often adopted that benefits the poor.

Table 6.11 Existing alternative solutions for water supply (Lyonnaise des Eaux, 1998)

Type	**Description**	**Average water supply (l/c/d)**	**No. of households supplied**	**Advantages**	**Disadvantages**	**Examples**
Communal open wells	A bucket attached to the end of rope is used to draw water from a well.	10–50		In principle, cheap source of water (but can be fairly expensive, taking into account the cost of ropes, buckets, etc.)	High risk of contamination	Dhaka
Tube wells/bore-holes	Pumps are attached to the top of a pipe which is sunk into an aquifer.			Clean water is available.	Reliability of pumps. Can be expensive to sink.	Dhaka, Jakarta
Water vendors	A vendor delivers water to the home.	5–50		Water is brought to the home.	Expensive quality of water supplied.	Jakarta, Maroua, Ho Chi Minh
Public tanker trucks	Trucks deliver water to the home.	5–50		Water is brought to the home or close by.	Expensive	Phnom Penh
Water kiosks	An operator sells water from a kiosk, from where users have to carry it home.	5–20	100	Clean water is available. People only pay for water they use. Very high payment rate.	Expensive. People have to walk and carry long distances.	Port-au-Prince, Ouagadougou
Communal standpipes	A tap shared by many homes	10–50	25	Cheaper installation than other systems - low bills.	People have to carry their water. Different methods of payment.	Jakarta, Ouagadougou, Delhi

Table 6.11 (continued).

Type	Description	Average water supply (l/c/d)	No. of households supplied	Advantages	Disadvantages	Examples
Yard or roof tanks	Water is drawn from tanks installed at each home; the tanks are filled daily.	30–50	1	Water is always paid for. People can judge the amount of water they use.	Tank may empty if consumption is high.	Durban
Yard taps	Service pipe delivers water to tap situated in the yard.	30–100	1	Water is available on-site.	High investments.	Kliptown
House connections	Service pipes are laid to supply water direct to taps inside the home.	40–250	1	Very convenient	Water use escalates. high investment costs.	City centres

6.5 DRAINAGE SOLUTIONS

6.5.1 Drainage paths

Whereas piped water supply is generally under pressure, drains flow part full is by gravity. Alignment is therefore important as well as freedom from obstructions. It also means bigger pipes and margins for ingress of stormwater are bigger, i.e. pipes run part full. In fact pumping systems are discouraged in drains as they require sumps and disruption of flow.

There can be three different streams of water to be drained off a low cost township, as opposed to two (sewage and stormwater) from a developed town:

6.5.1.1 Stormwater is often the largest stream, albeit irregular. Drains for collecting and conveying stormwater may therefore be the most costly from the point of view of construction. They will be associated with a multitude of inlets and can occupy considerable width of roadway, especially during construction. Underground stormwater pipes will be preferable if surface space is at a premium, but they are the most costly to design and install, and maintenance is difficult. Manholes are required at regular (- 100m) intervals and at bends and junctions.

Surface channels are cheaper but take up valuable space and can be dangerous to traffic and humans. However, if roads and paths are unsurfaced, removal of grit is easier from surface drains.

Pollution of stormwater is endemic in poorer communities. Sources of pollution include overflowing toilets and sewers, surface deposits, discarded solid and liquid wastes, and air borne solids (smoke, wind, thrown). Characteristics of the pollution vary widely. Suspended materials can be plastic bags, household wastes, building wastes and biological matter.

Cleaning by means of screens at the outfall is the minimum for runoff from poor areas where solid waste is discarded on the ground and collection is minimal. Regular planned cleaning of screens and disposal of the solids is therefore required.

Settling may remove inert solids, but requires open basins and more sophisticated solids removal, e.g. attenuating basins or grit channels, and spading. Neutralization of BOD (biochemical oxygen demand) requires aeration, which can be done in the receiving waters, but the reaches of river thus rendered sterile require consideration.

Septic tanks and soakaways are an alternative or supplement to aeration. They may be safer in some circumstances but the seepage path needs tracing and perhaps protection.

Removal of solutes is too expensive for informal treatment works such as above. Likewise disinfection is seldom performed on storm runoff.

Stormwater drains can be designed as twin systems, i.e. minor and major systems. The underground pipes take minor storms, even up to 5 year storms, and the major storms overflow to roadways,. The proper drainage design of roads is expected and this may not be the case in poorer townships. So alternative stormwater management procedures are necessary.

Detention and retention basins require substantial land area, which may not be possible. Infiltration zones and disconnection of impervious areas are also seldom viable. Careful road layout and grading is one of the few ways of avoiding catastrophes. Care is also needed in redirecting stormwater from uphill catchments. Unstructured townships may develop their own flood warning system, but flooding and flood plain management remain serious problems in most poor townships.

Detention storage is also not as viable in extreme climates as in temperate climates. The variation in rainfall rates and volumes is large. So a detention pond capable of attenuating say a 5 year hydrograph by 50%, may only have a 5% effect on the 20 year hydrograph, and a 1% effect on a 100 year hydrograph. The value of the storage is solely to reduce the frequency of flooding, not the extent of large floods, i.e. it is a convenience structure, not a life-saving measure.

6.5.1.2 Grey water (or *sullage*) is household wastewater from baths, showers, basins and other wash water, but excludes toilet water and waste. The main pollutant is soap but some bio-matter and even grit can be expected to flow down the road and join the stormwater system. Where sanitary sewers are installed, grey water is automatically fed into the sewers. There are cases where the grey water is separated for use on gardens, especially in arid areas, but the possibility of human contamination should always be considered.

Where grey water is directed into stormwater drains (combined system type 1), separation at the downstream end and treatment of the concentrated stream is achievable. This may be only a simple treatment, e.g. skimming, settling and/or screening. Separation by gravity (diversion weir for heavy solids and skimming weir for floating solids or oils) is the simplest. Vortex separation using centrifugal action is also effective for removing denser solids, and the centrifuge usually occupies less space than a settler. The concentrated solids stream should be diverted into a sewer or to a decanting basin for regular emptying.

Separate conveyance of grey water to sewage treatment works is seldom considered. However, it does not require pipes the size of sewers. Minimum deposit velocities need not be maintained as for sewage, and blockage and rodding to clean the pipe is not of as great a concern. Household screening, e.g. in plug holes is, however, a prerequisite. The possibility of smaller pipes than sewers, e.g. 50 mm should be considered and few manholes are necessary.

In providing for sewerage of grey water only, the planner commits the community to non-waterborne sewerage. That is, the community will have to use pit or chemical toilets or hand transport of sewage. In some communities, the sewage is separated and used for composting. Upgrading from small bore to large bore sewers is difficult and expensive, so careful thought and discussion is needed as to future plans.

6.5.1.3 Sewage should be conveyed in closed underground pipes and preferably not combined with stormwater. Whereas combined systems still exist in many older cities, in the case of developing communities, the problems of combined systems are exacerbated. More blockages are likely from surface waste disposal, and alternative cleaning methods may result in grit and other matter entering the sewer. In addition, many developing communities are emerging in tropical climates, or in other extreme climates, where extreme or infrequent storms make separation difficult. The sewage is very diluted when a storm occurs so it is difficult to separate. Stormwater drainpipes have to be large compared with separate sewers, so in dry weather conditions, sewage velocities are low, resulting in separation and stranding of solids. Odours are worse in hot climates and the possibility of disease more prevalent.

Conventional sewerage in most cities is based on certain minimum pipe sizes and gradients. This (a) minimizes blocking and facilitates cleaning by rodding, and (b) ensures transport of solids with the flows usual in the circumstances. A minimum sewer pipe diameter of 100 mm is common for house connections, and 150 mm diameter for common street pipes. These may be compared with water pipe connections of 12 mm for houses and 25 mm in streets (allowing for no fire hydrants though). Thus sewers are much bigger and therefore more expensive than water pipes, notwithstanding the internal pressures in water pipes. Manholes also add to sewer costs.

Alternative sewer systems could thus be considered. One system would be to exit houses or toilet blocks via a septic tank or holding tank. Solids could settle or digest before the liquid is decanted into smaller sewer pipes. Regular cleaning of the tanks makes this alternative unattractive, but if the full financial impact is imposed on the communities, the perspective could change. Cleaning the tanks could be by truck and siphon, or manually, with solids buried or further treated for composting.

The use of septic tanks and soakaways is seldom viable in dense urban settlements. Not only is their insufficient space, but contaminated seepage poses a health problem.

The discharge of raw sewage to the sea is common in coastal towns. Saturation and oxidation are required to be adequate if dilution is greater than 1/500, but floating matter can be a serious health hazard as well as unaesthetic, especially when washed ashore.

6.5.2 Treatment and reuse

A broad study of the alternatives for economical treatment of wastewater from urban communities, and possible reuse, was made by Rose (1999) for the IDRC.

There is a differentiation between wet and dry sewage treatment. The scope for separation of solids and liquids depends on the level at which the sewage is intercepted. The dry techniques are applicable more at a household level where water borne sanitation is not installed or used for excreta.

The advantages of reclaiming wastewater and sewage for irrigation include:

- Increasing crop yields
- Decreasing use of artificial fertilizer
- Soil conservation and regeneration
- Reducing pollution by discharge of wastewater to streams
- Protection of freshwater sources
- Reduced cost of sewerage and plumbing

The disadvantage is the necessity for care with regard to handling waste and resulting crops. Health hazard due to disease transmission is however not as severe if wastes are properly treated. The problem is still rather social acceptance, and integration of the land area for disposal of wastes.

Dry sanitation was customary in many countries before water toilets become fashionable (reported by Niemczynowicz, 1996).

In the case of rural pit latrines the waste is allowed to decompose and soak into the ground. In more dense communities, the waste should be collected either by buckets or at less frequent intervals by tankers. Then a certain amount of treatment is permitted to occur. Asano and Levine (1996) indicate shorter recycling loops are possible as the decomposition and reuse become better understood and applied.

Composting toilets are examples of micro treatment and reuse. Double vault batch toilets facilitate reuse. Each vault needs to be adequate for 1 year's storage to ensure adequate composting. The compost can then be used for local gardening or transported to controlled agricultural projects. Local application can enable liquid (urine and grey water) to be evaporated.

Reed bed filters may accept untreated or partly treated sewage but are not designed for agriculture, just treatment of the waste.

Dry Alkaline Family Fertilizer (DAFF) latrines have soil, lime or ash added to maintain optimum moisture content. It is reported to produce a drier sludge, but concern about the retardation of stabilization exists.

Removal of night soil and deposition in pits to stabilize improves space and aesthetics, particularly in dense communities. The alternative is to convey the night soil to sewage treatment works.

The re-use of sewage wastes is more common in Asia where population densities are high than in Africa, where there is also less control.

Agriculture is a growing industry in peri-urban areas. They may use derelict tracts of land, inundated lands or seek land further out. Production may extend beyond subsistence and farming as a source of income with sales in the city. It also is a way of disposing of wastewater. However, health considerations need attention in this case, i.e. proper handling and treatment prior to application. As an example, 18% of China's large cities are self-sufficient in vegetables.

6.5.3 Practical treatment systems

Conventional sewage treatment works are relatively expensive and urban land is at a premium, energy input for aeration may be high. Simple systems, also less mechanized, exist.

Anaerobic digestion reduces solids volume and produces rich humus for land application. Upflow anaerobic filters, and biogas reactors are concentrated systems but more economic on a community scale rather than for individual households. Soil aquifer treatment is potentially economic but requires primary treatment and suitable aquifers.

Water based treatment technologies are more common but only suitable for larger communities and where suitable sites exist. Considerable success is being achieved with constructed reed beds to free water surface (FWS) wetlands.

Floating macrophyte plants capable of being grown include water hyacinth (*Eichnorniacrassipes*) and duckweed (*Lemnacea* sp., *Spirodella* sp.). Duckweed is particularly nutritious for stock and fish but results in hot climates have been not as good as in temperate regions. Careful cropping is important.

Sewage fed aquaculture systems have been used for centuries but there is a movement away from them in China, Japan and Taiwan. The wastewater must be treated in stabilization ponds (1 day anaerobic + 5 days facultative is considered adequate) before discharging into fish ponds.

6.5.4 Sewerage: the Cinderella service

In the year 2000, there are 23 mega-cities with a population of over 10 million. Eighteen of these are in the developing world (Black, 1994). Three hundred million of the urban residents have no access to sanitation. Urban drainage problems are accelerating ahead of water supply problems to these areas.

The cost of sewerage, ranging from typically \$100 for pit latrines to \$1000 for conventional water sewage systems, exceeds the cost of comparative level water supply. Flush toilcts consumc 20–40% of domcstic urban water, and cvcn more in poorer suburbs. UNCED (1992) emphasized the problems in Agenda 21.

The solution propounded by Rose (1999) is to solve sanitation problems and provide agriculture with compost and water simultaneously.

The most appropriate type of sanitation depends on (Alaerts *et al.*, 1993):

- Availability of sewerage system and water supply
- Site specific conditions, viz. Population density, pattern, soil permeability, slope, climate
- Groundwater and surface water pollution threat
- Institutional capabilities for maintaining, financing
- Social factors, customs, attitudes, health education
- Economics and financial viability.

Onsite sanitation occurs at household level and offsite sewerage and treatment at city level. There are also intermediate level options, e.g. neighborhood-level, which may allocate appropriate technologies to different zones.

Sanitation is a private function and often part of a developed custom, so changes in mode can be unacceptable socially. Table 6.12 shows a wide variety of different modes. New technology should not replace existing practice but possibly adapt it. Initial consultation is a pre-requisite. Cost is also a problem, particularly if people are unaccustomed to paying for performing necessities.

Customs that should affect the selection of sanitary means include whether communities prefer to sit or squat, and whether there should be separate facilities for the sexes. The method of cleansing could be wiping or washing. Smells, aesthethics and privacy also need consideration.

One of the most important factors in selecting treatment and disposal means is health. In particularly, if wastewater is used for irrigation, it should contain less than 1 intestinal nematode egg per litre, and less than 1000 faecal coliforms per 100ml (Mara and Cairncross, 1989).

Table 6.12 Existing alternative solutions for sanitation (Lyonnaise des Eaux, 1998)

Type	Description	Average flush water volumes (litres)	Advantages	Dis-advantages	Examples
Ordinary pit latrines	A pit with a seat, in a shelter.		Basic enough for people to construct themselves.	Often badly built. Problem with flies and stench.	Dar-es-Salaam, Tanzania
Bucket sanitation systems	A bucket is placed under a seat in a privy.			Problem with flies and stench. Bucket soon fills when many people use the system.	
VIP latrines	Reinforced pit with concrete cover and seat. Air vent with screen and anti-mosquito baffle.		Easy to construct. Cheap. Hygienic.	No good if ground is rocky or groundwater table is near surface.	Zimbabwe
Aqua-privy with on-site disposal	Water enters digester to be processed by bacteria. Liquid effluent soaks away.	1	Cheap. Easy to install.	Water tank needs frequent filling. Digester requires periodic emptying. Effluent can contaminate the surrounding ground.	India

Table 6.12. (Continued).

Type	Description	Average flush water volumes (litres)	Advantages	Dis-advantages	Examples
Septic tank	As aqua-privy with onsite disposal, except that this is a full flush on-site disposal system.	10–20	Can be installed where there are no sewers.	Expensive to install. Reserved for large sites. Cost of emptying. Sludge disposal.	
Aqua-privy with solid-free sewer or simplified network	Waste enters a digester to be processed by bacteria. Liquid effluent is evacuated by sewer pipe.	1	Small value of water needed for flush. No soakaway required.	Digester needs periodic emptying.	USA, Australia, Brazil
Intermediate flush sanitation (with sewer)	Similar to a full flush system but uses less water. All waste goes to a sewer	3–6	Full flush system is convenient. Uses little water. Smaller pipes.	Needs to be designed and installed correctly. Conservancy tanks costly.	
Full flush sanitation (with sewer)	Full flush system with on-site sewer to transfer waste to main sewer	10–20		Most expensive system to construct. Uses the most water.	Developed cities

Table 6.13 sets out the possibilities, in order of increasing cost.

Table 6.13 Successively more expensive solutions

Facility	Removal	Treatment
Behind trees	None	Decompose
Pit in ground	Burial	Decompose
Privy and bucket	Carry	Drying beds/compost
Privy and tank	Truck	Chemical
Privy and septic tank	French drain	Anaerobic
Pour flush and tank	Small bore pipes	Reed bed
Cistern	Sewers	Works
Hydraulic valves	Sewers	Works (settler, filter/aerator)

6.5.5 Payment

The willingness to pay for sewers is generally less than for water supply. Wright (1997) quotes several installed sewer lines where only about 10% of the design number of households connected up their houses. Sanitation is not seen as a life necessity as it has a longer-term effect on health than lack of water. On the other hand, subsidization of poor people's sanitation by richer city dwellers is probably more pressing, so sanitation may be more likely to become an institutional project than water supply.

Some utilities add a surcharge on water bills to cover sewerage. In fact, the cost of sewerage generally exceeds water supply and under-recovery of costs is common, resulting in disintegration of the administration.

6.6 CONCLUSIONS

6.6.1 Community participation

The concept of community participation in development started on a small scale in the 1950s. The concept then broadened and grew during the 1970s, stimulated, in part, by the failure of many of the large development projects of the previous decade; today it has become a significant factor in development planning. The early models of development, in the immediate post-colonial era, were purely economic and were based upon the "trickle down" hypothesis, whereby (in simplistic terms) economic growth or benefits in one sector (e.g. urban industrial development) would eventually spread to other sectors, such as the rural economy. This model has proved to be invalid and, if anything, the reverse has happened, in that while a small elite has prospered, the large majority of the rural population has become poorer.

There are exceptions to the above of course, and undoubtedly there have been cases where large scale investment in specific geographical areas (generally those with good agricultural potential) have benefited the majority in the area. Thailand is one such case. However, the Thai experience also showed that there was no spread of benefit outside the particular geographical area being assisted and that the net result was in fact an overall increase in the gap between rich and poor nationwide.

Public involvement programs usually have four primary general objectives:

- Identifying local needs, preferences and priorities
- Identifying all impacts of alternative and recommended programs.
- Promoting understanding and support for the identified objectives and solutions proposed
- Providing opportunity for those affected by water resource development to influence decisions regarding development.

Public views are probably both more important and more difficult to obtain in rural areas in developing countries. The difficulties stemming from a lack of communication are primarily due to:

- The dichotomy between national and local priorities
- Much of the water resource planning affecting rural areas is conducted by urban personnel of the national government
- Much of the water resources planning is conducted by or jointly with foreign agencies and foreign private consultants.

6.6.2 Affordability

Although it has been generally assumed that rural populations in developing countries are too poor to pay for adequate water supplies and sanitation, Churchill (1987) indicates otherwise. The ability to pay and the desire for good services are underestimated. People in some situations will pay up to 10% of their income. The number of villages that are on or below subsistence level is diminishing and there are signs of wealth practically everywhere. This may appear in the form of portable radios or bicycles and clothing.

In many situations, a scheme has failed because they have only provided communal standpipes at some distances from individual dwellings, whereas the people would have been prepared to pay for connections in their own houses. A careful examination of the wants and means of the population is therefore warranted.

Appropriateness of the technology of a water supply scheme is important to ensure its continuity. High-tech systems including pumps, windmills and even hand pumps will deteriorate and unless there is a source of spares and a means of obtaining them, the entire project could be rendered useless in a few years. Therefore gravity systems and those that will not exhibit less efficiency if left unattended need to be earmarked. Alternatively, arrangements need to be made for maintenance on a regular and ad hoc basis.

6.6.3 Economic reform

In an attempt to 'economically' reform the management of water in economic terms, Briscoe (1997) of The World Bank sets out lessons learnt. He suggests:

- Change should only be initiated when there is a great need
- A clear strategy and effective information is required for dealing with all interested parties
- Sensitivity is necessary for different institutions and environments
- Allow water markets for effective management
- Start with pilot projects in new environments
- Compensate economically and efficiently third parties affected, e.g. by lowered water tables.

Urban water supply is characterised by high supply cost and low opportunity cost (Figure 6.8). Accordingly, priorities for economic optimisation relate rather to the engineering side, which dominates the supply cost rather than the trading of water as an economic commodity. Briscoe cited examples in India where the reform process floundered because of this inflexibility, yet in California and Australia reform happened.

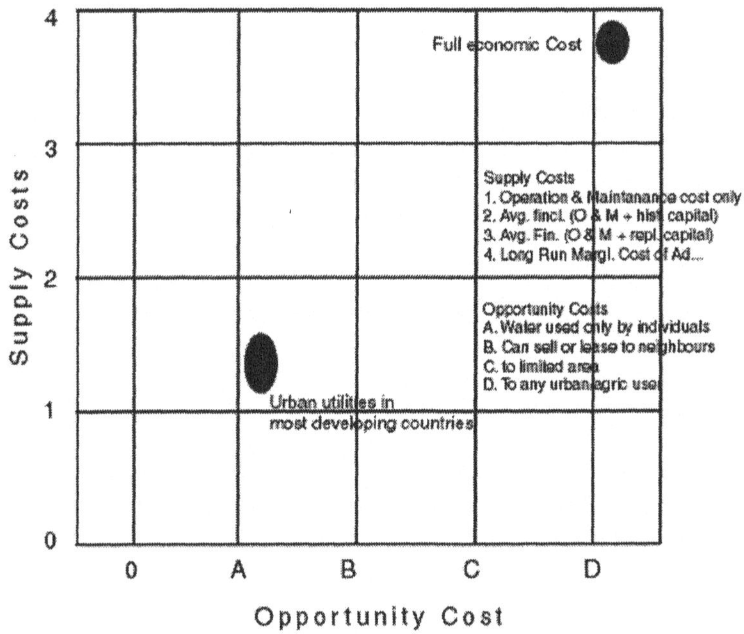

Figure 6.8 Supply cost, opportunity cost and full economic cost for urban water supply (Briscoe, 1997)

Hanoi, the capital city of Vietnam, exhibits an exceptional case of abandonment (Le Van Duc in UNESCO, 1997). The sanitation network covers only one seventh of the city. In addition, 70% of piped water is unaccounted for. This demonstrates that many developing countries' cities are in a state of crisis management, and emergency planning is needed to catch up with servicing the entire city in addition to natural expansion.

A further problem arises with uncontrolled urban growth in arid countries. The water resources may not be adequate or extremely expensive to source, e.g. desalination, or pumping from far away.

6.6.4 Policy

To summarise, the focus must be on developing institutional arrangements that provide services at least at cost and in a way that is responsive and accountable to beneficiaries. Where public finance is scarce, significant additional resources can often be mobilised within local communities. Attention should be paid to ascertaining from the poor what level of service it is that they want, discuss what is feasible and then to providing a range of affordable levels of service.

There needs on the part of governments worldwide (not only in developing countries) to be a changing emphasis from single focus projects (e.g. a focus on technology, or on preventative health, or on hygiene education), to a more integrated approached to the provision of infrastructure and the improvement of health.

This change of emphasis needs to be accompanied by other shifts in emphasis, particularly:

- From top-down approaches, to approaches that are a judicious mixture of top-down and bottom-up
- From focus on construction costs of facilities, to lifetime costs of facilities (i.e. including operation and maintenance) and
- Recognition of the need to ensure the financial and environmental sustainability of projects.

Conclusions relevant to WSS provision in the urban areas of developing countries include (Walls, 2000):

- Services suppliers must understand that they are selling a product, and this involves regarding the beneficiaries as customers and understanding their circumstances. Community contact and beneficiary education are essential
- It is better to improve an existing system in ways that are affordable, upgradeable and financially and environmentally sustainable
- Water resources must be managed better, taking account of scarcity, quality, economic efficiency and sustainability
- Those "private" services that people want and are willing to pay for (including water supply and the collection of excreta, wastewater and solid wastes) should be provided at full cost with due consideration to social factors. Scarce public funds should only be used for those services (specifically, treatment and disposal of excreta, wastewater and solid wastes) that provide wide communal benefits
- Flexible and responsive institutional mechanisms must be developed in order to supply these services, with a larger role for community organizations and the private sector where appropriate
- Supply of services has often been hampered by the conflict between levels of service and financial constraints. It is often not acknowledged by

governments, politicians and beneficiaries that this conflict reduces the potential for services to those not yet supplied

- There is a need for integration of all efforts to reach the same project objectives. Thus, in order to improve community and individual health, water and sanitation must be integrated with the provision of solid waste disposal, roads, stormwater drainage, flood prevention, education in water and sanitation use, primary health care, education in general health care, shelter upgrading, nutrition improvements and opportunities for earning income
- Attempts to improve any aspect of service provision must adequately address all elements, viz. technology, finance, institutional reform and social
- There is a need for greater political will to address the above, and to make difficult and possibly unpopular decision on issues such as management, capital investment, subsidies and prices, and on trade-offs between competing demands for limited resources
- There are no set policy, project and programme designs that are the "right" way to approach any urban water supply and sanitation need/situation. How actual implementation should take place would vary from situation to situation, with scope for tailored differentiation to appropriately cope with diverse individual and dynamic circumstances.

Options need to be kept open. Changes in population growth rates, migration and depletion of resources could result in wasted investment. Bu the plight of billions of poor people, health and threats and overflow should be sufficient to attract attention to the needs of the urban poor. The correct way to invest is the only question.

6.7 REFERENCES

Alaerts, G.J., Veenstra, S., Bentvelsen, M. and Van Duijl, L.A. (1993). Feasibility of aerobic sewage treatment in sanitation strategies in developing countries. *Water, Science & Techn.*, **27**(1), 179–186.

Asano, T. and Levine, A.D. (1996). Wastewater reclamation, recycling and reuse: Past, present and future. *Water, Science and Techn.* **33** (1) 1–14.

Bhatia, R. and Falkenmark, M. (1993). Water Resources Policies and the Urban Poor: Innovative Approaches and Policy Imperatives. UNDP-World Bank, Water & Sanit. Prog.

Black, M. (1994). Mega-slums: The coming sanitary crisis. London, *Water Aid.*

Briscoe, J., Feachem, R.G. and Ahaman, M.M. (1986). *Evaluating health impact: Water supply sanitation and health education.* UNICEF-ICDDR-IDRC.

Briscoe, J. (1997). Managing water as an economic good: rules for reformers. *Water Supply*, **15**(4), 153–172, Blackwell Science Ltd.

Cairncross, S., Carruthers, I., Curtis, D., Feacham, R., Bradley, D. and Baldwin, G. (1980). *Evaluation for Village Water Supply Planning.* John Wiley and Sons, N.Y., 179 pp.

Churchill, A.A. (1987). Rural water supply and sanitation. World Bank discussion paper, Washington, D.C.

Gleick, P.H. (1998). The Consequences of Water Scarcity : Measures of Human Wellbeing. *Proc. Intl. Conf. World Water Ress. at the beginning of the 21st Century – Water: a looming crisis*. UNESCO, Paris 3-6 June 1998. Addendum to IHP-U Techn. docs in Hydrology No. 18, UNESCO, Paris.

Habitat-UNEP (1996). Proposals for a multi-agency programme for integrated water resources management in Peri-Urban areas. Discussion paper for sub-com. on Water Resources.

Ilemobade, A. (2000). Private communication.

I.R.C. (1980). Public standpost water supplies – a design and construction manual. International Reference Centre for Community Water Supply and Sanitation, The Hague, The Netherlands, Technical Paper No. 14, 91 pp.

IRC (1981). Small community water supplies. International Reference Centre for Community Water Supply and Sanitation, The Hague, The Netherlands, Technical Paper No. 18, 413 pp.

Khouri, N., Kalbermatten, J.M. and Bortane, C. (1994). The reuse of wastewater in agriculture: A guide for planners. UNDP-World Bank. *Water and Sanit. Prog.* Washington DC, the World Bank.

Lagos State Government (1997). *Urban Services Improvement: A Willingness to Pay* Study. Ministry of Environ. And Phys. Planning.

Lyonnaise des Eaux (1998). *Alternative Solutions for Water Supply and Sanitation in Areas with Limited Financial Resources*. Paris.

Macleod, N. (1997). Tariffs and payment for services – The Durban Story, *Civil Engg., SAICE*, June.

Mara, D. and Cairncross, S. (1989). *Guidelines for the safe use of wastewater and excreta in agriculture and aquaculture*. Geneva, UN Environmental Programme/WHO.

Mara, D. and Pearson, H. (1998). *Design manual for waste stabilization ponds in Mediterranean countries*. European Investment Bank, Leeds, Lagon Technology Ltd.

Marshall, K. (1996). Water, Welfare and Institutions. Care challenges for Africa. *EDI/WUP Conf. Instl. Options in the Water Supply and Sanitation Sector. Private Sector Participation.*

Mwabu, G. (1999). *Health development in Africa* African Development Bank, Economic Research paper 38.

Mwanza, D.D. (1997). Case study on institutional options for urban water and sanitation services in Africa - The Zambian case. *Proc. Intl. Consultations on Partnerships in he Water Sector for cities in Africa.* CapeTown.

Narayan, D. (1994). The contribution of People's Participation. Evidence from 121 Rural Water Supply Projects. *UNDP-World Bank Water & Sanit. Prog.* TWUWU & Social Policy and Settlement. ENVSP, World Bank.

Niemczynowicz, J. (1996). Megacities from a water perspective. *Water Intl.*, **21**(4), 198–205.

Parikh, H. (1999). Slum networking, an alternative way to reach the urban poor. Keynote lecture, *Proc. Int. Conf. Urban Storm Drainage*, Sydney.

Petersen, M.S. (1993). Water for Africa, the Human Dimension. *Proc. Water – The Lifeblood of Africa, IAHR*, Victoria Falls.

Price, T. and Probert, D. (1997). Role of constructed wetlands in environmentally sustainable developments. *Applied Energy*, **57**(2/3), 129–174.

Pritchett, L. and Summers, L.H. (1996). Wealthier Is Healthier. *J. Human Resources*, **30**(4), 841–868.

Rose, G.D. (1999). Community Based Technologies for Domestic Wastewater Treatment and Reuse: Options for Urban Agriculture. Cities feeding people. IDRC report series CFP 27.

Sarah, J.M., Gam, M. and Katz, T. (1998). Some key messages about demand management approach. *Water Supply & Sanit. Conf.* UNDP- World Bank Water & San. Prog. World Bank.

Sen, A. (1995). Mortality as an indicator of economic success and failure. Intl. Child Dev. Centre, Innocenit Lectures, Florence.

Stephenson, D. (1999) Water demand management theory. *Water SA*, **25**(2) 115–122.

Stephenson, D. (2000). Management of urban runoff in South Africa. *J. Hydrol.*

Tchobanoglous, G. (1991). Land based systems, constructed wetlands and aquatic plant systems in the US. An overview. In Elnier, C. and Guterstam, B. (Eds), *Ecological Engg. for Wastewater Treatment.* 110–120. Gothenburg, Sweden. Boskagen.

Tejada-Guibert, J.A. (1998). Water Supply in Small Settlements and Peri-Urban areas – Selected Aspects. *Proc. Intl. Symp. Suitable Water Management and Technologies for Small Settlements*, Polytechnic University of Catalunya, Barcelona.

UN Conference on Environment and Develpment (UNCED) (1992). *Agenda 21, Programme of action for sustainable development.* Rio de Janeiro, Brazil.

UNDO (1998). *Community Water Supply and Sanitation Conf.*, Washington D.C.. UNDP-World Bank.

UNESCO (1998). *Proc. Symp. Water, the City and Urban Planning* Paris.

Van Damma, A. (1973). Water Supply to Developing Communities. *WHO Intl. Ref. Cr. for Community Supply.*

Walls, K. (2000). Private communication, Johannesburg.

WHO (1982). Activities of the World Health Organization in promoting community involvement for health development, Geneva.

WHO (1996). Water Supply and Sanitation Sector monitoring. Report 1/46. Sector studies as of 31 Dec 94, WHO Water Supply & San. Collaborative Conf. & UNICEF, WHO/EOS/96 - 5/Geneva

World Bank (1988). The challenge of hunger in Africa. Intl. Bank for Reconstruction and Development, Washington D.C.

Wright, A.M. (1997). Towards a Strategic Sanitation Approach: Improving the Sustainability of Urban Sanitation in Developing Countries. *UNDP-World Bank Water & Sanit. Prog.*

7

Economic and financial aspects

Terence Lee, Jean-Louis Oliver, Pierre-Frédéric Teniere-Buchot, Lee Travers and François Valiron

7.1 FINANCIAL REQUIREMENTS OF WATER RESOURCES INFRASTRUCTURE

Water management in urbanised areas includes three main aspects:

- Drinking water supply for domestic and municipal (highway maintenance, markets, etc.) use. It includes service activities and craft and industrial use
- Domestic and industrial wastewater collection and their treatment before discharge in the natural environment. The core problem of the fight against pollution
- Natural or canalised water management, environmental protection and its development (recreation and tourist use, navigation), prevention against flooding.

Edited by Čedo Maksimović and José Alberto Tejada-Guibert. ISBN: 1 900222 76 0

Without going back over the description of urban water infrastructure that was already mentioned in previous chapters, concerning water supply, as well as rainwater and wastewater control natural and canalised water management. Environmental protection and development, a general distinction may be drawn with respect to urban water infrastructure concerning its geographic location: whether it is inside or outside the urbanised area.

In fact, the water supply network and the wastewater and storm drainage networks are always inside urban areas. On the other hand, the source for the available surface and ground water must be searched for farther and farther outside because of the expansion of towns. It is the same situation for the structures for river pollution control and intakes or for protection of groundwater threatened by upstream discharge or by intensive farming. All of this has numerous and considerable consequences for the financing requirements, since the structures of water conveyance and those for upstream pollution control are more and more often outside the built-up area, like those necessary for controlling river high water and low water and those supplying the town and the canals that serve it. This leads to tensions between the towns and their upstream area. It is the same for the impact of downstream discharge. To avoid conflicts, a clear communication channel must exist between the water managers in the built-up area and those in charge of the upstream and downstream areas by the joint management of water resources in quantity and quality.

Wastewater disposal is often more expensive than drinking water supply and must be rapidly implemented to avoid the degradation of conditions of hygiene and the resulting epidemics.

The first consideration of a wastewater conveyance system should be public health. It can prove to be disastrous for environment if no sewage treatment it is included. Indeed, the whole pollution of a town is often brought to one point at the end of a sewerage system. This is another example of the relations existing between the inside (generally located upstream) and the outside (downstream) of a built-up area.

An additional expense, generally for several purification facilities must therefore be incurred in order to protect the natural environment. Those investments are often a burden for the tax payers, but the operating costs they create have an even more important economic impact. As we will see, those expenses are normally covered by either the water rates or from the municipal budget.

The sludge from sewage treatment plants constitutes a concentrated by-product that large towns must incinerate, store or apply to agricultural land. Once more, relationships with the rural areas are sometimes delicate and additional expenses may again result . It is indeed impossible to contemplate urban water management without taking into account the region around the town.

Similarly, the protection and the development of the ground and surface water resources depend economically on the situation of the urban area and on the external natural conditionsit affects.

Additionally, the investments required for infrastructure for navigation and protection against flooding can reach important amounts. As opposed to the previous case, in this case, it is not usually the municipal water rates that finances these, but other national, regional and local budgets. Nevertheless, they should be kept in mind since their implementation may constitute a driving force to develop to a global urban water policy.

The economic policy for water in urban areas – and its complement, the political economy – is the art of considering all the interrelations it involves and to do this at many levels:

- Resource: ground waters (plural, since there can be several ground aquifer systems), surface waters (fresh, brackish and sometimes saline for towns located on the shoreline), as affected by extreme events (low waters, floods, exceptional episodes);
- Competing uses (irrigated agriculture in the outlying areas of the towns, industrial and port areas especially) and the resulting national planning (balanced management of these economic activities development) ;
- Institutional system of decision making (Who is able to decide, finance and undertake, and for what amounts?) and local funding capacity (what financial sacrifices are the public and private political decision-makers ready to make for water?);
- Infrastructure implementation capabilities (management of international tenders, control by the owner, project management and foriegn or local companies) as well as the form (public or private) chosen to operate the infrastructure.

These last points are certainly the most challenging. To assess the resources requires skills but relatively few funds. For that reason, this kind of information is usually available, at least in part. Making a list of the alternative and competing uses of water is rather more difficult. This task suffers most of the time from the lack of knowledge of the volumes used and of the quality requirements for some of them, and furthermore enters the political arena in planning their future development. Here, we go from study activities necessary to obtain the information (usually insufficient) to make a concrete decision to management activities which have implications on the political balance of power.

Internal costs

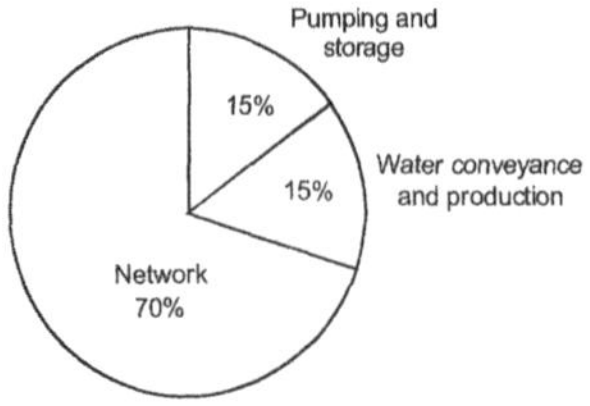

With 700/800 $ for
2.7 to 3 inhabitants/house
180 to 210 m3/year
out of external costs and out of private connection

External costs

Dam	0 to 10 %
Pipes and dam	10 to 50 %
Groundwater protection	0 to 5 %
Interconnection	0 to 10 %

Connection	10 to 20 %

Water supply costs

Wastewater

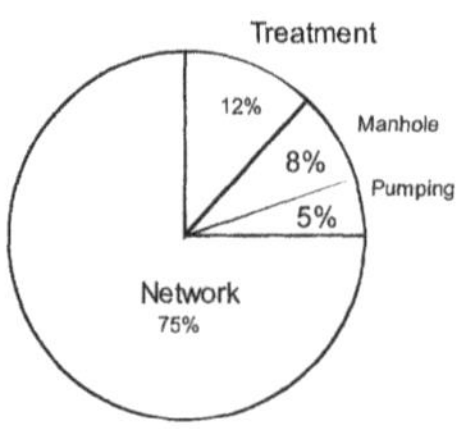

Cost : 700/900 $ (out of external costs)

Stormwater

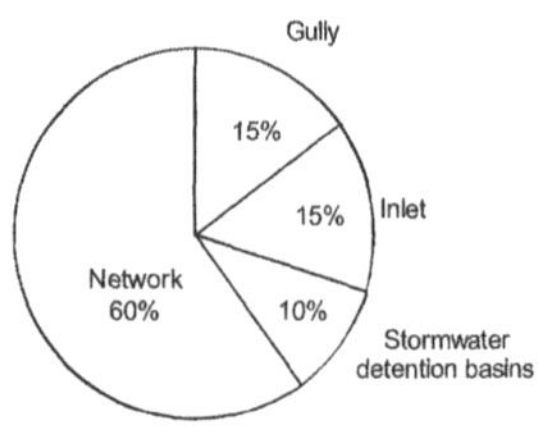

Cost : 650/700 $

Costs sharing for separate sewer systems
(out of external costs)

Internal costs

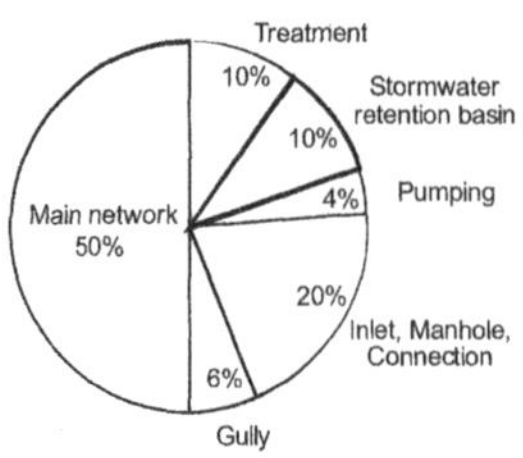

Cost : 1000/1300 $/inhabitant (out of external costs)

External costs

Environment protection	300 to 400 $/inhabitant

For separate and combined sewer systems

Connection	300 $/inhabitant

Costs sharing for combined sewer systems

Figure 7.1: Cost of water supply and wastewater disposal in a fully supplied and connected system. Cost sharing

On the other hand, the institutional analysis of the regional political and financial capacities requires governance skills where the sometimes contradictory wills of the people and private interests (industrial, at times agricultural), of the main opinion leaders (politicians, media, associations and pressure groups), and of the financial institutions, must be taken into account.

It would be ill-advised to think that the political will to undertake the necessary actions matches the demands. The obvious lack of water, in quantity as well in quality, does not imply that a financial priority will be accepted, or even encouraged. Water works, whatever their nature, are usually not very attractive compared to other more visible investments, profitable in the short-term and in the end better accepted by the population.

Water managers (whether for of a city or of the river basin to which it belongs), must keep in mind that their capacity for acting is more often constrained more by the availability of funds than by the funding requirements to cover water demands, which nearly always exceed the possibilities and therefore, are rarely satisfied.

The willingness to pay (as related to the water rate structure and taxes obviously constitutes an important parameter to ensure the essential balances that have been mentioned. The considerations below aim at facilitating and developing, in actual practice, such balances.

7.1.1 Overview of infrastructure costs

Figure 7.1 indicates the average costs per inhabitant (domestic connections not included) in dollars, comprising the whole necessary infrastructure for supplying and draining wastewater and stormwater, separating internal and external costs. These costs correspond to an annual water consumption of 180 to 220 m^3.

The investments required for industries connected to the network are analogous to those for housing, on a partially sliding scale according to the annual volumes used. As regards wastewater disposal, these connected industries are responsible for meeting the standards for domestic wastewater that is discharged into the network by means of a pre-treatment unit. The investments required for non-connected industries are in a lower range than they would have to pay if they were connected.

Part of the external costs, particularly in the case of dams, is used to finance multiple purpose works, thus reducing the investments that need to be made for other uses, such as flood control, irrigation or the canalisation of rivers. For example, in Paris (France), around one third of the cost of dams, which maintain the water levels and thus control floods, is included in the price of the water. This is also the case in countries such as Morocco, where the dams are used both for water supply and for irrigation. This contribution is invaluable since these dams are often very costly and their financing depends, to a large extent, on the government. The same is true for external investments linked to *wastewater* disposal, which contributes to reducing pollution in rivers, thanks to water purification installations.

7.1.2 Impact of various parameters on these costs

Costs can be classified into three categories, according to whether the investor has the possibility of acting on them or not:

- Those that depend on local geographical, physical or economic conditions,
- Those for which the investor has only a restricted range of action,
- Those on which the investor can act.

7.1.2.1 Defined parameters

As regards water supply, the climate influences the type of resource, it location, its importance (underground or surface water), and its variability. The topography influences the transfer of resources, the necessity of creating backflow, and also the *wastewater* drainage system where the amount of rain has an effect on the stormwater network and the combined network (structure of the network, storm water basins). Finally, it may be necessary to thoroughly purify the water (eutrophication) and treat rainwater discharges in order to protect the downstream quality. The maximum possible fluctuations are indicated in Table 7.1.

7.1.2.2 *Partly defined or adaptable parameters*

These include **the size of the town**, and the unit cost, which decreases in importance between 1,000 and 200,000 inhabitants; beyond 300,000 habitants, it no longer decreases. Finally, socio-economic conditions play a determining role through the density of occupation of housing and the unit consumption.

Urbanisation policy also has a significant effect on the cost of networks, which account for 60 to 70% of the costs of land management. If the water manager has correctly analysed the effects, he can then help the political decision makers to understand the importance of the choice of sites to urbanise, and properly control the habitation density (choice between multiple or individual housing), the way the land is used (the lay out of the road networks and buildings and the influence of surface sealing) and the drained area (see effects in Table 7.1).

7.1.2.3 *Adaptable parameters: service standards*

These objectives are fixed in order to ensure the correct operation of the service and to meet the requirements of the users:

- First, adapting the network to the wastewater discharge requirements of the users by providing additional capacity, ΔC, so that it fulfils these new requirements when it is put into place. The difficulty is in correctly calculating ΔC, since over-capacity that is not used has an effect on the financial equilibrium of the service,
- Second, ensuring the security of the network via interconnections and reservoirs, while maintaining sufficient pressure everywhere. However, the pressure should not be too high, as this generates losses that should be avoided,
- And third, acting on the water quality, its taste and its odour,
- In the case of stormwater networks, choosing the level of protection against overflows using storage facilities and increasing the diameter of the pipe work, since it is recognised that flow rates increase from 20–25% to 50–55%, if the return interval goes from 5 to 20 years,
- In the case of combined systems, the sewers are not changed but the spillways have to be over dimensioned.

The other decision that needs to be made as regards wastewater disposal is the level of discharge purification required to protect water quality in rivers.

The effects of these choices can multiply the cost by 1.5 (see Table 7.1).

Table 7.1 Summary of the possible impact of the various parameters, shown in terms of a multiplication factor.

	Climate and geographical situation	Size	Type of urbanisation	Service standards
Water supply	1–1.7; underground water 1–1.8; surface water	1–2	1–1.25	1 to 1.5
Purification				
- *wastewater*	20%	1–2	20–25%	20%
- rain water	45%			30%
- combined system	25%			20%

These different fluctuation factors are maximum values that cannot be automatically multiplied by each other. A maximum in one parameter does not necessarily correspond to a maximum in another parameter. An overall reduction of one third gives the maximum overall fluctuations that are usually encountered. This leads to the data shown in Table 7.2, which indicates the values observed in developed countries with ecologically reliable services.

Table 7.2 Infrastructure costs of different sanitation systems (in dollars per habitant)

Service	Sub-group	Min. / capita	Multiplying factor	Max. / capita
Water supply		450	4	1800
Wastewater treatment	Wastewater	650	2.2	1400
	Stormwater	970	3	1250
	Combined system	900	2.4	2200

As a conclusion to this section, several ways of minimising costs may be defined:

- First, correctly analysing the water demand and programming the investments so that they meet the requirements as closely as possible
- Second, correctly analysing the effects of certain parameters, in order to make them known to the decision makers and demonstrate the financial consequences of their decisions as regards security, urbanisation and flood protection
- Finally, employing efficient and economic techniques.

This *order* is economic and acts as a reminder that although the technical aspects are important, they are not the determining economic factor.

7.2 WAYS AND MEANS OF FINANCING

7.2.1 The ways

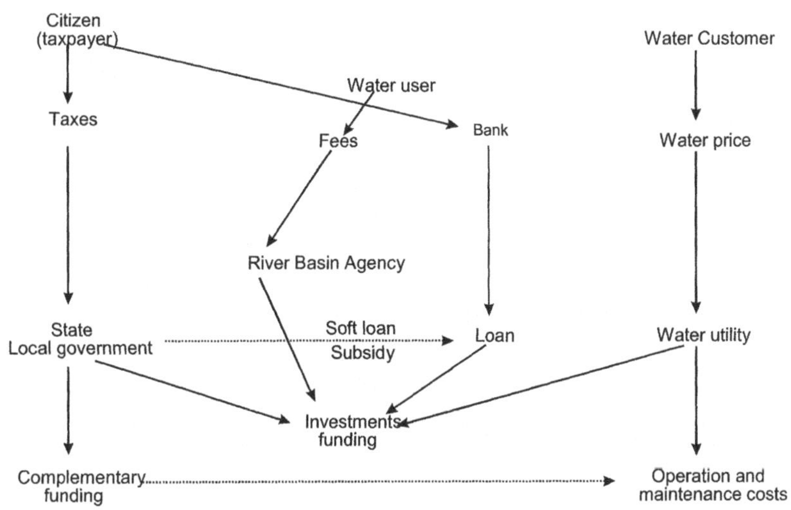

Figure 7.2 The financing channels

Figure 7.2 shows the links between the five ways of financing: taxation, fees and levies, capital contribution, self financing and borrowing, which concern the citizen, the direct end user and the user of the service. It shows that, in the final instance, these three categories bear the financing costs. Many people are affected on two or three accounts, depending on whether they pay:

- Taxes to the State or the community – in this case they belong to the category "citizens"
- Specific water taxes, such as disposal fees – in this case they belong to the category "water end user"
- Their water bills – in this case they belong to the category "users of the service".

Three channels are shown in this figure:

- The shortest channel, via the price of the service, is the most logical and equitable in that there is a lack of equipment for measuring storm water drainage and the fact that the costs have to be reduced for the poorest
- The channel via the direct end user is also logical since it links the tax and the benefits provided
- Taxation, whatever it is indexed to (income, consumption, production or assets), is convenient, but it is not equitable since it is not linked to the benefits.

7.2.2 The means

7.2.2.1 Subsidies and interest rate rebates

These have a tendency to disappear when they come from taxation, in the case of works that the user can support, such as water supply and *wastewater* discharge, and are then transferred to the financing of storm water discharge systems and for major irrigation infrastructures, flood protection or canals. This is not the case in countries with low incomes, although they have perverse effects, since they can lead to over-dimensioned structures being constructed.

7.2.2.2 Contributions from beneficiaries

The first route establishes fees that are borne by the users as a function of the discomfort and inconvenience that they cause to others. These are the "polluter pays" or "taker pays" fees that have been set up, particularly in France, with Water Agencies. The funds collected in this way are redistributed, in order to finance works with a common interest through subsidies or loans, and these then become mutualisation plans.

The second route requires part or all of the necessary investments to be borne by those responsible for the investments. This is the case for taxes per m^3 that are required for constructing or building service water networks. It is also the case for taxes on the area of land that generates run off and storm water pollution, amended according to its surface-sealing factor. These are used, particularly in Sweden, Switzerland and Germany, to finance stormwater discharge systems.

7.2.2.3 Borrowing and investments

This system is widely used with banks and specialised organisations. The duration and rates applied depend on the works that have been financed and the status of the borrower. Periods vary between 10 and 20 years, and rarely more than 25 or 30, unless the lender is an international organisation. The rates are normally 1 or 2 points above the rate of inflation, again except in the case of certain international organisations.

An interesting variation is used in the USA with " bonds " sold to the public by the operator. These are paid back over a period of 20 or 30 years at a fixed rate, and are guaranteed by the price of the water.

7.2.2.4 Self financing

When the services can apply a water price that is above the financial equilibrium price, this constitutes an endowment that is intended for self-financing purposes. This method of financing is of interest because of savings in the costs linked to interest payments. This reduces the cost for future users but present users are penalised with an overcharge. This disadvantage is eliminated if the self financing only applies to the renewal of works.

7.2.3 Equity

Table 7.3, which shows the effects of the various types of financing on past, present and future costs, makes it possible to determine if the costs are borne by present or future users.

Table 7.3 Effects of the method of financing on past, present and future end users

Means of financing	Effect on the past or present cost	Effect on the future cost	Remarks
Borrowing or capital contribution	0	↗ ↘	Cost passed onto future end users
Self financing	↗	↘	
Surcharge	↗	↘	Cost born by past or present end users
Fees	↗	↘	
Subsidy or interest rate rebate	0	↘	Cost borne by present users

7.3 THE COSTS OF WATER SERVICES

To obtain water, humans have always had to dig wells, dig ditches and construct water reservoirs for producing, conveying and storing the water. Unlike electricity, which is easy to transport but difficult to store, water is a local natural resource, although often erratic, which is expensive to transport and purify but easy to store.

As we have seen in section 7.1, the costs of water services can vary considerably:

- For physical and geographical reasons; they depend on the location and quality of the resource used: underground water is, in general, less costly than water extracted from rivers, since it involves lower production, transportation and treatment costs as a result of it being normally of better quality. Since the source can be a long way from the place of use, it is often necessary to constitute reserves of raw or treated water in order to compensate for fluctuations in the resource or the requirements. The transportation and distribution costs are also dependent on the geographical dispersion of the users. When the resources are insufficient, costly works have to be constructed, such as long distance water conveyance structures or multiple purpose dams – reservoirs, which can, depending on how they are exploited, reduce flooding (providing they are empty), compensate for water shortages or produce energy (providing they are full), or create water tracts for leisure purposes (at constant level).
- For financial reasons, the most important of which is the age of the networks, whose cost represents, on average, ten times that of treatment plants. In fact, the construction of a network is often financed through borrowing over a period of around 20 years, while the operating life of the networks is around 100 years. Compared to new networks, old networks therefore involve significantly lower financial costs, especially since the depreciation provisions are often not set out in a very strict manner. Another financial factor is the diversity of the subsidy schemes from the public authorities.

The cost of the water services is therefore the objective, and in principal measurable, consequence of the works required to provide the water for multiple use.

7.3.1 Theoretical considerations

Traditionally, one distinguishes:

- internal costs, which are located within the management unit (investment and operation)
- external costs, which are not located within the management system, and concern local governments, as a consequence of water use. It is said that external costs are "internalised" when these are introduced into the price of the water borne by the user.

7.3.1.1 Breakdown of the costs of the water service

The cost of the water service is made up of the following four components:

- The capital costs and the cost of repaying loans, which consist of loan instalments with part of the repayment and the interest being repaid. As regards the capital, a fraction has to be provided for each year and an appropriate remuneration has to be added when the service is in financial equilibrium. These costs have an obligatory fixed nature, and are independent of the number of service units.
  The *operating costs*, which include:
  - Labour, management and maintenance costs (a),
  - Supply costs (electricity, treatment chemicals) (b),
  - Direct maintenance costs (replacement parts) (c)
  - General costs (premises, management, vehicles) (d)
  - These are divided up into fixed costs (a and d) and variable costs (b and c).

    A reduction in the fixed costs, like a reduction in the variable costs, has effects that are more or less long term; the right balance is difficult to determine by the operator and depends on his experience
- Renewal costs enable the necessary replacements to be made in order to maintain the installations in good condition, thanks to provisions that are financed without resorting to borrowing.
  One requisite of good management is making sure that the works are in their original condition at the end of their renewal periods; which implies taking into account inflation, which increases the cost

- Taxes, levies, fees, benefits: these are the obligatory costs that are set by the public authorities; only the benefits, once they have been paid by the lease holder or concessionary, are at the discretion of the operator.

7.3.1.2 Cost structures

The overall cost **Cg** includes a fixed part **Cf** and a variable part **Cv,** proportional to the number **U** of service units : **Cg = Cf + Cv**

Cf can vary between 70% and 90% of the overall cost. In order to reduce the cost per service unit:

$$\mathbf{Cu = Cg/U}$$

The installations must operate at a level as close as possible to the nominal capacity for which they have been designed.

- Breakdown into specific costs
  The overall cost **Cg** is the sum of the energy costs, the product costs, the capacity costs and the general costs.
  The customer cost is born by the connected user, independently of the demand, while that of the income varies with the service units.
  The capacity costs are those that correspond to the demand, while the general costs do not depend on either the use or the number of connections.
- Marginal cost and opportunity cost
  The marginal cost is the cost of the last delivered unit; the short-term cost corresponds to existing works and the long-term cost reflects the cost of any necessary extensions or reinforcements.
  The opportunity cost is the sum of the costs for mobilising the resources, the cost of impoverishing the natural resources and the deterioration costs. It represents the costs required to restore the service to its initial condition, in terms of resources, quantity and quality.
- Costs for the subscriber and for the habitant
  These costs are obtained by dividing the overall cost **Cg** by the number of users or habitants supplied.

7.3.1.3 Evaluation of costs

- The book, or apparent, cost reflects the figures from the water service accounting system and differs from the *"real cost"* in terms of items that have not been taken into consideration in the financial accounts, as is sometimes the case for maintenance charges, opportunity costs and certain external works.

The difference can reach 100%, with renewal accounting for 35%, costs carried forward to the local public body 20%, opportunity costs 20%, and connection costs 25%

- Costs depending on the type of service
  First, this involves making a distinction between water supply costs and water purification costs, which can include sewage and storm water, either separately or together. This is also the case in the grouping together of communities or wider areas, such as hydrographic sub-basins as in England.
- It is necessary to compare the *cost variations* against inflation in order to highlight real increases.
- The cost variations can range from 1 to 15 or more, but on average from 1 to 5, as a function of the combination of differences between the investment costs and those for the methods of financing and maintenance, as well as differences in operational costs .

7.3.2 Water supply

We will not return to the financing costs that depend on investments, the variations for which are shown in Table 7.4 below. According to an investigation by the IWSA – the International Water Supply Association - the operating costs, expressed in equivalent terms, can vary by a factor of 5.6 to 13.8 and from 0.45 to 1.3 for 1 000 habitants.

The overall cost, in dollars, for 100 m^3 / year varies as follows:

Table 7.4 Water supply – costs (for 100 m^3 / year)

Financial costs	\$0–40	on average \$85 for 100 m^3 per year
Maintenance costs	\$0–45	
Operating costs (of which 30% are labour costs)	\$15–60	
Taxes and miscellaneous costs	\$3–15	

In the U.S.A., the cost per m^3 varies from \$0.2 to \$1.2. In France, the average price in 1990 was 0.7 dollars / m^3 with a range from 1 to 10.

Given fixed costs of around 40%, the diagram below shows the effect of supply yield against the percentage use of works; it demonstrates the importance of reducing leaks and fully exploiting works, and therefore the necessity of having efficient (?) equipment (Figure 7.3).

Figure 7.3 Relation between costs and use of the equipment

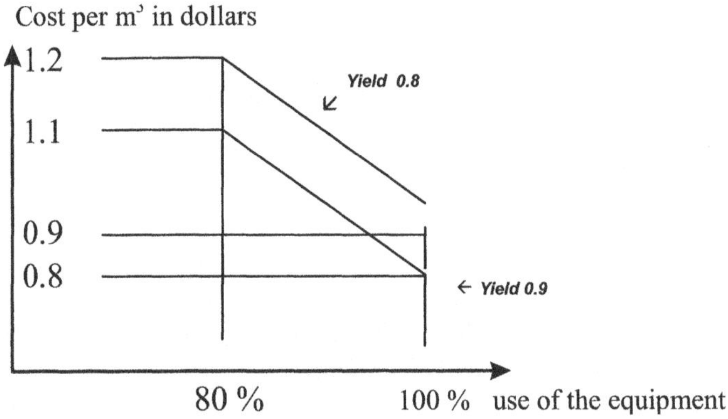

7.3.3 *Wastewater* disposal

The remarks made regarding the financial and maintenance costs in the previous chapter also apply, in general, to *wastewater* disposal. Table 7.5 shows the costs, in dollars, for 100 m^3 / year in a town with 100,000 habitants for all of the equipment needed for combined systems or separate systems (*wastewater* / storm water):

Table 7.5 Costs of wastewater disposal (in $ for 100 m^3 / year)

Type of cost	Combined system	Wastewater	Stormwater
Operating costs	35 ± 15%	25 ±40%	15 ± 20%
Maintenance costs	13–25	8–15	5–13
Financial costs	15–25	10–16	9–15
Other costs	4	2.5	2
Total	60 (80) 95	35 (50) 60	25 (35) 45

About 60% of these costs correspond to labour costs, 20% to electricity and water treatment products costs, and 20% to equipment and replacement parts. The average cost for larger towns is 20% to 30% lower, but almost double for small towns. In these costs, pollution treatment represents 35 to 40% for the combined system and the waster water system, and 20% for the storm water system with storm basins. Most of these costs (80 to 90%) are made up of fixed costs; this leads, as in the case of water supply, to the unit cost being highly dependent on the number of cubic metres sold.

Figure 7.4 Relation between cost of wastewater treatment and volumes sold

7.3.4 The effects of the service organisation

Since they are located nearer to the areas and populations concerned, local governments have to assume the responsibility for local public services such as drinking water supplies or *wastewater* discharge: this is the application of the subsidy principle.

To meet these responsibilities, local governments, whatever their size, have a prime interest in setting up an *additional budget*, which itemises all of the financial operations of the service and provides necessary accounting information, information on tariffs and management indicators.

Local governments, or larger associations, can then either set up a specialised *autonomous public company*, with the status of a corporate body and financial autonomy, or a *mixed enterprise corporation*. Both solutions have the advantage of greater flexibility in organisation, planning and financing : in fact, as urban development continues, it is then statutorily or contractually easier for them to extend the area covered by the service. In addition, the shared management of water supply and wastewater discharge services also offers the following advantages:

- Economies of scale
- Similar techniques
- Better coordination
- Greater diversity and flexibility in labour management.

These benefits vary, depending on the size of the urban area and become irrelevant beyond certain extension, or network complexity, thresholds.

In keeping with the trend that is developing for all types of infrastructure networks (railways, telecommunications, etc.), one solution consists in dissociating part of the *ownership of installations,* which remain in the public domain and which continues to be managed as an asset in the traditional way, and *operation* , which can then be managed more appropriately for mobilising the technical and financial capabilities of the private sector, within a contractual context.

7.4 WATER CONTRACTS

Local governments can, using different types of contract, hire a private operator for more or less complete responsibility, as shown in Table 7.6.

In practice, the management methods are not as clearly defined as . In fact, each service is unique and has to be treated with a lot of pragmatism.

The responsibility for water services is often incumbent on communities or groups of communities with average populations between 1,800 and 10,000 inhabitants in Switzerland, France, Spain and Germany, and from 30,000 to 50,000 inhabitants in Sweden and Austria, with, sometimes, specific entities, as for example in Belgium or the United Kingdom (170,000 to 10,000,000 inhabitants).

This delegation can reduce general costs and make it possible to have improved labour management, and provide the required technical criteria (use of the same resource, interconnections, etc.) or topographic criteria, in the case of *wastewater* discharges.

Regionalising water supply and *wastewater* disposal can lead to economies, thanks to a shared general management structure, a common management structure for customers, etc. This is normal in the United Kingdom, and sometimes, found in Germany and Switzerland, where water, electricity and gas supplies are combined in a single municipal corporation, although *wastewater* disposal remains under the responsibility of the community.

The legal status of the service, whether public or private, also has an impact on costs. Public management, either direct or autonomous, controlled by local regulations, or a more flexible structure such as a mixed ownership corporation with delegated management or even a B.O.T. (Build Operate Transfer) structure, where a private operator is contractually responsible for the management, or with total privatisation, where the local public body nevertheless keeps control of, and regulates, the services provided.

Table 7.6 Responsibilities entrusted to private operators

Type of contract	Concession	Leasing	Public ownership with private management	Management	Services
Who finances new works	franchise holder	local government	Local government	local government	local government
Who finances trading capital	franchise holder (participation)	lease holder	local government	local government	local government
Who sets the rates paid by the users	Public authorities via contracts	public authorities via contracts	local government	local government	local government
Body to which the users are contractually linked	franchise holder	lease holder	lease manager	local government	local government
With remuneration of the private operator	included in the rate	included in the rate	0% of costs plus a productivity bonus for the parameters	fixed rate, as a function of physical parameters	according to the contract
Cover for the costs of local government	surcharge	surcharge	income	income	income
Responsibilities of the private operator	very high	high	average	average	low
Financial commitment of the private operator	very high	high	average	average	low

The advantages and disadvantages of these various structures are summarised in Table 7.7 below. They depend to a large extent on the means employed for control and regulation that have been put in place when the private sector is called upon, whatever final structure is chosen.

Table 7.7 Advantages and disadvantages of delegated management contracts

Type of management	Advantages	Disadvantages
Public management	No benefits	Rigid organisation Difficult to mobilise specialists Financing problems
Delegated management or B.O.T.	Specialists can be called upon Flexible organisation Setting up financing	Remuneration of the service providers, including a benefit
Private management	Specialists available Possible grouping together Flexibility in Financing	Benefits born by the cost

The control exercised by the public authorities not only involves the terms and duration of the agreements and the opening up to competition during renewals, but also ensuring that the terms are respected while the contract is in force, with the possibility of terminating the contract if the terms are not respected. In addition, the conditions for reviewing prices may be strictly controlled.

Delegated management is widely used both in France for water supply (80%) and *wastewater* disposal (50%), and in other developed countries, particularly the United Kingdom (100%) and Spain (40%).

Vision 2025, presented at the Hague Forum in March 2000, forecasts an intensification in this trend, with private financing, on an international scale, increasing from 5–6% to 24% in 2025, with the investments themselves increasing significantly, depending on the requirements, in a very capitalistic sector.

7.5 WATER AND WASTEWATER PRICING

The last two sections analysed the cost of providing network water and sanitation services to urban people. This section begins to explore some of the key pricing problems connected with those services. Most of the discussion addresses pricing of a water monopoly—which dominates most urban supply and sewerage. Government regulation plays a large role here. But our emphasis on network pricing overlooks many other ways that people pay for water. When faced with low network water quality, urban people invest in home filters or buy bottled water. In poor countries quality may be so low that households must buy much more: water storage jars or tanks, pumps to boost pressure, and filters, fuel to boil water, or vended or bottled water. In fact, the urban poor often lack network water or wastewater service altogether and rely totally on private or mixed public/private solutions such as a neighbour's tap, a public standpipe, or a water vendor for water. Their personal hygiene may rely on public facilities, dry latrines, or soak pits. These special features of developing country urban water and sanitation will be addressed in the following section.

The remainder of this section discusses briefly the economic theory behind water and sanitation pricing, then turns to the practical problems of pricing. It concludes by highlighting some growing challenges in service pricing.

7.5.1 Pricing theory

Economic theory provides reasonably simple guidance for water and sanitation pricing: charge customers the marginal cost of supplying them with water and a sum equal to the marginal damage caused by their wastewater. The motivation for the theory is equally simple: people charged less than those amounts consume too much water or discharge too much wastewater; those charged more consume too little for their own or society's benefit.

Good pricing practice acknowledges, yet deviates from, this theory. Water and wastewater present three principle challenges to full use of the theory: first, water and sanitation have a social aspect much stronger than that of most other goods or services; second, implementing the theory requires effective consumer metering and full knowledge of marginal costs; and, third, other needs, such as adequate and stable revenue and simple, easily understood tariff systems outweigh pure economic efficiency as goals. These challenges play a large role in shaping good international pricing practice, as will be discussed below.

7.5.2 Pricing practice

7.5.2.1 Metering

Without metering a utility and its customers cannot monitor consumption. When cities move from non-metered to metered consumption, consumption falls. And it can fall with little impact on consumer welfare. Only a metered household can easily know whether they have leakage problems in buried pipe. Only a metered household can see the virtue of using a water-saving toilet or showerhead, or of turning off water immediately. In cities with initial domestic consumption above 200 l/c/d[1], metering can easily lead to reductions of 25% or more. Yet meters and meter reading are costly, and meters can be tampered with. This leads to some simple rules: meter large users first, then work down to average households. Meter sections of the network to know where usage is greatest, and meter there. Metering allows price adjustments for seasonal consumption, helping to hold down consumption peaks that otherwise require expensive additional capacity. Falling prices for remote and continuous meter reading will allow pricing structures that further ease demand fluctuations, for example by lowering prices to large water users who refill private storage during night hours. These will make sense first in high water price markets in the higher income countries, but offer future promise to all.

7.5.2.2 Water pricing

People place a high value on water consumption. A value high enough that virtually everyone, no matter their income level, quite willingly pay a price at least as high as the cost of service from an efficiently run urban utility network.

[1] 200 litre/capita/day equal to about 75 m^3/year

Poor people paying that price may not consume much water. WHO guidelines recommend consumption at no less than 20 l/c/d for metabolic and hygiene needs—in households without a flush toilet. Where water is very expensive, primarily rural areas where people must walk long distances for water, but also some urban areas in arid climates where people rely on vended water, consumption may fall to about 10 l/c/d, but these people face prices at least five times as high as water network supply prices. In urban network supply, the cost of a connection rather than the price of water causes the greatest access challenge. And, because they need to control water bills through careful consumption, the poor will suffer more than the non-poor from a tariff dependent on fixed charges rather than volumetric charges.

To meet efficiency goals, water prices function most effectively when based on volume consumed and, in all but exceptional situations, at levels that cover the cost, including capital cost, of efficient operations. Where the cost of service, including capacity costs to meet peak demands, varies among classes of consumers, then water prices should likewise vary. The efficiency dimension –to constrain consumption by charging full prices – has been accepted in principle by many countries, but still fully met by few. For example, among OECD countries, the U.K. prices water at over 90% of full costs, while Greece and Spain charge only 25 and 33%. Cities can charge less than full cost and still maintain a viable utility because government grants have been used to finance capital investments, because they enjoy tax exempt borrowing privileges, or simply because, for public agencies, they do not seek a positive rate of return on investment. However, some cities in advanced economies and many in the less advanced economies hold down prices by holding down maintenance, accumulating a service deficit that will be paid by future generations (see Table 7.3 above).

Water endowments in a country can make a big difference in what full cost prices mean, within the OECD the range runs from US$ 0.8 in Canada to US$ 2.24 in Germany[2]. Whatever the full cost of current supply, cities facing the need to develop new water sources often face huge cost increases such as shown in Figure 7.5.

If consumption is increasing at current prices, necessitating large new investments in the near future that will considerably increase costs, then the policy challenge will be to understand whether accurate pricing would cause customers to reduce consumption enough to avoid or delay the investment. The market test is to begin raising tariffs in anticipation of the investment and observe reactions.

[2] "full" cost is here limited to water supply. It does not include wastewater (sanitation) cost, protection against flood or other nature protection expenses which (at least) double the current water supply price.

7.5.2.3 Protecting the poor

The importance of water for life often leads people to advocate the use of water tariffs as an instrument of social policy. Even many advanced industrial economies feel the urge to supplement other forms of social insurance with discounted water bills. This poses little threat to utility viability if it can be limited to the poor, or if the protected consumption is held to the WHO guideline of 20 l/c/d. Many cities mandate "lifeline" water rates. These provide lower tariffs for some small first block of consumption. Once that level is exceeded, tariffs to a level above costs to make up the revenue deficit. If the advanced industrial economies use water pricing as an anti-poverty measure at all, they generally manage eligibility well enough to keep utilities well financed. By contrast, cities in poorer countries tend to suffer because of overly generous initial consumption blocks (in some cases exceeding the average consumption in a northern European city) that apply to all domestic consumers. When this happens, the non-poor, who are the ones with water-consuming appliances, capture most of the benefit.

A "lifeline" tariff is a form of increasing block tariff. This is one possible financing approach (although one that works well only when the initial block is sufficiently small). A second popular is to charge industrial and commercial users enough above cost to offset the domestic income loss. The second is to use increasing block tariffs designed to that same end.

7.5.2.4 Cross-subsidies

Using industrial and commercial cross-subsidies to offset financial losses in domestic supply can succeed. But success depends on industrial and commercial users lacking the ability to opt out of the network supply. They will opt out when self-supply costs fall below network costs. That, in turn, occurs most in areas with abundant surface or groundwater. Many examples of this behaviour can be cited. Regulations prohibiting self-supply are often ineffective in stopping the practice. Where avoidance is not possible, cities may find that such cross-subsidies discourage new industry from setting up in the city. Not only is new industry discouraged, the utility itself may be discouraged from making new domestic connections, since each new connection can lead to a net revenue loss.

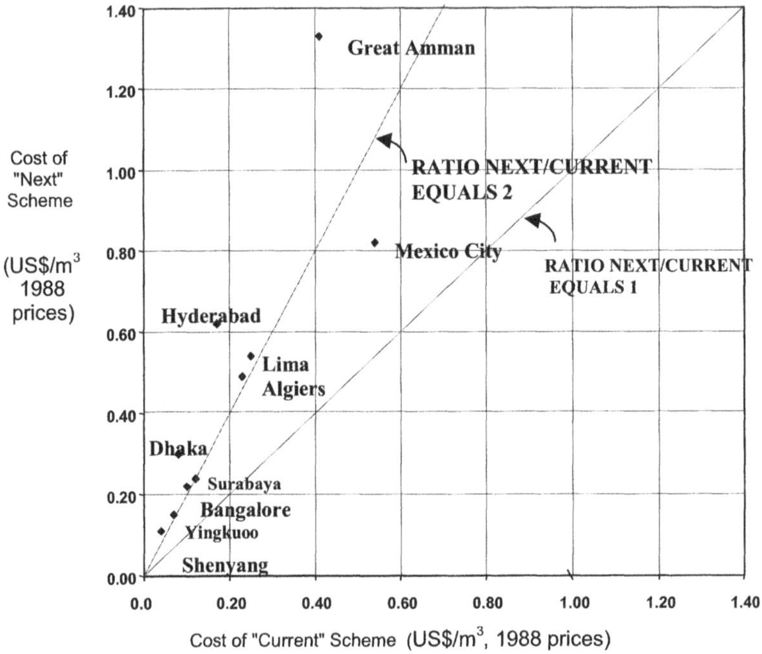

Figure 7.5 Impact of development on water supply costs

The increasing block tariff faces a different sort of problem in cities in developing country . There, few individual households will fall into high use categories. Instead, block tariffs tend to harm the poor. The poor often have markedly larger households e than the non-poor, so even where the poor have individual household metered supply (itself a rare thing), they are more likely than the non-poor to have total basic consumption which pushes them into a higher consumption bracket. When the poor lack their own connections and they buy from other households, those households will find their total consumption pushed into higher price brackets, hence charge buyers that higher price. The same thing happens if several households share a tap on a communal basis.

7.5.2.6 Direct government subsidies

Direct government subsidies continue, even in OECD countries (see Figure 7.2 above). The consumer efficiency loss from this has already been discussed, but loss can also occur on the producer side. Producers depend on certainty of subsidy payments. Without timely payment, they will reduce maintenance and other long-run efficiency measures. For that reason, if given at all, subsidies should be directed toward capital, not operating and maintenance costs. In poorer countries, to make subsidies pro-poor they should go first to connection costs for poor households or to direct payment of that portion of a poor household's bill that covers the minimum consumption norm.

7.5.2.7 Wastewater pricing

Unlike water, wastewater pricing ideally responds to damage caused, not the cost of conveyance or treatment. Households willingly pay to make wastewater disappear, but studies consistently show a reluctance to pay the cost of purifying wastewater for the benefit of downstream users. Decisions to invest in wastewater treatment therefore tend to be joint political decisions, in which everyone agrees that all will share the burden by cleaning up their own waste streams. Perhaps to make the political bargain easier to understand and implement, most countries settle on fixed treatment standards rather than relative damage assessments when mandating treatment. The financing, hence pricing, problem thus departs far from the theoretical basis of offsetting environmental damage.

The cost of a mandated level of treatment may be well above or below the damage untreated wastewater would cause. This deviation is worth worrying about, especially in poorer countries. In the advanced industrial economies, wastewater treatment typically costs more than providing potable water. Adding secondary wastewater treatment can therefore more than double a water bill.

Experience in all countries has given clear guidance for wastewater pricing. Too many wastewater treatment plants operate badly or not at all due to lack of income. That is less likely to happen when consumers pay directly for wastewater treatment than when plants depend on government subsidies from general revenues. So tariffs should cover at least full operations and maintenance costs (see Table 7.5). And to encourage conservation, which will allow smaller wastewater treatment plants (with very large cost savings), tariffs should be based on the volume of wastewater produced. Since it rarely makes financial sense to meter wastewater directly (the exception being large industrial enterprises), billing can be done on the volume of water consumed. To reflect uses, such as gardening, that do not end up in the sewer, some cities base wastewater bills on winter water use, when outdoor and other water uses are at a minimum.

Although pricing may not follow theory, experience proves the value of paying attention to the principles. Cheap water gets wasted and wastewater treatment costs are high. Water and wastewater services priced to cover the *full cost* of efficient operation offer the dual virtues of supporting high quality service and minimizing environmental costs.

7.6 CITIES IN DEVELOPING COUNTRIES

The financing of urban water management is a perennial problem in cities all over the world, but particularly in cities in developing countries. The issue of financing water services is becoming increasingly important, as the population of the world is now predominantly and increasingly urban. Particularly in poorer countries, the rate of urbanisation is still rising rapidly. Urban agglomerations of population and economic activities are emerging of a size previously unknown. 17 of the so-called megacities are located in developing countries (World Resources Institute, 1998–99). It is not surprising to find that the magnitude of the new urban populations is having serious negative consequences for the environment, in general, and for the water resources, in particular. One distressing example of the consequences arising from these huge concentrations of population is the continuing wide occurrence of waterborne disease. The spread of epidemics is undeniably aided by the deficient sanitary infrastructure in many cities.

The growth in the urban population has made cities increasingly important as users and consumers of water. However, despite the increase in the importance of urban water use, institutional structures have not evolved to manage the water resources compromised by urban growth or to ensure the efficient provision of water-related services. On the contrary, in many countries, the existing institutional systems tend to act as obstacles to effective urban water management due to over-centralisation, bureaucratic inefficiencies and lack of sustainable finance. Many institutions seem to passively allow the problems to increase and consequently fail to create spaces for the consideration of solutions.

7.6.1 Financing water services

The financing of water services is not only a question of financing the initial capital investment, but also of generating the funds necessary for the operation and maintenance of the services provided. Moreover, the financial demands continue to grow as urban populations increase, water sources grow more distant, pollution of water bodies worsens and there is increasing occupation of areas subject to flood and other water-related natural hazards. The financial issues are complicated and exacerbated by a widespread lack of awareness of the economic importance of effective and sustainable management of urban water resources. The discussion here will centre on water supply and sanitation services, but the arguments used apply equally to other water services.

7.6.1.1 The failure to appreciate the economic value of water

The failure in many cities and societies to appreciate the real economic value of effective, efficient and sustainable management of the water resource is one of the main contributing factors to the lack of financial sustainability of urban water management services.

The first issue to be addressed in this discussion, therefore, must be the reasons for this failure to appreciate the value of good management or to grasp the scope of the damage caused by the lack of it - the costs of poor management so prevalent in so many cities. There is a marked trend of accepting long-run economic costs of catastrophes as inevitable, whether natural disasters or epidemics, when many times they are the result of poor management. There are also further current costs associated with poor management including endemic water-related disease and the restrictions on water use imposed by pollution. These too are taken as part of the natural order of things.

7.6.1.2 Pricing water

The result of the failure to perceive the economic value of water leads to the adoption of tariff policies for water services that fail to cover costs. The price of water services, water supply and sewerage in particular, must reflect the total cost of provision so as to meet the costs of operations, debt service and capital expansion, including a reasonable return on capital. It is patently clear that water services cannot be provided in the amount and quality required for sustainable urban development without satisfactory tariff levels (World Bank, 1996).

Tariffs should also encourage efficiency in the use of resources; and many would argue, on grounds of equity, that the overall tariff structure should make services affordable to all, including the poor. Reconciliation of all these objectives is rarely achieved, but it need not be impossible.

There is a tendency for tariff rates to have two components: a fixed charge and a volumetric charge related to consumption (see above 7.3.1). The first is often intended to cover the fixed costs of the utility and the second the variable ones. In addition, the volumetric charge usually increases as consumption increases. There is also a common tendency to discriminate among customers, charging higher rates for commercial and industrial customers and not charging public agencies

Rate discrimination is often the result of the decision to provide a cross-subsidy from some groups to others. More recently, rate discrimination has been claimed to promote water conservation. Whatever the argument, tariff discrimination introduces economic distortions in the use of water. Lower tariffs for the poor can discourage utilities from providing service and the failure to charge realistically is one factor contributing to high levels of unaccounted for water.

Problems arise because the tariff levels and structures tend to be arbitrarily determined. They tend neither to reflect the real costs of providing the service or the defined goals either for service coverage or quality. For example, fixed costs may be so high as to negate any cross subsidy or the costs of connection to the system may be beyond the means of many households (Artana, 1997).

The solution to these difficulties is to have one price for water with some form of direct subsidy for the poor. In the poorest countries, the subsidy might take the form of low cost service through public taps while, in the less poor, direct subsidies for bill payment seem to be more appropriate.

7.6.1.3 The ability to pay

Recently, following the adoption of policies by the World Bank to encourage private participation \in the provision of water supply and sanitation, there has been strong negative political reaction. Most striking was the revolt in Cochabamba, Bolivia, against the transfer the operation of the local water supply system to a private international consortium. The main argument used for the rejection of "privatisation" is that the price of water will rise (see Figure 5). This is far from certain, even if the tariff for water supplied through the system does increase.

The argument that the poor cannot pay for adequate service is not substantiated. In many cities the water supply system only extends to part of the urban area, for example, in Cochabamba to less that one-third of the population, in Buenos Aires it did only to 60 percent prior to privatisation, in Dakar, Senegal, only to the old core of the city. This always means that it is the poor that have no service and consequently pay anything from at least 10 to over 100 times the tariffs paid by households fortunate enough to be connected to the system (Lee, 1992).

The argument that the poor can pay is strengthened by the fact that the poor consume less water, only a fraction of that consumed by higher income households, less than a quarter in Santiago, Chile, for example. The consumption of water has been clearly shown to be income-elastic (Hirshleifer, 1960; Lee, 1969). This is largely because water consumption and use is dependent on having incomes that permit the enjoyment of facilities such as bathrooms, domestic appliances, swimming pools and so on.

7.6.2 Factors influencing the viability of institutions providing water services

At the United Nations Conference on Environment and Development in Río de Janeiro in 1992, the governments adopted Agenda 21. In Chapter 18 there is a section on water and sustainable urban development. In the discussion emphasis is put on the urgent need for better water management for urban use so as to contribute to the reduction of poverty, to the improvement of health and to the control of pollution.

Statements made in documents originating in major international conferences have to be taken with a degree of scepticism and the poor state of urban water management cannot be blamed on the failure of governments to meet the commitments made at international meetings. The issue is much more complex. There has been a general, almost universal, lack of serious attempts to establish adequate policies. In turn, institutional arrangements, despite the importance of a clear and effective distribution of responsibilities, are not the key to better urban water management. It has been observed, in another context, that the differences in performance of water institutions does not just depend on institutional structure – although it is better to have open, transparent and representative institutions – but on the *dynamism of management.* Dynamic management can only come with adequate financing. This is the crucial step. Adequate finance is the best indicator of a real commitment to resolve urban water management problems both by government and by users.

Of course, there are elements in institutional structure that can contribute to better management (see the section 7.3.4 above). The first, that each individual management task be assigned clearly to a particular agent, is very obvious, but nevertheless not always achieved.. For example, when the water supply and sanitation sector was restructured in Chile in the late 1980s, responsibility for storm drainage was taken away from the sector and not assigned to any other agent. The result was more than ten years without a co-ordinated policy or administration of storm drainage and little or no investment in storm drains.

Second, the evidence of many studies indicates that the clear delineation of responsibilities between institutions responsible for the management of the water resource and institutions responsible for the management of water services can make a major contribution to good management and adequate finance. Transferring the responsibility for the management of water services to the private sector can be a valuable instrument for successfully making and maintaining this distinction. Privatisation is not a requirement, but it is more difficult to maintain a clear distinction between the two tasks if the regulator and the operator are in the same institution. So separation is important through decentralisation to autonomous operators, whether within public or private sectors.

There is also evidence that greater user participation does produce better water management. In consequence, irrespective of specifically how, means must be found for strong and committed public participation in urban water management. This probably can only be achieved if an important role is given to municipalities. For too long, the municipality has been pushed aside and all services provided from central governments. Fortunately, there is now a reconsideration of the municipal role, as a way of reducing the burdens on the central governments of the provision of many public services, particularly, health and education. Responsibilities for water services have also been transferred to the municipality, but this may not always be the optimum solution. However, the municipalities can, and perhaps should, take on the role of acting as the focal points for the discussion of priorities and the establishment of overall policy frameworks for urban water management.

No form of institutional innovation can *per se* resolve the problems of urban water management unless the users, as well as the polluters, pay adequately and justly for the services they enjoy from the water resource. This is a further argument for extending the role of the private sector in the provision of water services. In sum, the provision of water services in urban areas must be, as far as possible, *self-financing*. If it is not, then it can be expected that the water resources in the vicinity of the large cities of the developing countries will continue to be degraded as they have been in the past.

7.7 SOME RECOMMENDATIONS

Economic and financial costs of water services (drinking water supply, sanitation, eventually water resource protection, and the alleviation of water related disasters) invites realism especially in regions where financial resources are limited.

Realism consists first in not dreaming about free water for all or to having nightmares about the impossibility of reaching solutions for water problems in the poorer countries.

Realism also consists in giving due consideration to the principles listed through this chapter:

- To take into account all costs (global cost, total cost) in particular if current tariffs do not allow cover them all
- To adopt transparent and openly accessible financial practices. This facilitates debate, choice and augments confidence

- To permit users of water services toparticipate in the decision making process, especially at the local level for domestic, industrial and others water uses
- To transfer to water prices the costs imposed by these users ("water user pays")
- Take into consideration the solidarity that must be built between the haves and have-nots in order that economic and social development comes together with water availability. About that issue one can refer to the recommendations of the Social Charter for Water discussed at the World Water Forum in March 2000 in The Hague
  - Finally, take care that the options related to water development and management be considered closely with respect to urban planning and vice versa, in technical, institutional and, of course, economic and financial areas.

7.8 REFERENCES

Artana, D., F. Navajas and S. Urbiztondo (1997) La regulación económica en las concessiones de agua potable y desaguas cloacales en Buenos Aires y Corrientes, Argentina, Interamerican Development Bank, Washington D.C.

Hirshleifer, J., J. C. De Haven and J. W. Milliman (1960) *Water Supply, Economics, Technology and Policy* University of Chicago press, 1969.

Lee, T. and A. Jouravlev (1992) Self-financing water supply and sanitation services, *CEPAL Review*, N°48, December, 1992.

Lee, T. (1969) Residential *Water Demand and Economic Development* University of Toronto Press, Toronto.

World Resources Institute United Nations Environment Programme, United Nations Development Programme and the World Bank, *World Resources(1998-99)* New York, Oxford University Press.

World Bank (1996) Water and Sanitation Division of the Transportation, Water and Urban Development Department, Water and Wastewater Utilities, Indicators 2nd edition, World Bank, Washington D.C.

8

Social, institutional and regulatory issues

Jan Lundqvist and Anthony Turton, with contributions from Sunita Narain

8.1 PURPOSE OF THIS CHAPTER

8.1.1 Urban dynamics is a test case of sustainability

Cities and urban areas in general are rapidly becoming the most significant arena of human activity, hope, environmental decay and frustration. A nexus between strong social forces, technological and financial constraints and environmental issues, create significant challenges for institutional and regulatory arrangements. Managing the urban dynamics is a test case for the possibilities to achieve sustainability. Water plays a crucial role in this regard, both in terms of its close relation to social indicators but increasingly also in terms of its basic importance for economic and environmental stability.

Edited by Čedo Maksimović and José Alberto Tejada-Guibert. ISBN: 1 900222 76 0

The purpose of this chapter is to discuss a strategy that could contribute to an improvement in living conditions and, at the same time, reverse the ongoing degradation of water systems and, generally, the environment. A review of the prevailing system of management is made to elucidate the need for a change of the institutional arrangements concerning water management, which could bring the necessary changes about. The centralised system of management, run by a monopolistic operator, and with "big pipes" supposed to be extended to cover the entire city, was initiated during the 19th century and became widespread during the 20th century. It was naturally tailored for the characteristic features of the cities of those days (Newman 2000). Today, the urban setting and the urban dynamics in most parts of the world look significantly different.

An assertion of this chapter is that the hardship and tremendous challenges that haunt the inhabitants of urban areas, especially in the South, will not be effectively tackled unless civil society initiatives are takes seriously and groups of important stakeholders are actively involved in management. This is not a new argument. Involvement of stakeholders and, generally, the water users, sometimes referred as the "beneficiaries", has been propagated for quite some time. Although an involvement of various stakeholders in management tasks is of vital importance, the conventional understanding of this assumption falls short of addressing two of the perhaps most basic challenges.

8.1.2 Civil society initiatives need serious consideration: not harassment

One challenge refers to the need for recognition and support to civil society initiatives from the Government system. A successful involvement of civil society in management presumes institutional arrangements through which the users and various interests groups are active and which the community accepts. But in the absence of a political support, it becomes very difficult for these institutions to be effective, apart from work on a small scale and in a piecemeal manner. As will be discussed below, water problems cannot be effectively addressed if rigid rules and regulations prevail. Monopoly control over water and land policies in urban areas, for instance, are often used capriciously to deter entrepreneurs from the community to provide water and sanitation services, even if such services are not organised by the formal water agency or its concessionaire. This is, for instance, reported from a ten-country study in Africa (Collignon and Vézina 2000). Regrettably, public sector officials tend to harass people who are performing essential services to the community (*ibid.*). Agarwal (2000) charges government officials for corrupt practices, churlishness and incompetence, especially against the poor. Tropp (1998) found that a typical feature in the behaviour of the state representatives is to keep NGOs and CBOs (community based organisations) at arms' distance.

At the same time, it is inconceivable to imagine well functioning management arrangements for large and complex urban systems without a formal regulatory system. Urban water management is a particularly complex task in a river basin context where the upstream and downstream relationships have to be taken into consideration. In this regard, it is not less Government that is needed but better governance, where the roles and rules of public sector and civil society organisations should be clarified and adhered to (Lundqvist 2000). Civil society organisations and initiatives are also of quite varied quality and scope, which is another important reason for a multiple set of formal and informal institutional arrangements.

Water services in an urban context include a wide array of tasks. Water provision, for instance, includes arrangements for withdrawal, conveyance, storage, treatment and distribution to various sectors, groups and individual users. Bulk supply obviously requires other management skills as compared to distribution to individual households. Similarly, the management of wastewater and effluents refer to many inter-linked tasks. A proper policy and management strategy in this regard should include efforts to reduce waste generation, i.e. reduction of pollution load generated through production processes and consumption in society. It must also include the actual treatment and disposal of the waste, sludge, etc. In order to effectively deal with the various tasks associated with water services, it is imperative to include the various actors and stakeholders in society who are playing an active role in the respects just indicated.

With this integrated view of water resources management, attention is naturally also directed to the other basic challenge in urban water management. It refers to behaviour among water users and the driving forces in this connection. An excessive and careless use of water among segments of the population is unfortunately a recurrent phenomenon. Misuse of water in one part of the society, will deprive people in other areas and other social strata the opportunity to get easy access to water and to water that is safe. Both water users and officials occupying various positions in management institutions contribute to the water crisis. To put it simply, the problem with water is not water, but society. In this regard, the problems will not be effectively tackled by focussing on hydrological or technical aspects alone.

It is absolutely vital that the connection between human and societal development and its consequences on life support systems is recognised and that people from all walks of life learn to deal with it. Management of water must stimulate and facilitate human behaviour and social resources in a direction where the dynamics of water and other natural systems is duly considered (FAO 2000). A combination of educational efforts, incentives and sanction to stimulate the most worthwhile use and stewardship of water in society is crucial. If management is solely associated with technical and formal organisational issues, it is unlikely that the human potential to adjust to changing social and environmental conditions and to develop solutions that are sensitive to new circumstances is given adequate attention.

8.1.3 Three types of management systems

The conventional heavy reliance on a combination of technical solutions and a supposedly benevolent state that arranged for a heavily subsidised supply of water is no longer a feasible strategy. Yet, the expectation that the Government shall provide water remains a common attitude, despite the fact that Governments at the same time face dwindling support in many countries. It is mind boggling that, while a considerable proportion of people tend to take water services for granted and expect that they shall continue to be heavily subsidised, i.e. through Governments –, the more difficult and the more expensive it is to secure such services, especially in big cities. Expectations that the public sector or some other benevolent external agent should make sure that growing demands are met have been stimulated through promises by politicians who tend to see members of community as vote banks, but also by other groups. To the extent that these promises are not fulfilled, which is common, the frustration in society is likely to increase and the need for alternative approaches increases.

Water use and the role it plays for social and economic development and for environmental change, may be seen as the outcome of three types of management systems (Figure 8.1). The predominant system has been one where the public sector plays a major role, whereas currently the importance of the two other systems is growing. In this chapter, the focus is on the growing importance of the civil society initiatives in management and the institutions and organisations that may grow out of these. The notions of institution and institutional arrangements refer to the "rules of the game" and the structural nature of institutions (Kemper 1996). 'Organisation' in the everyday language is often used synonymously with institution, but in this text it is used to specifically refer to a concrete social entity.

As shown in Figure 8.1, the management tasks carried out within the civil society presume a facilitating formal institutional and regulatory framework and thus a significant role for the Government system. Today, this link is weak or characterised by a patronising attitude of the government. It is also implied from the figure that water use and the behaviour of people are related to what kind of management system(s) that operate. This, in turn, will have implications for the social and economic benefits that are derived from resource utilisation as well as water resources and environmental parameters.

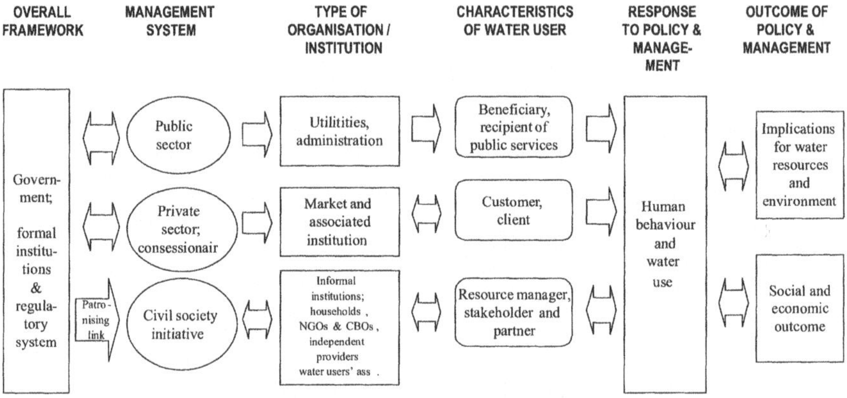

Figure 8.1 Three types of management system and the context in which they operate. Arrows show hypothetical direction of relationships and response

Poor coverage of water and sanitation services is an indication that the prevailing policy and management system does not match the tremendous challenges, which exist particularly in water short regions. Figures on coverage are uncertain and vary between source of information, but a recurrent observation is that a considerable part of the urban population does not have access to water and sanitation services through public sector arrangements (Table 8.1).

Although the number of people served has increased during the period 1990 - 2000, it is also shown that the number of people who did not have access to water services increased in Africa, Asia and Latin America and the Caribbean during the last decade of the previous century (Table 8.1). According to UN statistical information, the sanitation services improved in Asia and Latin America and the Caribbean while the number of people without access to sanitation services increased in Africa. Figures do not permit interpretation of which agency is instrumental in providing the services nor the quality of the service rendered. Information in other reports (see, for instance, Gleick 1998; Collignon and Vézina 2000), suggest that the figures presented in Table 8.1 are inflated, or simply that they indicate total coverage, i.e. arrangements through public sector, private sector entrepreneurs and civil society organisations.

Table 8.1 Urban water supply and sanitation coverage by region, 1990 - 2000
(Population in millions)

REGION / SECTOR	1990				2000			
	Population	Number Served	Percentage Covered	Number Not served	Population	Number Served	Percentage covered	Number Not served
Africa WS*	197	168	85.0	30	297	253	85.1	44
Sanitation	197	164	83.1	33	297	249	83.8	48
Asia / WS*	1029	971	94.3	59	1352	1254	92.7	98
Sanitation	1029	678	65.9	351	1352	1047	77.5	305
Europe/WS*	—	—	—	—	545	542	9.5	2
Sanitation	—	—	—	—	545	537	98.5	8
LA and Caribbean/ WS*	313	283	90.4	30	391	353	90.3	38
Sanitation	313	247	78.9	66	391	335	85.7	56
North America WS*	—	—	—	—	239	239	100.0	0
Sanitation	—	—	—	—	239	239	100.0	0
Oceania WS*	—	—	—	—	21	21	98.5	0
Sanitation	—	—	—	—	21	20	96.0	1
World /WS*	—	—	—	—	2845	2662	93.6	183
Sanitation	—	—	—	—	2845	2427	85.3	418

* WS = Water Supply, LA = Latin America

A dash (—) indicates a lack of comparable data; Totals may differ from the sum of other columns because of rounding. **Source:** The table is a condensed version of a table that is presented in Neto and Tropp (2000) and based on information in UN (2000).

The figures do, however, probably give a reasonable comparable picture of what has happened during the 1990s and how the level of service varies across the major regions of the world. In the absence of sufficient capacity of the State and its affiliated bodies, the only conceivable approach is that civil society organisations, and ultimately the water users themselves, become partners in urban water management. Involvement of NGOs, independent providers and the private sector as a complement to public sector management is already a noticeable sign of change in a large number of countries.

Lack of sufficient capacity in the public sector is only one argument for an active and broader involvement of organisations from civil society and the water users themselves in management. Another argument refers to the difficulties of a centralised bureaucratic system, which is bound in its operations by formal and often standardised regulations, to handle diverse and dynamic social systems, notably in the unregulated squatter areas, which constitute a large part of many cities. Flexible and informal institutional arrangements are better suited to handle urgent tasks, which fall outside the purview of the formal management system. Such solutions may also prove to be comparatively cheap as compared to solutions executed through Government agencies.

In addition, it is relevant to mention that a large part of the challenges that arise in water management are generated through activities in civil society. Water pollution, in particular, is related to the use of water and other resources and not so much to supply, which is the typical public sector activity. While shortcomings and irregularities in the public sector could amplify the challenges, serious water problems emanate from production processes, transport systems, and human behaviour at large, i.e. activities in civil society. Certain cultural practices and the evolution of consumer society are together with the large and growing number of people per unit of water important driving forces behind the water crisis. People are consequently contributing to the evolution of the challenges and they must be part of the solution.

The inclusion of civil society organisations and initiatives in policy formulation and management tasks will facilitate an improved awareness of the challenges that have to be faced. Such an approach is likely to strengthen skills, which are necessary to solve essential problems. The opposite is even more plausible. If various groups of stakeholders are not properly involved in management, their effort to improve overall water use is likely to be insignificant and even counterproductive to what is required.

8.2 THREE PRINCIPAL TASKS

The institutional and regulatory framework must be arranged so that three major tasks can be carried out. The primary task is to bring water *to* people and to various activities of society. Arranging for adequate supplies of water to growing urban centres is a substantial technical task and it entails considerable financial expenditure. The level of investments in the water sector and related services will have to increase substantially, at least twice the current level, if the goal of water security, as defined by the World Water Commission, is to be achieved by 2025 (WWC 2000). Water security refers to a much wider task than urban supply, but this component makes up a large part. Commitments made by the Government of Mexico illustrate that the magnitude of the task can be staggering. To meet the demand in the three most important cities in the country will cost more than the total national investments in the entire public sector including education, health and social security, agriculture, livestock and rural development (Tortajada 1998). With the easily accessible sources already exploited, an additional supply of water, conveyed over long distances and often at another elevation as compared to the city, is progressively difficult to arrange in technical terms and in terms of energy requirements with escalating costs as a consequence.

Another principal task is to arrange for the safe and proper treatment and disposal of water *after it has been used.* With an intensified use of water, particularly in a dense urban setting, the volume and the quality of wastewater and effluents create a serious threat the urban sustainability. Many of the polluting substances that affect the population and the environment through cyclical and cascading effects in the basin, have a time scale for clean-up of up to 100 years (Peters and Meybeck 2000). The tendency in most countries to concentrate on water provision with scant attention to what happens to water after it has been used will therefore result in the build-up of a significant environmental debt. Simple inter-generation equity considerations do illuminate that this is a matter that needs urgent attention.

Already today, the disposal of untreated sewage and effluents in small rivers or water bodies easily convert a fresh water source into an unpleasant and unhealthy sink, which should not be used for any purpose. Even if it is not used in the conventional sense, the degraded water will affect the functioning of aquatic and man-made ecosystems and, hence, also society (WRI 2000). Apart from the impact on human health and the functioning of ecosystems, which is bad enough, it is important to note that many industrial processes require water of high quality. Pollution, which to a large extent is generated through the growth of the industrial and related urban sectors is consequently contributing to the development of a situation where industry is having problems to secure water for its continuous operations (Blomqvist 1996; Hertsgaard 1998).

There is a third task that has been given surprisingly little attention in conventional water management discussions. It refers to the actual use of water in society and what kind of institutional arrangements, incentives and sanctions that could best facilitate the most worthwhile distribution and use of an increasingly scarce and vulnerable resource (Kemper 2000). Even if an augmentation of supply to urban centres is technically and financially possible, it increasingly conflicts with growing water demands and water use in areas from where water it is to be abstracted. Management of the demand and use is therefore not only referring to the urban system as such but to a growing support area since local surface and ground water sources are invariably insufficient to cater for the required increase in supply in the urban centres. A judicious use and exploration of re-use options are examples of urban water management that would reduce the reliance and the consequences for support areas outside the city. Overdraft of groundwater in the city is another issue that warrants attention (Foster *et al.* 1998) and for which there is no technical solution.

Linkages between the three tasks are important. The use and re-use of water in the city determines the amount of water that needs to be supplied, and thus withdrawn from somewhere, including the amount of groundwater that is lifted. Use of the resource will determine what happens at the other end of the system in terms of the amount and characteristics of the wastewater that is generated. The distribution and use of water refer to social, institutional and political judgements.

The three tasks are schematically portrayed in Figure 8.2. As shown in the figure, they require different types of skills and approaches. To arrange for the supply of water to the various users basically requires technical competence and, of course, adequate financial resources to build the structures required. Similarly, the challenge to deal with pollution and environmental degradation is to a large extent a financially and technically demanding task. In addition, it is necessary with a proper understanding of biological and chemical aspects.

Management of water distribution and use requires a very different type of competence as compared to water supply and treatment and disposal of wastewater and effluents. Sanctions against inefficient use and pollution become important together with efforts to meet the needs in the urban area as a whole. In many cases, water services are mainly directed to one part of society while other segments of the population do not enjoy the opportunity of easy access to water nor to sanitation services. As usual, it is the poor and the environment that suffer. The highly variable access to water and sanitation services between various social segments and settlements and the degradation of water and environment quality in the urban area as a whole are two crucial issues for urban sustainability.

As indicated in Figure 8.2, this part of urban water management presumes interdisciplinary competence, institutional and social adaptive capacity and a political will (Olsson and Lundqvist 2000). The combination of a supportive political system and an active involvement of stakeholder groups in management apparently represent a much-needed system for management. There is no blueprint for such a system, but a transparent system with checks and balances are necessary instead of the previous non-transparent, hegemonic and centralised decision-making system (Turton and Meissner 2000).

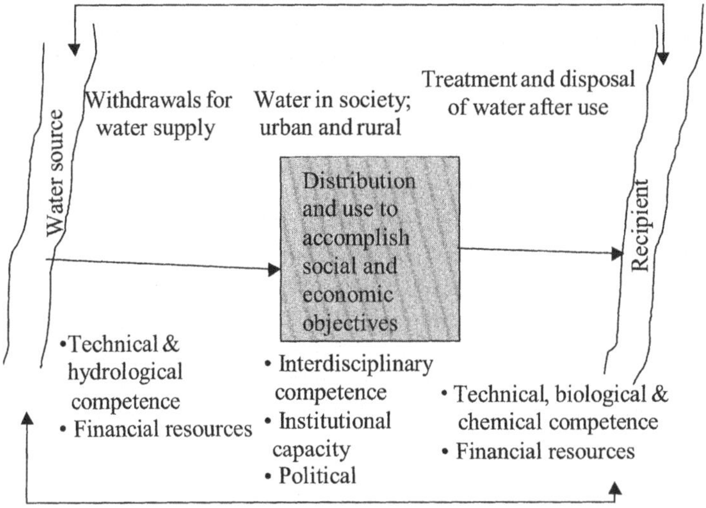

Figure 8.2 Three principal tasks in water management and the competence and resources required for their implementation

A pre-occupation with the first task and a relative neglect to take care of wastewater and effluents is understandable in the face of urgent needs and growing demands for additional water supply. In addition, the handling of wastewater is quite costly. On average, the cost to supply a unit of water is much lower as compared to the cost to treat a unit of wastewater (Serageldin 1994). The omission to manage the actual use of water and the lack of appropriate and effective institutions that are able of fulfilling these kind of management functions requires another type of explanation. One aspect in this regard is the narrow range of perspectives and skills that have conventionally been associated with water management. As elaborated below, chapter 8.3, large-scale supply-side engineering efforts have dominated the perception on water management.

Today, it is widely recognised that water management is an undertaking, which must include other aspects than the engineering ones. Needless to say, these remain of basic importance, but the social, institutional and regulatory dimensions are assuming significant proportions. Some of the pertinent questions are: *who* should provide the various water services, as indicated above, and *on what terms* should they be delivered? The answer has traditionally been: the public sector should deliver water services and they should be subsidised or free. Surely, there is a need for a new answer.

Currently, a new set of questions needs to be addressed. How can services be arranged that are *affordable and acceptable*? To what extent do organisations of civil society and the users themselves, have the *required competence* to take care of water management? What is the *coverage* of civil society management, in terms of area and population served? And how can civil society initiatives be properly utilised?

There are, of course, conflicts within and between groups in civil society and there is no intrinsic management culture in civil society that makes it automatically fit to take over management tasks that formal management arrangements are handling poorly. Many civil society organisations are beset with imperfections like, for instance, lack of transparency and foresight, poor attention to succession of leadership and limited technical skills. It is therefore of interest to discuss how the work of civil society organisations can be co-ordinated and how the relation between civil society organisations, professionals and the public sector might develop.

8.3 THE NEW HYDRO-SOCIAL CONTRACT

To better understand the role of the public sector and civil society's involvement in water management, it is useful with a short introduction to the historical evolution of water management principles. The prevailing order in the current water management system developed at the end of the last century, or in some cases even before that, and that relies on an unwritten contract between the public and the Government (Turton and Olsson 1999).

Initially, the members of a community had to arrange for the water services on their own. When the individual is no longer capable of mobilizing sufficient water for its own personal survival and development, government gradually takes on and executes this responsibility. Historically, this process has been evolving in most societies for a long time, but it is during the 20th century that the Government gradually becomes the main agent and executor of water services (Turton and Meissner 2000). This hydro-social contract acts as the basis for institutional development, and also determines what the public deems to be fair and legitimate practice. Indications are that there are at least two, and possibly more, forms of hydro-social contract.

Historically, the first variant is known as the Hobbesian form of hydro-social contact (*ibid.*), which mirrors the thinking of Thomas Hobbes. One of his key ideas concerns the need for the establishment of a strong central power and an almighty ruler – a Leviathan – to whom the citizens hand over the authority over their affairs. In return, the ruler or the State is supposed to be able to assure them prosperity and to avoid chaos. Under this variant, "the hydraulic mission of society" (Reisner 1993; Swyngedouw 1998) is established which gives rise to the development of a distinct group of professionals, who is charged with the noble responsibility of mobilising water in order that society may grow and prosper. Thus the era of "heroic engineering" (Platt 1999) is born, at which time the hydraulic engineers perform a series of "hydraulic miracles" (Reisner 1993) or engages in the "production of nature" (Swyngedouw 1998). The politically and socially sanctioned strategy is to get more water from increasingly distant sources. Under these conditions, water provision is seen primarily as a technical challenge. Water is treated as a commodity rather than as a finite and vulnerable natural resource, which is part of the hydrological cycle and which fulfils basic functions not only in society but also in the landscape. To the public, water becomes something that comes from a tap, is expected to be clean and healthy, and is expected to be delivered at no cost or at a subsidised rate.

From a hydropolitical perspective, this also becomes the era of what Homer-Dixon (1994) has described as "resource capture". Cities, for example, can only continue their growth and expansion by capturing water and other resources, which are available in areas further and further away from the city itself. Highly sophisticated, costly long-distance transfers of water are found in many countries, including the US, China, South Africa, India, etc. Capturing resources through other means than legitimate administrative arrangements and agreements have implications for social and political relations in society at large. Increasing water provision to expanding cities is not possible without formal control over the sources of water and/or the consent of the communities who reside in areas from where the resources are to be abstracted.

With growing need and demand for water across society, the Hobbesian form of management system is increasingly difficult to pursue. Gradually, a second type of hydro-social contract evolves which may be referred to as the Lockean form. It refers to the idea of John Locke, who argued for the sharing of power through a "social contract" between the Government and the public (Turton and Meissner, 2000). The principal of mutuality, i.e. that the validity of the contract hinges on that each party fulfils his/her obligations is important in this connection. This new order is the outcome of a reaction to the centralised and technically dominated strategy of water management. One of the first clues of the evolution of a new hydro-social contract lies in the emergence of interest groups in society who start to question the prevailing hydraulic mission as the dominant principle behind water management. A change in the perception of water management is stimulated by a lingering and often growing inability of the Government to provide the needed and promised services. Events in terms of, for instance, droughts or serious environmental threats serve to mobilise public opinion for new arrangements. Central to the emergence of a consciousness for the need of a review of the prevailing water management system is the cost, in social, financial or ecological terms, of constantly mobilising more water from ever-distant sources without any involvement of communities in the policy decisions and without satisfactorily solving major socio-economic problems.

A new relationship between government and the public is forged, based on the notion of checks and balances to hegemonic decision-making. Government recedes from its conventional role as the sole and dominant agent for water management. NGOs and CBOs are becoming more important and privatization of utilities becomes an accepted order (cf. Figure 8.1). This does, however, not mean that the Government should withdraw from management responsibilities in general. With new actors assuming management tasks, the regulatory arrangements become crucial. Without political support and consent, it is most difficult and often impossible for operators from civil society to carry out many of the basic water management tasks. Investments in equipment and structures for water provision in non-regulated areas, for instance, are risky since the Government or municipality authority may formally take over such structures, often without compensation (Collignon and Vézina 2000).

Parallel with a shift in management arrangements, the perception on water changes and widens. Deliberations in connection with the Second World Water Forum in The Hague, in March 2000, illustrated that the perceptions on water currently include a recognition of both the scarcity value of water as well as a strong plea that water should be seen as a fundamental Human Right (WSSCC 2000; Gleick, 1999). Under the Lockean form of hydro-social contract, man's part in nature is redefined, with a strong element of public opinion suggesting that Man should no longer view her- or himself as being above nature, but should instead seek to live within the parameters that nature has created. A manifestation of this is the emergence of a desire to rehabilitate degraded rivers, and reintegrate them into the urban landscape. It also implies a growing awareness that natural resources have to be used in a manner, which takes a rapid growth of population into account and the need to alleviate poverty.

It is interesting to note that the World Water Vision that was announced at The Hague in March 2000, has as one of its central themes, the relationship between Government, Civil Society and the Private Sector (WWC 2000, 7). This tri-partite alliance, which seeks "more community participation [in order to] ... provide a sense of ownership and empowerment to local stakeholders" (Cosgrove and Rijsberman 2000, 64), is nothing less than a manifestation of the Lockean form of hydro-social contract. The World Water Vision is supported by a "social charter for water" that seeks to promote a new water policy for the 21st Century that is "designed by the policy-makers and their experts, in partnership with the citizens" (Agence de l'Eau 2000, 22). Similarly, the Collaborative

Council for Water Supply and Sanitation argues that "a shift in emphasis to local effort, local initiative and local leadership" (WSSC 2000) is important.

Regulation and support to activities carried out by civil society organisations is still a significant task for the formal governance system. It is not conceivable to have a well functioning management in the absence of a legal and regulatory framework. For this reason, there is an obvious need for a formal regulatory system, which, at the same time provides incentives and sanctions in agreement with social, economic and environmental objectives. In short, water must be "everybody's business" (Agarwal and Narain 1999; WWC 2000; Cosgrove and Rijsberman 2000).

8.4 HISTORICAL EXPERIENCE AND A NEW MILLENNIUM

It is, indeed, ironic that as we go into a new millennium, with all the sophisticated technologies and a tremendous potential in terms of accumulated knowledge and human experience, the key challenge will be the good management of the world's most basic resources – its water and land resources and air. The application of hardware and software in efforts to promote development has increased wealth generation to the detriment of the environmental situation. It also appears as if the combination of urban life together with technological progress and growing wealth have contributed to alienate people from Nature and to erode a sound and proper understanding of the non-negotiable character of the hydrological cycle and other natural resource phenomena.

Economic growth is, of course, very much welcome in areas where mass poverty prevails. As usual, the coin has a reverse side. During the 1970s and the 1980s, Southeast and East Asia grew at a rate that was unprecedented in human history and in a cross-country comparison. Today, this region is also the most polluted on Earth. Literally, most cities in the South are gasping for air that can be breathed and water that may be safely used.

Studies from different parts of the world illustrate that economic growth generates a pollution load, which is often growing relatively faster than the economy. In terms of GDP, the Indian economy doubled over the period 1975–1995 and during the same period the industrial pollution load went up by four times and the vehicular pollution load by eight times (Narain 2000). Partly as a result of a rapid economic growth and without the execution of an effective environmental policy, many rivers and small streams have been seriously polluted, virtually transforming them into toxic drains (Agarwal, Narain and Sen 1999).

Pollution load is not only emanating from production processes. With growing wealth, it is increasingly a consequence of consumption pattern and household characteristics. According to calculations done at the Asian Development Bank, the daily per capita generation of solid waste is between 0.4 to 0.7 kg in urban centres of low-income cities such as Karachi and Calcutta. The cities in higher income countries, e.g. Seoul, produce almost six times as much on a per capita basis (Shah 2000). By comparison, the amount of waste generated in Japan as a whole, i.e. not only the urban refuse, by the beginning of 1990s, is comparatively modest with about 1.12 kg per day, capita (Matsui 1996). To what extent the urban component differs from the national average in Japan, is not known. The challenge related to a growing amount of waste and polluting substances is a global experience. For OECD countries it is reported that the "... delinking of effluence from affluence generally remains elusive". In spite of 30 years of environmental and waste policy efforts in OECD countries, figures illustrate that waste production continues to increase in direct proportion to economic growth (OECD 2000).

8.4.1 Repeating historical experience or alternative path of development?

Expectations for improvements in standard of living and attempts to alleviate mass poverty are thus likely to increase pressure on water resources and natural ecosystems and generate huge amounts of waste and pollution, particularly in the short run. Most of these challenges will be concentrated to urban areas. For the very majority of the population, there is no acceptable alternative to poverty reduction and improvements in standard of living. The awareness of the importance of safe and healthy living conditions is presumably also on the increase. Improvements in socio-economic and other living conditions and safeguarding the environment have tended to be dealt with separately and efforts to improve the material standard of living have been given priority. As elaborated in a growing literature, the two objectives must be combined (see, for instance, Berkes and Folke 1998).

Historically, it has been observed that the connection between economic growth and environmental degradation is not necessarily reflecting the close connection between 'affluence and effluence' as indicated above. Environmental degradation tends to continue up to a certain level of economic development after which improvements have been recorded (Arrow *et al.* 1995). This historical relationship, which is only an empirical observation, is often portrayed as the inverted U-curve (Figure 8.3). It is, however, sometimes put forward as an argument or as an excuse to give priority to economic growth at an early stage of economic development and to postpone the development and implementation of an effective environmental policy. It has also been used as an excuse for not doing what could be done without major investments. Considering the acute and pressing demands from large segments of the population for improvements in the standard of living, it is not surprising that economic growth is given high priority in the South. Postponing an effective environmental policy is also convenient since the hazards for the environment from economic growth are generally less tangible as compared to the material components of development. In addition, the problems typically appear some time after the desired goods have been produced.

With regard to the historical experience, the environmental awareness and technologies, which are effective in reducing pollution load and its environmental impact, have developed comparatively late. Cities in Europe, US and Japan became quite polluted before concerted actions were taken to reverse the trend. In some cases, the efforts to improve the situation are quite recent. In Brussels, the capital of the European Union, it is only during recent years that systematic efforts have been made to build effluent treatment plants.

But 21st century problems should not be approached by using 20th century experience. For countries that are currently in an early stage of industrial development, the real challenge is to stimulate a development, which does not repeat the historical experience encapsulated in the inverted U-curve. It has been repeatedly warned that a "business-as-usual" approach is likely to result in "nasty" consequences (WWC 2000). Averting a looming hydrocide (Lundqvist 1998), i.e. a wilful neglect of significant degradation of water quality, is a vital challenge in many countries of the world, particularly in urban areas. Water policy for urban development, as well as for the country as a whole, must be a central and integrated component of an effective environmental policy. It must be linked to realistic policies for industrial and other development aspirations. A postponement of an environmental policy and its implementation in the hope that the economy will become strong enough to allow for a larger allocation of funds for that purpose is fallacious as illustrated by the recent economic recession in the Asian region. Substantial reductions in Government budgets have led to sizeable cuts in the environmental protection measures. The wastewater sector in Thailand, for instance, experienced a reduction in its budget by about 38% between 1997-1999 (Shah 2000).

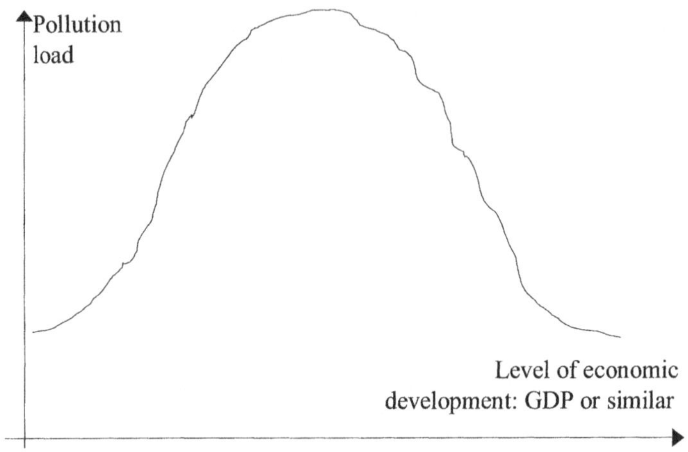

Figure 8.3 A schematic curve of a hypothetical relationship between pollution load and level of economic development

In efforts to create a better future for people in the South, a replication of what has been done in other parts of the world is consequently not an effective nor a realistic option. Considering the magnitude of the problem and rapid pace of urban degradation, the time frame must be short and may not be compared to the period during which the environmental policy evolved in the West. Identification and implementation of a viable development and environmental policy under these conditions is a tough task. A broader and more active involvement of various stakeholders in society is often suggested as a major pre-condition for improved water resources management. Newman (2000) argues that the new urban water management, in rich and poor cities, should be seen as a process where the total water cycle in a catchment perspective is recognised and where localised community processes are fully recognised and integrated in the management strategy. Attention to water users and to various stakeholders is a common denominator in the contemporary debate about strategies to tackle the water crisis (WWC 2000; WSSCC 2000) .

8.4.2 Walking on two legs: the experience in the West

It was in the post-World War II period that the world began to see what was then an unprecedented economic boom in Europe, Japan and North America. By the 1950s itself, cities from Tokyo and London to Los Angeles were choking under pollution. Stockholm, for example, was seriously affected to the extent that the waters surrounding the city were considered unfit for most uses. Today, Stockholm and the downstream archipelago is presented as a success story of environmental management. Reduction of pollution load and treatment of sewage and effluents have contributed to convert the state of the recipients from an unhealthy water pool into a resource where people may have a bathe, angle and even drink water directly from the stream in the centre of the city. It is, however, noted that the sediments in the area downstream of Stockholm contain huge deposits of heavy metals that have accumulated over centuries (Jonsson 2000).

Western society responded to the water and environmental threat with a combination of legal and regulatory measures, economic incentives and sanctions and an increasing investment in pollution control. In the 1970s, the implementation of this policy resulted in significant investments. In Japan, for instance, both public and private sector investments increased from a very low level to become a significant part of overall investments and GDP. During a couple of years around 1975, the big enterprises, i.e. not the entire sector, were annually spending about 1,000 billion Yen in pollution control measures. Recent levels of investments are about 20–30% of that level. For central and local Governments, the annual budget allocation for environmental conservation in the mid 1980s was also about 1,000 billion Yen and this has been a level of public investments since the beginning of the 1970s (Figures from Environment Agency, Government of Japan).

In Sweden, the introduction of environmental legislation in combination with a sharp increase in water fees at the end of the 1960s contributed to a significant improvement. Investments in water supply and treatment plants in Stockholm amount to an estimated US$320 (or about 3,000 SEK) per capita, while the investments in distribution networks for water supply and sewerage is roughly ten times higher, or about US$3,200 (30,000 SEK) (personal information Lars-Gunnar Reinius, 2000). To these figures should then also be added figures on investments made in industry in terms of new production technologies and other changes, which may or may not have been motivated by environmental considerations. As a result of these measures, water demand decreased drastically in industry and households, i.e. mainly in urban areas (Figure 8.4) and also pollution load to recipients. New production technologies in combination with treatment of effluents are key ingredients in this transformation.

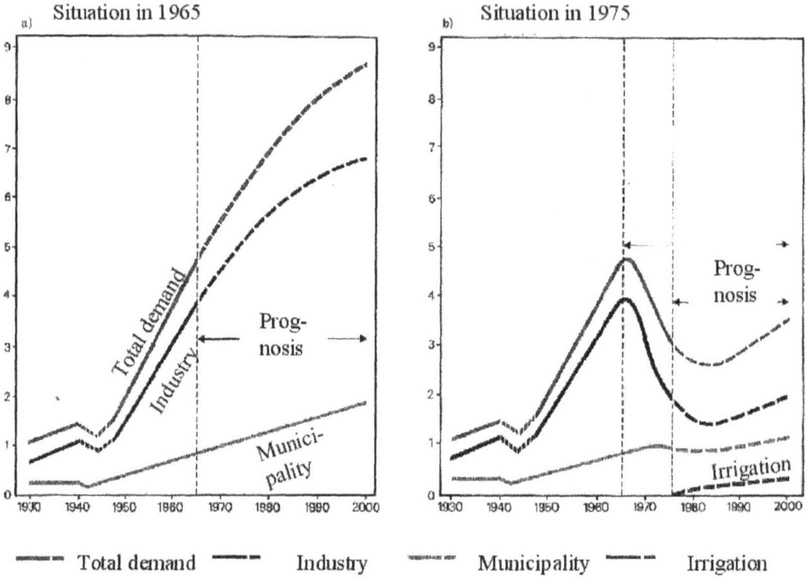

Figure 8.4. Water demand in Sweden from 1930: in 1965 (left) and in 1975 (right). Source: Swedish Ministry of Agriculture in connection with UN Water Conference in Mar del Plata (1977)

The apparent success of environmental policy in the West can, to a large extent, be described as walking on two legs in the sense that simultaneous changes were made through the public sector together with a change within the private business domain. The public sector arranged for end-of-pipe solutions. Many cities have gone through various stages including the addition of a third stage of the municipal treatment plants, i.e. chemical precipitation of phosphorous in addition to the previous mechanical and biological treatment. The private sector responded to incentives and regulations by modernising production technologies, thereby significantly reducing the demand for water input in production and reducing the amount of pollutants that were emitted through the wastewater and effluents to the recipients. Current water policy within the European Union, however, significantly increases the pressure on the countries to improve water quality with heavy investments as a consequence.

It is fair to say that the evolution of an environmental policy in the West, so far, has been a "walk" rather than a rush. Although environmental policy is usually associated with the period from the 1960s and 1970s and onwards, many countries – although certainly not all – had started to implement various types of environmental control measures long before that. In Sweden, for instance, three phases have been identified with an interval of 20 years between each stage. In the 1930s, mechanical treatment was introduced. Twenty years later, biological treatment became common and after another twenty years, the time was ripe to introduce the third stage, i.e. chemical treatment (Cronström 1986). Hence, the evolution of water and the associated environmental policy in Western countries progressed over a fairly long period of time. To a large extent, it has been orchestrated in a context of consensus of the soundness of the new environmental policy across various stakeholders in society. Opposition and reluctance to comply with environmental regulation has, of course, been observed. But, by and large, enforcement and litigation have not been characteristic features of the process.

8.5 URBAN DYNAMICS IN THE SOUTH: SOCIO-ECONOMIC AND ENVIRONMENTAL ASPECTS

For countries in the South, the context is different in many ways as compared to what has happened in the West. With a rapid change characterising the urban setting in the South and with a more turbulent history in terms of the evolution of institutional and political structures, the possibility to use the western experience is limited. One crucial question revolves around the financial and technical requirements to deal with the needs and demands. In the short perspective, countries in the South can rarely mobilise the resources that would be required to copy the western approach. The challenges are, however, also very much related to social forces and the effectiveness of the institutional and regulatory system.

8.5.1 Opportunities and hardship

The pace of urban growth is quite high in the South, both in countries with a booming economy as well as in countries where socio-economic development is slow. Urban centres in Asia and the Pacific grow at a rate of about 4% annually while some of the least developed countries in Africa experience an urban growth rate of about 5% per year (WRI 2000). The growth rate is declining on all continents, but in absolute figures the increase is estimated to be substantially higher during the first decades of this century as compared to the last couple of decades of the 20th century (*ibid.*) To a large extent, the growth may be related to a rural-to-urban migration. The opportunities in rural areas were comparatively better some decades ago in connection with the introduction of the green revolution, water development and supply programmes, community development programmes, etc. In other parts of the world, rural areas were the only option for a majority of the inhabitants since migration to urban areas was restricted and selective, for instance, under the colonial rule. In South Africa, the apartheid system did not allow families to move to urban areas but only "able bodied men" (Schreiner and Naidoo 1999).

Today, the opportunities are less tangible in rural areas whereas growth of the urban economy is significant. According to estimates made some years ago by the World Bank, about 80% of the future economic growth in the developing world will take place in urban areas (World Bank 1990). The Asian Development Bank arrived at the same figure, i.e. 80%, in their calculation of the urban contribution to overall growth in the Asian economy (Shah 2000).

Economic opportunities are, however, limited and uncertain for the great majority of people in urban areas. A large number of urban households do not have a regular income but have to rely on temporary and casual jobs. The possibility to pay monthly fees for water and sanitation services is therefore scant while small daily expenses are usually possible as reported by Collignon and Vézina (2000).

One of the results is a quite diverse character of urban growth and significant variations and segregation in social conditions. The expansion of the urban system is characterised by a growth of secluded areas for residential purposes of a wealthy middle class who want to escape crime and environmental decay in the old parts of the city. The withdrawal of trade, hotels, offices, etc. from parts of the city to suburbs and other locations, means that job opportunities and tax revenues will dwindle. With this kind of expansion of urban systems, the resources required to deal with decay in some parts of the system are not forthcoming. Not all people are able to move and those who are left behind will face rough times and many of them exacerbate the social disorder. Downtown Johannesburg is an example where the decay is noticeable and where even newly erected buildings have been abandoned and where formal and lawful institutions and organisations have great difficulties to operate.

Although the urban economy is booming in many countries, the magnitude of the task to plan for a rapid influx of people is enormous and the capacity is limited. The cost of new schemes is often 2 to 3 times higher in real terms as compared to the schemes now operating (Macy 1999). The public sector does not have, and can hardly be expected to generate, sufficient financial resources that would be required to augment supply at such a rapid pace that is needed in fast growing cities. It is estimated that the level of investments must increase quite substantially "at least twice the current level" if the identified challenges are to be met (WWC 2000). With tightening budget situations in most countries and with demands for increased and improved services in various parts of society, where the public sector has traditionally been involved, the possibilities for a larger contribution from Government sources are slim. The demographic situation where a large fraction of the population are young and with low or irregular income for most of the community, public revenue in terms of taxes is small in relation to the needs (Drangert 2000).

8.5.2 Authorities and unauthorised settlements

Apart from the limited capacity of the public sector, a large part of urban areas do not have access to services for other reasons. Growth of settlements and activities largely take place outside planned areas. For people seeking a place to settle the options are either an unauthorised area or trying to find a place in the existing regulated areas, which thus become more and more dense. The combination of high density and extensive unauthorised areas – where density could also be quite high – is a typical feature of rapid urbanisation in the South.

The proportion of the urban population who live in unauthorised areas is typically between 30–60% of the entire urban population. It is estimated that at the turn of the century the number of people residing in informal areas is a staggering 1 billion (Foster *et al.* 1998). In most cases, these settlements are not recognised and there are no or rudimentary public services provided. The formal premise for the public sector to work in these areas does simply not exist. Many Governments have gone one step further by prohibiting utilities to provide services to people residing in unauthorised or informal settlements, obviously fearing that such provisions would act as a "pull factor" and thus stimulate the growth of informal settlements. Worse even, the government system is often harassing organisations of civil society to provide the necessary services that the public sector is unable or reluctant to provide (Collignon and Vézina 2000).

Combination of informal settlement and congestion represents a major obstacle in efforts to extend water services. For the densely populated areas, conventional systems are particularly unsuited. For Dar es Salaam, it is estimated that maybe as much as forty percent of the population reside in areas where the density of houses, shops and other physical features preclude the possibilities to build conventional structures for water supply and waste management (personal communication Mark Mujwahuzi 2000). Practical difficulties also arise when people settle in flood prone areas or along steep hill slopes as reported for many cities in Africa and Latin America (Collignon and Vézina 2000). The lack of formal recognition compounds the problem.

With a rapid growth of informal settlements, the infrastructure for water provision, wastewater handling and also other components of the basic infrastructure required for a functioning society are consequently inadequate and often substandard as compared to what is required. For the people living in these areas, the situation becomes a vicious circle. Since the settlements are not recognised, people often have to solve problems by way of bribes and through political patronage (Hasan 2000; Tropp 1998). With a lack of title to the land on which people reside and with no official recognition, the inhabitants are also denied access to formal credit. They have no other opportunity but to rely on informal loans where monthly rates of interest of 10% or even higher are common (*ibid.*).

The unfortunate social and institutional situation, which characterises living conditions in the informal settlements is thus extremely costly in monetary terms for members of the community. Collectively, the community suffers from the decay of the environmental situation in the neighbourhood. At the same time, the urban economy as a whole is to a large extent driven by the labour force that people in informal areas provide (Figure 8.5). Wages offered to people who make up the informal sector are low and people can be hired temporarily or on an hourly or daily basis with no social security.

The vulnerability of people living in unauthorised settlements is, of course, most serious for themselves. Indirectly, a worsening of the social and environmental situation in the informal areas will affect the urban system and the urban economy as a whole.

Poor water and sanitation services in these areas should therefore not only be seen as a matter that concerns the fate of the people living in informal settlements but for the functioning of the urban economy as a whole (personal communication, Michael Mutale and Dennis D. Mwanza, 2000).

The existence of the informal settlements can not be overlooked or denied infinitely. Extending the area to be served without a corresponding increase in public resources will, of course, not solve the problem of poor public services. If the public sector is unable to mobilise additional resources and to deal with the problems on its own, the realistic option is to allow and support organisations in civil society that carry out necessary tasks. The public agencies must also provide the regulatory arrangements so that systematic solutions for the urban system as a whole can be identified and promoted. As shown from various parts of the world, independent providers and civil society organisations are carrying out many vital tasks but the degree of co-ordination is limited (Collignon and Vézina 2000; Hasan 2000; Drangert 1997). It is also important to note that the services offered in this way are widely accepted among the citizens. Often, but not always, it is a cheaper solution as compared to an approach where the Government tries to solve the problem, through contractors.

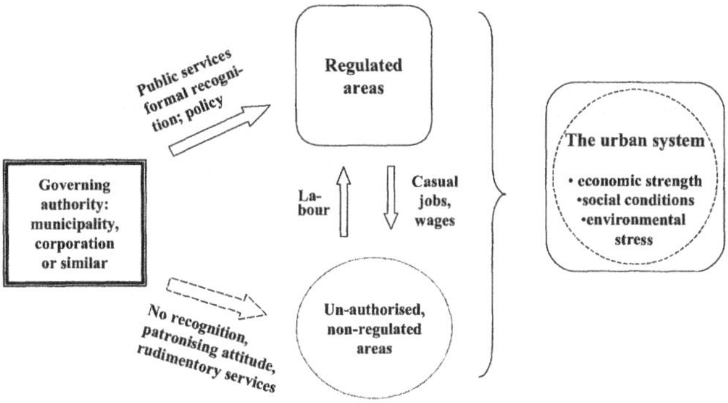

Figure 8.5 A schematic illustration of the combination of regulated and non-regulated areas in urban system and the lack of a policy for how to handle challenges in the non-regulated areas which are of significant importance for the fate of the urban system as a whole

The informal character is a significant feature in areas where poverty prevails but it is also common in areas where the economy is booming. The combination of rapid growth of informal settlements and mass poverty has contributed an extreme density of human population and a complex mixture of formal and informal social and economic activities. It is common to find industrial units, and especially small-scale enterprises at sites where the conditions are uncertain. The premises where the activity is housed are often on lease contracts where tenure over extended periods of time is not secure. Under such circumstances, the entrepreneurs are reluctant to invest in structures that would contribute to an improvement of the environment situation in the neighbourhood (Blomqvist 1996). It is also common that small-scale industries, are located in areas which are not considered suitable for industrial activity. For Delhi, the number of industrial units was estimated to about 126,000 (in 1996). Out of these, some 97,000 are located in so-called "non-conforming areas" which refer to residential areas or areas that are not designated for industrial activities in the Delhi Master Plan (Ahmed *et al.* 2000). Location in these areas is convenient for the labour force and the prices on land are comparatively modest. At the same time, many of these small-scale industries are extremely hazardous from an environmental point of view.

8.5.3 A new category of pollution

It is a matter of grave concern that water, which is a *sine qua non* for almost any kind of development and human activity, is clearly a victim of urban growth. Both in terms of overexploitation of readily accessible sources and in terms of degradation of quality of surface as well as ground water, the trend is alarming. The sufferings and problems caused by pollution of water in terms of impairment of human health has been a matter of concern for some time (WSSCC 2000). The perception of water quality problems in the South has tended to be associated with bacteriological contamination due to poor sanitation and hygiene behaviour and the threats to human health and wellbeing have been stressed.

Although attention to the connections between drinking water – sanitation – poverty – human health rightly remains a prime concern, it is important to recognise a new threat to water quality. It refers to the substantial pollution load from a growing industrial sector, from transport systems and from energy production. It is still related to low level of socio-economic development but not so much to poverty as such. This kind of pollution is very much the result of the structure of the economy, industrial growth, especially of the small scale units commented on above. A lack of systematic and stringent pollution control measures compounds the problem. The lack of policy or, rather, its proper implementation is an expression for the weakness of the prevailing institutional and regulatory system and an absence of social forces that could put pressure on the legal and regulatory system.

8.5.4 An integrated approach in a basin context

With an increasing number of people and economic activities in a small geographical area, it becomes increasingly important with an integrated approach in water management.

Integration refers both to the situation within the urban area as well as in a river basin context. Within the cities it is common that industry, commercial establishments and public institutions, e.g. schools and hospitals, make their own arrangements for water supply and, occasionally, that they are taking steps to reduce their contribution to environmental degradation. Also for individual households, the public supply is sometimes inadequate, for instance, in connection with house constructions, weddings, etc. For various reasons, ground water is often the most convenient choice. Foster (1999) estimates that " ... more than 1,000 million urban dwellers in Asia and 150 million in Latin America probably depend on well, spring and borehole sources" (p. 125).

If withdrawal of groundwater on a large scale is not part of an integrated plan for urban water management, a number of problematic consequences will appear. It will, for instance, result in the drying up of shallow wells. Another consequence is that the amount of wastewater is different, and usually larger, than the amount of water that the water utility plans for. The design of common treatment plants becomes a problem since volume of wastewater can hardly be projected and the possibility to charge for water pollution cannot be related to water supply. Moreover, heavy withdrawal of groundwater in coastal cities increases the risk of salt-water intrusion. Socially, it implies that people who have access to private wells and who have the economic means may pump water from deeper and deeper layers whereas people who rely on shallow wells do not have that option.

Another kind of integration refers to the need to relate urban water management to the basin within which urban centres are located. Cities located in the lower segment of a basin naturally face other problems as compared to cities located in the upper segments (Falkenmark 1997). Rivers are sources of survival, while at the same time, they are the recipients of waste and various by-products from society. Increasing withdrawals inevitably mean less water in downstream segments of the river, that is, unless the amount of groundwater is not making up for the difference. Increasing withdrawals also imply that each unit of water is more intensively used and that more wastewater and effluents are added to the recipients. The synergetic effect is obvious; more pollutants are added to the recipients while the dilution capacity is reduced.

For the big cities in semi-arid contexts, where the base-flow of rivers is small, the balance between the natural flow and the amounts of sewage and effluents that are channelled to the recipient streams are growing with city size. Rivers, which pass through the urban centres in the South are today literally turned into sewers from the waste generated by urban-industrial complexes. According to a high level panel associated with the World Commission on Water for the 21st Century, the world's rivers are in crisis "Some are dying, others could die ... more than one-half of the world's major rivers are being seriously depleted and polluted, degrading and poisoning the surrounding ecosystems, thus threatening the health and livelihood of people who depend on them for irrigation, drinking and industrial water" (World Water Council 1999; http://worldwatercouncil.org; Press release 29/11/99).

India is one of the countries where a large proportion of the people are facing a continuously worsening situation in urban areas and along rivers. A dramatic deterioration is reported from Delhi, which is located along river Yamuna. Even though the city constitutes only about 2% of the catchment area of the Yamuna basin, the contribution of pollution load from the area is about 80% of the overall pollution load in the basin (GoI 1997, 10).

8.6 STAGES OF A WATER USE AND POLLUTION ABATEMENT POLICY

It is unfortunate that the urgency to focus on what happens to water after it has been used is not universally recognised. Human ingenuity to handle waste problems and threats towards water quality degradation, which also implies environmental degradation, is far less developed as compared to progress in terms of exploitation of water and other resources. Strategies to prevent problems in these regards are generally poorly conceptualised and rarely implemented. It is, however, possible to identify three stages along which the environmental policy has progressed. A fourth stage, also discussed below, is currently less articulated, but it appears to be a vital component in the evolution of a comprehensive and effective pollution abatement policy.

8.6.1 Stage I: out-of-sight and out-of-mind strategy

At an early phase of industrial growth and urbanisation and with a limited environmental awareness, the strategy to deal with growing volumes of by-products and human waste is usually to get rid of it in a simple but deceptive manner. An "out-of-sight and out-of-mind" approach has invariably been practised and surprisingly seldom questioned. Wastewater and effluents are transported from areas where they are generated, i.e. in areas where people live and work and where they are creating a direct nuisance, to areas where they are not supposed to cause problems. The pipes that carry wastewater, and the chimneys through which smoke is channelled, are prolonged, literally and in practice.

If the pipes cannot be prolonged from existing units, the strategy may instead be to move the polluting industry to some other site where exposure to the pollution is reduced. In Delhi, for instance, an area of about 15,000 acres has been set aside at the outskirts of the city for industries which are currently contributing to a considerable pollution in the city (Ahmed *et al.* 2000; ENS November 27, 2000). Shifting polluting industries to new sites does not solve the problem and it is a costly and difficult operation.

The conjunction between out-of-sight and out-of-mind is, for natural reasons, most significant below the ground surface. Throughout history, a wide variety of substances have been emitted from activities in society and have gradually percolated through the ground and accumulated in soils, ground water and sediments. The environmental history that is recorded is these archives, illustrates that the threat can be found on all continents (Sampat 2000). Groundwater contamination is a common problem in the industrialised countries both underneath cities and farmlands (Stanners and Bordeau 1995). Pollution of groundwater is done in various ways and for various reasons. It could be direct or indirect and could be the result of deliberate action or ignorance. Even in societies with an advanced environmental policy and where technological resources are considerable, the use of the ground as a convenient sink is practised. In Florida, for instance, underground injection wells are often used to pump contaminants underground for disposal (Sampat 2000). It is therefore not surprising that similar practices are reported from countries having less technical and other resources at their disposal. Officials of the Andhra Pradesh Pollution Control Board in India, for instance, found a deliberate injection of toxic waste, arsenic-laced effluents, directly into borewells (Tiwari and Mahapatra 1999).

Since a part of the waste injected or otherwise dumped into the ground, consists of substances that are only slowly degradable, and which move very slowly, the build-up of an environmental debt is a reality. However, it could take a long time before the hazardous substances reach aquifers on which society depend. Although the lack of reliable data prohibits a detailed picture of the extent of the threat, a growing number of reports indicate that it is a mounting problem throughout the globe. In 1995, the Central Pollution Control Board in India reported that the groundwater pollution in a large number of sites in 16 States in India was 'critical' and found that industrial activity to be the primary reason for it (*ibid.*). Tests carried out at the Indian Institute of Technology in Kanpur on samples taken from eight places in three States, Haryana, Gujarat and Andhra Pradesh, showed traces of heavy metals in various combinations. The levels of mercury were 'dangerously high' in all samples with concentrations up to 0.2683 mg/l – to be compared to the permissible limit of 0.001 mg/l set by World Health Organisation (*ibid.*). Similar findings are reported from countries on all continents (Sampat 2000; Stanners and Bordeau 1995).

The deliberate dumping of toxic waste into the ground is a sign of a wilful neglect on the part of the polluters to avert a looming hydrocide (Lundqvist 1998). This kind of behaviour is probably based on a hope that dilution and natural processes in terms of dissolution, degradation and dilution of non-desirable substances will take care of the problem. With such a perception, ground is not only a convenient but also a "safe" place to store waste (see Figure 8.6).

Apart from deliberate practices, the daily life of people and also legally sanctioned practices result in the deposition of huge amounts of pollutants in waterways and onto or into the ground. Households without easy access to garbage collection and sewage system have no choice but to dump waste wherever possible and convenient. Vehicles emit substances, which to a large extent end up in the ground and in water bodies. Use of low quality gasoline is common in many countries, which increases the load of harmful substances that are emitted from traffic. Landfills are common in most countries and the sludge from treatment plants and sanitation waste must be put somewhere. Dumping sites can be found right in places where people live, but locations at some distance from urban residential areas are usually selected. These are, of course, still in watersheds and thus become a nuisance or a threat to other settlements nearby and in downstream positions. In an erroneous belief that the problem is thereby solved, a new one is created.

Over the centuries, substantial amounts of waste and by-products from society have accumulated in the ground and in sediments. Historically, most of the accumulated pollutants emanate from mining industry, i. e. from sites that are generally at some distance from major urban centres and later on the industrial and also agricultural sector as a whole has been important sources of pollution. Currently and for the future, the accumulation of various harmful substances will increasingly take place in urban areas, first in connection with the industrial expansion, transport and energy use, but gradually the pollution will increase from the consumption in society, i.e. from the citizens themselves. Urbanisation, industrial development and a growth in consumption in society will make the out-of-sight out-of-mind strategy completely unacceptable.

8.6.2 Stage II: end-of-pipe strategy

Gradually the problems related to water and environmental pollution become subject to policy intervention in terms of the creation of institutions through which pollution abatement efforts are supposed to be executed. Pollution control boards and other types of environmental agencies are created and laws and regulations are developed to deal with environmental issues. When waste water and waste management can no longer be solved through the prolongation of pipes to out-of-sight destinations or when deliberate dumping into the ground is no longer accepted, the strategy is usually to build effluent treatment plants (ETPs). Building of ETPs represents an end-of-pipe solution in the sense that the solution is applied at the end of the man-made water system, i.e. where the pipes otherwise would have delivered the waste into the recipients or other parts of the landscape. Building of effluent treatment plants is probably the most practised form of water quality management component in the South today.

It is usually the public sector, which formally demands that ETPs are built. Financially, it is a substantial task and the facilities are supposed to be financed through fees and taxes paid by the polluters, although subsidies in one form or another are common. In order to facilitate public control and monitoring, but also with regard to economies of scale, the preferred solution from society is usually to build common effluent treatment plants (CETPs) to which effluents from several point sources can be disposed and treated. From the industries that are to be connected to treatment facilities, the reluctance to comply is noticeable in many, if not most, countries in the South.

Delays and failures to respond to environmental regulations demanding connection to CETPs are frequent. In the first place the reason of the reluctance to comply are referred to as financial. Fees, which are supposed to contribute to the construction and running costs of treatment plants, could be quite high in relation to other investments for a company, particularly for small-scale industries who run their business with the help of simple technologies in simple premises. It is important to note that the building of CETPs presumes that pipes can be drawn from the various industrial units to the site where the treatment plant is located. Land tenure and existing physical structures in the city/town may, however, hinder such a solution (Blomqvist 1996).

Many industries prefer individual treatment plants (ITPs). From the industry it is argued that a common solution would invite free-riding and that it presumes a separate, bureaucratic management organisation and thus outside the control of the industry, although they are expected to contribute financially (*ibid.*). Treatment facilities attached to the industrial site would give the industry a much stronger position since they would then be in a position to do the monitoring and, in many cases, also decide about what quality standard to apply and how to interpret data from measurements.

The very slow progress of implementing an environmental policy in terms of building and connecting industries to effluent treatment plants is illuminated during recent policy exercises in Delhi, capital of India and considered to be one of most polluted cities in the world (Agarwal, Narain and Sen 1999; ENS 2000). The media, civil society organisations and Government officials have repeatedly reported and complained about the situation and demanded that steps should be taken to turn the trend. In a "White Paper on Pollution in Delhi With Action Plan", it is lamented that "It is widely perceived that the problem is threatening to get out of hand" (GoI 1997, 1). The most common suggestion to improve the situation is to build CETPs and/or ETPs. The policy is, however, neither stringent nor transparent. Apart from the stress on end-of-pipe solutions in terms of the building of treatment plants, the Government is pushing for a re-location of industrial units that are currently located in areas that are not designated for that purpose. For some of the industries who invest in treatment facilities, there is therefore no guarantee that they will be allowed to continue their operation (Ahmed *et al.* 2000).

Partly due to the lack of clear directives from the responsible authorities, the adherence to policy is poor. It is unofficially estimated that few industrial units have actually installed ETPs. And for those who have complied with court orders, the result in terms of an improvement of the environmental situation is not obvious and certainly not direct. Since the standard for the ETPs is not regulated officially, it is up to the industries themselves to design and install treatment plants according to their own judgements and to provide photographs and affidavits to the authorities as proofs that they are no longer polluting. Given the lack of a coherent and systematic enforcement of existing laws and regulations between the various official institutions involved, the compliance with the policy is poor and the much-needed improvements of water and environmental quality seem far away.

Concentration of attention to treatment plants as the only component in abating pollution could be questioned from another perspective. Its cost is very high, particularly if chemicals and pesticides are to be removed, both in terms of capital equipment and in operation and maintenance. Even if treatment plants were installed, their effectiveness is restricted due to lack of proper standard, lack of design with regard to volume and effluents to be treated, temporary or more frequent bypass arrangements by certain industries, etc. In a recent study conducted by Greenpeace in India, for instance, it was concluded that end-of-pipe trapping technologies, such as building of Common Effluent Treatment Plants (CETPs) are a waste of money and that they are basically non-solutions. Despite introduction of CETPs in 1996 in three industrial estates in Gujarat, areas continued to be contaminated with POPs and heavy metals from the industries (Labunska *et al.* 1999).

8.6.3 Stage III: clean production technologies

If treatment plants are inefficient in trapping by-products and pollutants from production and other activities in society, the attention must turn to what is considered to be the source of the pollution. A logical task in this connection is to change production technologies. Introduction of "clean technologies" which, by definition, reduce the usage of harmful substances and/or make it possible to contain pollutants in connection with production processes, represent a much more promising approach as compared to the building of end-of-pipe treatment plants. Clean production technologies are generally also resource efficient. The amount of water, but also energy and other inputs in production, can be reduced per unit of production output. The introduction of such technologies marks, however, a significant departure from the previous stage in terms of institutional and management arrangements and they have considerable social implications.

With regard to the institutional handling of the problem, the focus shifts from an activity, which is imposed by the authorities to become an internal matter for the industrial and business community, i.e. usually the private sector. The financial and management responsibility of the clean technologies will be part of the business considerations. The likelihood of an introduction of these technologies is, therefore, not so much related to environmental judgements but to conventional business considerations. The likelihood of an introduction of these technologies is, of course, also related to incentives and sanctions from the Government. If the use of "green" or "clean" technologies are seen as an attractive business option, the likelihood of having such changes accepted, or even promoted by the water users themselves, is naturally much better as compared to imposing solutions in terms of treatment plants which industrialists tend to consider "dead investments". Similarly, if incentives in terms of special loans, promotion campaigns are provided, the possibilities for an introduction and spread of clean technologies increase.

Depending on category of industry and the context where it is introduced, calculations show that introduction of clean technologies can be economically justified, especially if the externalities are considered. Initially, the capital and running cost are likely to be high in connection with a change of production technologies. However, if such technologies were not introduced, the expenditure to bring water to a particular site would increase. Similarly, the treatment of effluents presumes considerable investments and running costs. Cost for a re-location of polluting industries should also be part of the calculation. Apart from calculations referring to the industrial plants themselves, it is of great relevance to acknowledge the improved image, and thus market advantages for the industry that introduces clean technologies (personal communication, Hans Lundberg 2000). With an increasing awareness among consumers about environmental issues and with the promulgation of "green labelling" and similar trade regulations, the advantages with clean technologies become obvious to representatives of the business community.

In contrast to obvious environmental advantages and the prospects of economic rationality, the likely social implications from changes in production technologies are problematic. Depending upon sector and how it is organised, an introduction of clean technologies might have considerable implications on employment opportunities and structure of the industry, among other things. "Clean technologies" tend to be capital intensive and labour saving and they are usually fairly sophisticated to operate in terms of skills required. Adoption of such technologies are therefore associated with substantial social costs in the short run and add to the number of people out of work. An introduction of these technologies could amplify social stratification and income disparities. It is generally not possible to introduce environmentally appropriate technologies within the framework of the existing industrial structure. Small-scale industries make up an important fraction of the industrial sector in many urban areas in the

South. They operate with the help of family members and unskilled labour and the activities are carried out in premises and in areas, which are only occasionally intended for industrial activities. Changes of production technologies in these units are most difficult. These industries constitute, at the same time, a major threat in terms of pollution. The introduction of "clean production technologies" is necessary but considering the social implications, competent and strong political leadership and institutional arrangements are necessary.

8.6.4 Stage IV: modification of social and institutional responsibilities

At a fourth stage of water use and water pollution abatement policy the focus is, again, shifting. This time it is from production technologies to the citizens themselves and what role they can and must play to tackle the water and environmental crisis. The serious degradation of water and environmental quality in urban areas will not be effectively tackled by concentrating attention to end-of-pipe solutions and production technologies. Both production and consumption patterns and living conditions in terms of, for instance, sanitation are intimately linked to social as well as to environmental security and sustainability, or the lack of it. Level of consumption among the poor is low but the absence of sanitation and arrangements for waste management in informal areas contribute to environmental hazards. Parallel with widespread poverty, a consumer society is evolving in other parts of the urban areas where water demand, among other things, could be quite high.

Citizens are important in two respects. They are the actual water users and they are also consumers of energy and various goods, which have implications for the generation of wastewater and other types of pollution. In contrast to pollution from industrial production processes, the contamination from consumption is much more diffuse in character. Each individual and household contribute small amounts of waste and pollution in terms of bacteriological contamination, nutrients, heavy metals, chemicals, medical substances, for instance, endocrine disrupters, etc. Aggregate human consumption in large and densely populated urban areas result in substantial amounts of pollutants which are picked up by the water flows through society and which contribute to the pollution load that is generated in a city.

Citizens are also part of social networks. In this capacity, they are a potential force in the shaping or modification of policy. As will be discussed further below, the Government and the formal management system are still reluctant to recognise citizens and their organisations as a positive force in this regard. Citizens could, however, act as a pressure group and demand an improvement in public sector performance, they could act as "watch-dogs" and monitor changes, for instance, in environmental parameters. With strategic support from professional technical and managerial expertise, citizens could also build and manage structures for water and sanitation and control public works.

In the search for effective ways to deal with the water and environmental crisis in urban areas, it is necessary to strengthen or develop institutions and organisations through which the citizens could contribute to an improved water management in the two respects indicated above. One crucial pre-condition in this regard is a political recognition. Another obviously important circumstance refers to strategic support in terms of technical and managerial skills. A third crucial input refers to education and awareness campaigns. Water quality problems, for instance, are hard to understand without appropriate scientific training.

The four stages of a pollution abatement policy are portrayed in Figure 6. The conventional, historical approach for a gradual improvement in water and environmental management has been to climb the "stairs" from lower left step to higher steps (see arrow in the lower, left part of Figure 8.6). From what has been discussed above, the conventional approach is not likely to be an effective strategy for countries in the South, simply because adequate treatment facilities presume levels of investments which few countries can afford. Furthermore, it is likely that this approach would take a substantial period of time before noticeable improvements may be recorded.

A strategy that emphasises that the problem must be approached from the other end represents quite another approach. It is radically different from the historical path, but it is not an alien component in the current context. As described above, a Hobbesian form of management system, where the Government and its public sector agencies have the sole responsibility to arrange for water and related services, is gradually giving way to a system where institutions and organisations in civil society assume important management tasks under a Lockean form of hydro-social contract.

In terms of technical solutions, there is no evidence to suggest that it is necessary or even rational to build treatment plants instead of using the same resources for changes in production technologies. Due to financial constraints, most countries will obviously not have the resources to develop all possible solutions at the same time. Choices must be made. With regard to what might be the most effective pollution abatement strategy, it is conceivable that the third stage in the ladder, i.e. the introduction of clean production technologies, should precede or, if circumstances permit, be parallel with a development of the second stage, i.e. the building of treatment plants (Figure 8.6).

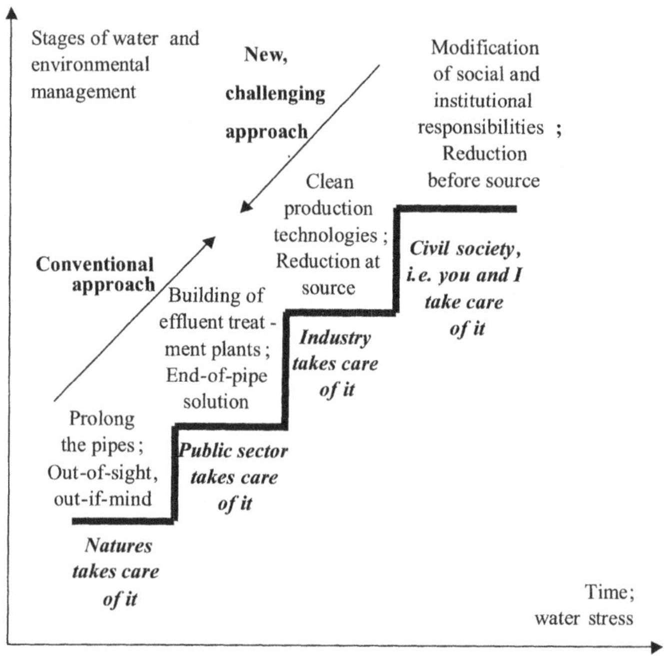

Figure 8.6 A schematic illustration of important stages in water and environmental management

However, the two approaches indicated by the two arrows should not be seen as alternatives. By stimulating the involvement of civil society organisations in water management and environmental stewardship, i.e. the top stage of the stairs, a development of the stages that are in the middle of the stairs is facilitated.

8.7 RESPONSES OF CIVIL SOCIETY TO WATER AND ENVIRONMENTAL CHALLENGES

8.7.1 Support from community but not from Government

Involvement of civil society in water management often starts as a reaction against the official policy and/or as an attempt to address significant social needs and environmental problems as experienced by the community. Typically, initiatives of civil society are stimulated by crisis situations, i.e. when problems appear acute and when the prevailing policy does not deliver the services. It is relevant to distinguish between two types of institutions or organisations. One type is the outcome of a process where a number of people decide to join forces and work for a common cause. The growth of NGOs, CBOs and, more generally, social movements are examples of organisations and institutions that belong to this category. Another important water management organisation is made up of what Collignon and Vézina (2000) refer to as independent providers of water and sanitation services. This group consists of entrepreneurs with a strong social affiliation to the community in which they work and who provide water and sanitation services to households and other customers wherever there is a demand for such services.

A characteristic feature of both types of institutions and organisations is that they adjust to local social and environmental contexts. For independent providers a high degree of flexibility in terms of response to customers' preferences is necessary. For organisations that involve a group of people or segments of a community, i.e. NGOs and CBOs, the importance of charismatic and visionary community leaders is obviously important in mobilising support and to draw up a programme for the kind of activities that are required. A common denominator for both types of organisations is also that they are seldom, if at all, recognised by the formal government or municipal system.

With these characteristics, civil society initiatives and the ensuing organisations tend to be in opposition to Government policy and they represent a threat to the legitimacy of the formal institutions and their officials. In addition, the work of civil society organisations often enters into conflicts with powerful economic interest groups of society, which play an important role in the emergence of problems for which a solution is wanting.

For civil society organisations to be able to achieve desirable objectives, the support from community members is critical. Competition among independent providers is often tough and unless the customers are satisfied with their services and pay for them, they are no longer able to operate. For the other category, financial support is also important, but the survival of these organisations hinges more on the number of supporters and proponents, who are able to defend the programme and who could contribute with voluntary labour and other support in kind.

In the absence of a formal constituency and with no direct access to project funding, organisations of civil society have to balance between interest groups to seek necessary support. Divergence of interests within the community is common, something that tends to be overlooked in assumptions about the potential of civil society organisations.

For both categories of civil society organisations, it is moreover very difficult or even impossible to solve the water management challenges on a larger scale without the consent, support or collaboration of Government policy and formal institutions. Most civil society organisations are comparatively small with a limited geographical coverage and in view of the complicated and large-scale technical challenges, which characterise urban water supply and sanitation systems, it is necessary to collaborate with authorities. They have the capacity to organise bulk supply of water and they often control the land on which the disposal of waste from households, including the sanitation system, can be disposed. In addition to the technical issues, the legal and regulatory system is of prime importance for the operation of civil society organisations.

The balancing between the various tasks is delicate since the legitimacy and strength of civil society organisations may easily be eroded if contacts with government are too close, particularly if the outcome is not satisfactory in the eyes of the groups that support the organisation. On the other hand, the possibility for civil society organisations to organise and execute water management projects in an urban context is limited. The degree of collaboration and mutual trust between formal, government institutions and stakeholders of civil society varies from a relatively high degree of harmony to mutual distrust and protest. Cases from Asia and Africa illustrate that an increased role of civil society institutions in urban water management is a complex and arduous challenge. It is also obvious that the relations between civil society organisations and the formal institutions are of crucial importance for the outcome of water management efforts

8.7.2 Communities as pressure groups—examples from India

In terms of number of people's movements and level of activity within civil society in general, India is one of the most prominent countries in the world. Social movements have had a strong influence on the political level and their significance in the daily life and in the cultural sphere is apparent. During recent years, a large number of civil society organisations, i.e. NGOs and CBOs, have become engaged in attempts to do something about the serious water and environmental problems. A strong concern about a rapid deterioration of the urban environment is noticeable. With pollution growing and the authorities seemingly unable, or even reluctant to control this trend, people living in urban centres, as well as in rural areas, have become engaged in various initiatives to change the situation.

In the gigantic Ganga River basin in the northern part of India, a large number of cities and towns of various sizes are located along the banks of the river and its tributaries, which are hold in reverence among major segments of the population. One of the important tributaries is Yamuna, which flows through Delhi, the capital of India, and Agra, two of the more famous cities. Altogether some 57 million people living in these and other urban centres rely on Yamuna as their main source for drinking water. At Allahabad, Yamuna joins Ganga and the waters of this river continue to flow through a number of important industrial and commercial centres with a large and rapidly growing population who are highly dependent on this river.

Even though the big cities and major rivers by their sheer size represent challenges of significant proportions, the communities living along and depending on smaller rivers may be relatively more acutely affected. The river Betwa is an example of a comparatively small river, which has been heavily affected by industrial developments.

Betwa flows through Madhya Pradesh, into Uttar Pradesh and then joins Yamuna some 400 kilometres downstream of Delhi. Along its journey it provides water to many towns and settlements. One such town is Vidisha, with a population of over 100,000 inhabitants, which has its main water supply from the river. A worsening of the water situation for the inhabitants over many years has stimulated the growth of a strong movement among the members of community. Pollution had been growing in the river for several years with recurrent signs of severe environmental consequences from the end of the 1980's. A flashpoint of the problem came in June 1996 when thousands of dead fish were found floating in the river at the water uptake point of Vidisha. Massive amounts of algae were found in the river and in spite of ample volumes of water flowing through the river, the water was declared as unfit for use (Agarwal, Narain and Sen 1999, 84–85).

With thousands of fish dead in the river and collected in the small dam where the water intake to the city is located, the people of Vidisha were outraged. Exposed to the evidence of the degradation of the water in the river, they decided to come together to protest saying "nobody has the right to pollute the Betwa river, our source of drinking water." A "think-tank" of the movement was created and a planned campaign against the State Government was launched which was supported by all political parties and also by the business community. A team of citizens travelled upstream to identify each source of pollution that entered the river. With this information they began campaigning in the town, stopping people in the streets asking them to take a pledge to clean the river. They led processions through the town and in December of 1996, a novel method was employed to draw attention of the authorities. They closed shutters and the entire town went on strike. The shopkeepers joined the movement saying that the river concerned us all.

With people's anger mounting, the state government was forced to take action. The minister for housing and environment was relived of his portfolio. Within days, the chief minister intervened. He ordered closure of the Som Distillery, one of the leading liquor manufacturers in north India and located upstream of the town, was seen as the main culprit. But this distillery is only one of the many distilleries, mills and other industries that contribute to the degradation of the river waters.

Within a few months the distillery was reopened. An independent organisation, the National Environmental Engineering Institute (NEERI), had been asked to examine the premises and production technologies and to come up with suggestions for how to prevent the serious pollution from June 1996 from recurring. The impact from the social protest was thus comparatively modest and the level of activity of the popular movement has toned down considerably after the initial response (*ibid.* 82).

As in many other cases, the initiatives of the local community were stimulated by events that affected the entire community. With the major threat to the city identified as coming from activities outside the town, the mobilisation of a movement that had active and loyal support across broad segments of the community was possible. The protests of the movement were obviously necessary for the responsible authorities to act, but since monitoring is poor, enforcement weak and the sustained effort of the popular movement lacking, the effect of the protest is limited and the river is still subject to considerable pollution (*ibid.* 82). The capacity of the civil society organisations to continue to push for an improvement in the water and environmental policy is apparently limited.

Apparently, the Hobbesian form of hydro-social contract is seriously questioned by the community. But the pressure from civil society is still not strong enough to result in a lasting shift in policy. The initiative is still to develop into a viable management alternative, which has the capacity to tackle the tremendous social and environmental challenges that are so common in India and many other countries.

There are many similar cases where the water supply in the cities and towns is seriously affected by activities in upstream areas. The general picture is that the protests from groups of civil society are only leading to temporary relief and piecemeal changes. Narain (1999) describes a number of similar cases from urban centres of India where a struggle for clean water to drink has been going on for decades, but where the prospects for a significant improvement are still to be seen.

8.7.3 A choice between public health and income opportunities?

The example described above, showed that mobilisation of civil society is often stimulated and facilitated if the threat comes from some identifiable source located outside the urban centre. If pollution and misuse of water occurs inside an urban area, the difficulties for a strong and united community effort increase. Since a reduction of pollution load and improvements in water quality presume changes in industrial activities, transport system, etc. there is an obvious risk that an improvement in water management is at the expense of jobs and income opportunities, at least in the short run.

The conflicting interests between the need for job and income opportunities, on the one hand, and measures to reduce the ongoing degradation of water and environmental systems, on the other, has been an unresolved and most challenging issue in Delhi for quite some time. The situation has become critical and the difficulties to execute a policy that would reduce the problem have been exposed through a number of reports and media coverage during the last years (Ahmed et al. 2000; ENS 17/1, 27/1 and 22/11, 2000). For the last five years or so, the Supreme Court has ordered that polluting industries, which do not build or connect to effluent treatment plants (ETPs) should be closed. The order to seal the industrial units has been iterated at several occasions, but the adherence to the court order has been quite weak and the date for compliance has repeatedly been postponed.

Contrary to the firm action taken by the legal authority to arrest pollution, the government control system is weak and a large number of industries are reluctant to respond. In some cases, the polluting units have been allowed by Delhi government to reopen on a trial basis based on claims that they have installed treatment plants. In other cases, units have installed treatment facilities as prescribed, but they are still not allowed to operate their industrial unit, apparently because they are located in areas that are not intended for industrial activities (Ahmed *et al.* 2000).

Contradictions in the public policy and its implementation have been illuminated recently. The high court issued closure notices to 1,384 industrial units by November 1999. A few months later, in January 2000, the Delhi Government ignored the legal order and allowed 372 units to start operations again. The decision by the Delhi government to allow the units to start operations again was based on ".. photographs or affidavits that they were no longer polluting". The validity of these proofs was, however, questioned. According to information and comments made by officials " .. a large number of polluting industries have furnished the exact same photographs to prove installation of the treatment plants" A prevailing perception of the industrialists is that treatment plants represent dead investments and they do therefore find ways and means to circumvent the administrative order (ENS 17/1, 2000).

Events took a violent turn in November 2000 when the Supreme Court ordered that about 7,000 polluting industrial units located in the residential areas in Delhi should be closed. Workers and owners joined forces and took to the streets were mass demonstrations brought the city to a virtual standstill. The heated situation opened up for confrontations between the police who opened fire on what was described as violent mobs and hooligans who targeted public property. At least three people were killed and the city was in a chaotic state (Ahmed *et al.* 2000; ENS 22/11, 2000).

The events in Delhi reveal two important circumstances, which influence water and environmental policy. One is that the formal institutions, in this case the Government of Delhi and the Courts, are pursuing contradictory policies. Representatives of the Court are "disgusted" by the poor performance of the Government, while Government officials are obviously unable to execute a policy that conforms with existing rules and laws. The hesitant steps they are taking do not give them support from any group. The court orders, which deliberately give top priority to public health even if it is at the expense of jobs (ENS 22/11, 2000), are also meeting a strong reaction among the public. The events have shown that there is a considerable public worry related to the prospects of losing jobs and that this might be felt as a more immediate and bigger threat as compared to the effects of a worsening environmental situation.

Again, the example illustrates that policy and the actual management system is in a deadlock situation where the Government through a Hobbesian type of central, monopolistic control is not able to handle the situation and deliver the necessary services. With a mutual distrust between various institutions and the associated contradictions in policy and with goal conflicts both among groups in society and at policy-making level, the management system is in disarray. Modification of policy is urgent. The most polluting industrial units and other polluting activities must be modified in terms of production technology or in operations or closed down. It is a tough option in the short run, but a prolongation of the current contradictory policy increases frustration while environmental degradation is worsening. Government has a prime responsibility that the necessary decisions are taken while implementation will be most difficult if civil society organisations are not made partners in this endeavour.

8.7.4 Scaling up operation without recognition in policy: challenges for the Orangi Pilot Project, Pakistan

Mobilising the community and provision of technical support to carry out self-help schemes is, of course, a basic task for civil society organisations. With increasing scale of operation it gradually becomes crucial to establish links with Government and the formal agencies and to become an integrated part of policy. While work at a local level and on a small scale can be carried out independent of Government support or recognition, it is increasingly difficult when the scale of operation increases. Both water supply and sewage systems in urban areas have to be part of systematic and integrated solutions. Water supply has in many cases to be arranged from sources far away and bulk supply can hardly be arranged by civil society organisations. The distribution of water within the city or town can, however, be organised by other entities than the bulk supplier. Similarly, the sewage system and the system for disposal of sludge requires an approval of the Government or public sector agencies in agreement with a urban plan and political decisions. Physical structures can, for instance, not be drawn over Government land without permission. Lack of permission implies a risk that the structures are confiscated or obstructed.

The interaction between the formal Government or municipality system and the civil society organisations thus becomes a critical test for the possibilities to scale up activities of the civil society organisations. With growing size and complexity of work that is carried out by civil society, the interaction and coherence within these informal organisations also becomes a crucial factor for the degree of success. One interesting example of a civil society effort where community members have become involved on a large scale to solve their most pressing problems is the Orangi Pilot Project (OPP) in Pakistan (WWC 2000, 37). Established in 1980, by Mr. Akhtar Hameed Khan, the renowned Pakistani social scientist, the founding idea behind OPP was to develop solutions that people could manage, finance and build (Khan 1996; Hasan 2000).

The Orangi Township has a population of over a million, which is a little over 10 % of the population of Karachi. Most of them live on state land, which has been informally subdivided. To cater for the requirements of the entire city of Karachi it is estimated that 80,000 housing units are required annually, but formal permits are only issued for about a third of this number. Similar proportions are valid for Pakistan as a whole (*ibid.*). The rapid urban growth has thus generated a demand for housing and infrastructure far in excess of what the Government system is able to meet. Lack of public finance and institutional constraints are two of the major shortcomings in this connection. With less than a third of the demand satisfied through the formal process, the growth of the unauthorised settlements and the application of alternative solutions has become necessary.

One of the most keenly felt problems by the community in Orangi Township concerned lack of sanitation. Through social and technical research carried out by the founders of OPP in the area, essential aspects of the living conditions in the area were revealed in terms of both sociological and cultural characteristics of low-income communities and also in terms of physical features of the settlements. From the very beginning of the OPP, it was understood that improvement in sanitation is a complicated task, which required the involvement of technical competence and managerial skills at a level, which was not available within the community. In the absence of such skills, the solutions tend to be substandard. Possibilities for communities to put a pressure on contractors and government authorities do also hinge on professional support, which has to be given from outside.

A vital component in the work was therefore to create an interaction between professionals from outside the community and members of the community. Since professionals tend to come from social and cultural contexts which are quite different from the conditions that prevail in the unauthorised areas and since social contacts of the professionals are usually closer to bureaucrats than to the communities that are supposed to be served, the creation of mutual trust and functional links between the professionals and communities in poor and unauthorised settlements is, in itself, an achievement.

The results so far from the work of the OPP are noteworthy on many accounts. Progress in terms of coverage in Orangi is impressive. Out of a total number of 7,256 lanes in the area, with their individual lane organisations, 6,082 have built their sewage systems. It is perhaps even more important that OPP has been instrumental in the building of these systems for a fraction of the cost that would have the case if the Government had done the job. On average, the investments paid by the members in the lane organisations, is US$16.50, which is calculated to be about one seventh of the costs that would have been incurred if the government had done the job (Hasan 2000, 4). A low unit cost is of significant importance not only for saving scarce financial resources or to get more work done for a given amount of money. It will also make the project affordable to people in low-income areas, one of the pre-requisites for sustainability and community commitment.

Special challenges refer to the scaling up of activities. Contrary to most other community-based activities, the OPP has been successful in mobilising communities in a number of places and in supporting activity in these areas. Apart from work in Karachi, which now involves 46 settlements, efforts are made to replicate the project in seven other towns in Pakistan and also abroad. The model developed for sanitation has been taken up in projects supported by the Asian Development Bank in Pakistan. According to Hasan (2000), the use of the model developed by OPP has reduced the cost of the ADB financed project from an estimated 1,300 to an amazingly low level of 36.2 million Rupees.

The expansion of OPP is not part of the official programme and its growth has increased conflict with existing government policy. Lack of Government recognition is of growing concern with growing size of operation. Scaling up operations are also hampered by the difficulties to provide adequate professional technical and managerial support. With growing size, close personal links between people fulfilling various functions are impossible. At the same time, the size of activities and flow of money increases. As a result of these new circumstances, the risk that community leaders are being lured to manipulate large contracts and turn into mafia agents increases (*ibid.*). Checks and balances are obviously required both at the community level within civil society as well as between civil society and the formal government system.

The OPP illustrates that a viable alternative to the central Hobbesian type of management system can be developed out of a civil society initiative, which has the capacity to handle complex technical tasks and, at the same time, be socially accepted. A major feature of the OPP is that the solutions have been developed at a much lower cost as compared to conventional solutions under Government contracts. With a low cost, the solutions are affordable to the local community, which is critical to its acceptance and functioning. Together with a management system that is based in the community, the prospects for sustainability are much better as compared to expensive schemes, designed and run by agencies which are not affiliated with the social, cultural and environmental context surrounding the scheme. Yet, this alternative is not officially recognised and it is still not part of policy (Hasan 2000).

8.7.5 Establishing a working relation between civil society and Government: experience from Cebu City, the Philippines

A common characteristic of the examples on civil society initiatives and activities described above is that they have not been officially recognised. The Cebu Uniting for Sustainable Water (CUSW), Philippines is an example of a civil society that has attracted a great deal of attention locally and internationally. It is an initiative that has developed into an institution that is formally registered and which functions as a meeting place for a large number of stakeholders from civil society and the Government system. The orientation of its work is urban and watershed planning and management. It is described as "a forum facilitator-convenor-mediator where stakeholders can voice their opinions, as well as become educated on the various water resource problems and discuss possible solutions". Formed in 1995, the CUSW seeks to develop collaborative, consensus-based solutions to complex problems (Nacario-Castro 1997). The following discussion is largely based on this publication and other report from the CUSW.

The core of the problem, which CUSW addresses is a very rapid population and economic growth concentrated to a relatively small area with a limited resource base on the island of Cebu. The population of about 1.4 million is concentrated in a small stretch of land between the coast and a rugged mountainous area that makes up most of the watershed. About half the population resides in Metro Cebu City on an area of only about 7,000 hectares. Unfavourable climatic conditions contribute to make the water supply situation critical. With limited availability of surface water, a heavy and largely uncontrolled abstraction of ground water has resulted in salt-water intrusion as far as about 4 km from the coastline. A transformation of land use from low input subsistence to intensive cultivation of high value crops in the catchment has increased the use of fertilisers and chemicals and created another threat to water quality.

With the rapid growth and high population density concentrated in a small piece of land, the water resources in the vicinity of the city have become inadequate. At the same time, the uplands of the catchment in which Cebu City is located, has become an attractive area for exploitation. As a natural consequence of an increased demand for water and land resources, a complex set of conflicting interests between various interest groups has developed. A key feature in the conflict and the way it stimulated a civil society initiative is the lack of a policy and a plan for the resource management in the catchment. A group of NGOs, tired of a lingering deficiency of the Government to formulate and execute water resources planning and management, reacted to the development plans of a major real estate corporation to transform part of the catchment for housing and a golf course by organising a meeting at the end of 1994. The idea of a broader participation of citizens in water watershed planning and management was taken up. Through the organisation of a number of discussion-field exposures and through headlines in the media, public awareness about the water problems increased and the CUSW was created in early 1995.

The major objective of the CUSW is to work for a "more sustainable approach toward the protection, management, and development of Cebu's water resources" (Nacario-Castro 1997). A formulation and institutionalising of an integrated resource management and land use master plan for the Metro Cebu area, with special consideration to the water resources, is seen as an important step in this regard.

CUSW has made considerable progress in terms of creating a forum where different interests groups meet. Initially, the CUSW met a lot of scepticism from representatives of civil society who felt that their interests were threatened or not taken care of by CUSW. Political leaders and people working in various Government departments were reluctant to support the organisation. With a strong commitment to build a coalition where consensus on the problems facing Cebu can be reached, the organisation has managed to become accepted by a wide range of interests groups. More than 150 members are stakeholders in CUSW, representing a wide range of government agencies, NGOs, private interest groups, professional and civic groups, religious groups etc. As such it has become an important channel for increasing a general awareness of the water and environmental challenges that the Cebu community as a whole have to face and tackle. It is also a recognised force in the local policy making process.

The main objective as formulated at its inception was the formulation of a master plan. This is a considerable task that has not been achieved as yet. The lack of success in this regard is partly explained by the technical complexities involved. It is, however, also important to note that in spite of the success to build a coalition of interests, it is also observed that "... many key offices mandated to pursue more effective environmental management actively ignore or oppose stated policy in response to other bureaucratic and political interests. Among the groups hostile or indifferent to CUSW's mission are both the powerful and the numerous, such as politicians, business, landowners and small farmers".

8.7.6 Relative harmony between Government and civil society: Southern Africa case

Socio-economic and political change has been quite strong in Southern Africa. The establishment of a democratic government in South Africa in 1994 has created a completely new context for institutional arrangements with significant opportunities for the involvement of civil society organisations in the management of water and other resources. The current development stands in sharp contrast to the previous system. A review of important features in water management in South Africa shows that radical changes can occur in relatively short periods of time.

The Witwatersrand area of South Africa, centred as it is on Johannesburg, owes its origin largely to the discovery of gold in 1886. The name "Witwatersrand" translated literally means "ridge of white waters", which was an apt description of the area at the time gold was discovered (Turton and Meissner 2000). These cascading waterfalls, the result of an abundance of springs and small rivers that existed above a major watershed, were the characteristic features at that time. They exist no longer, a lamentable fact of rapid industrial development. In 1887, barely one year after the discovery of gold, the population was around 10,000 inhabitants. Typhoid became a serious problem because no sanitation existed in the ramshackle mining town, and the local supply of water was soon either depleted or so polluted as to be unsuitable for human consumption. Between 1886 and 1903, Johannesburg ran out of potable water, resulting in a condition of prevailing water scarcity. This first-transition saw the birth of the Hobbesian form of hydro-social contract, with the Rand Water Board being constituted in 1903 (Bath 1999) and charged with the responsibility of mobilizing a secure source of water on which economic and social development could be planned. The problems of water supply became increasingly complex over time, with a steady growth in water demand of around 5% per annum for the next 80 years. To sustain this growth, the Rand

Water Board, in conjunction with the South African department of water Affairs and Forestry, extended its search for water over vast distances. Inter Basin Transfers, some of which are the largest in the world (the Lesotho Highlands Water Project being the best example), became the order of the day (Turton and Meissner 2000). By 1999, the Rand Water Board was responsible for supplying water to around 10 million people (Bath 1999).

By the time that the democratic transition occurred in 1994, approximately 75% of the total South African population lived on just 13% of the land (Abrams 1996), mostly in apartheid styled self-governing "homelands" or Bantustans. The consequences of this were far-reaching for water management. Not only were there an estimated 12–14 million people without any access to formal sanitation services, but there are also environmental effects of poverty which impact on the water resource base of South Africa today. The impact on the urban environment was large, with massive informal settlements, mostly without clean running water and sanitation services, geographically located alongside extremely wealthy suburbs with lush sprinkler-irrigated gardens and swimming pools. The best example is found in Johannesburg, where Sandton – the exclusive and upmarket northern suburb – exists within sight of Alexandra – one of the so-called "black spots" under apartheid and arguably one of the most underdeveloped portions of the Witwatersrand (now given the African name of "Gauteng" meaning place of gold).

The second-transition in South Africa coincided with the democratic elections in 1994, which saw the emergence of a Lockean form of hydro-social contract (Turton and Meissner 2000). This is manifest in the National Water Act (Act 36 of 1998) and the Water Services Act (Act 108 of 1997), which together provide the legislative framework. Underpinning this is the Constitution of the Republic of South Africa (Act 108 of 1996). These various pieces of legislation make the role of civil society much more powerful than before. It should be noted however, that civil society played a major role in the democratic transformation in South Africa, with NGOs in particular, championing the cause of human rights, environmental abuse and other social maladies.

It is therefore interesting to note that it is civil society in South Africa, and not Government, who are championing the notion of Water Demand Management. At first glance this may seem to be illogical. Why would civil society seek to limit its own consumption of water? The answer is peculiarly South African. Under apartheid, resource capture was so endemic that the minority had amassed the overwhelming balance of hydropolitical privilege in society, so by the time that democracy came, NGOs with a strong human rights and political agenda, championed the cause of the ecologically marginalised majority, calling for reduced consumption of water and the introduction of notions of ecological sustainability. This experience may be unique and may not necessarily be reflected elsewhere in the world.

8.8 WHAT ARE THE OPTIONS FOR THE FUTURE?

The options for the future will be a combination of many factors. Some of the key aspects have been discussed above. Technical and financial aspects are, of course, of significant importance. But in the absence of a policy where the social conditions and the links between the socio-economic system and water and the environment are duly recognised, improvements in water resources situation are likely to remain a distant dream. Changes in official policy are slow to come in most countries. The build up of viable and effective institutions within civil society is also a considerable challenge. It is clear though, that while the formal government must rule, this is not always effective from an social, equity point of view and the effects from prevailing management on an environmental count, shows a most worrisome trend. The role of civil society is becoming a key interceding variable. The situation is dynamic in nature, and it is important to capture the essence of hydro-social interdependencies to enable analysts and planners to understand these relationships better. It is hard to see any realistic alternative to a management system where civil society, in a broad sense, and government agencies are carrying out various tasks in tandem (cf. Figure 8.1).

The examples presented on the contribution of civil society to improved urban water management reveal a considerable variation. In South Africa, the Government and civil society is in relative harmony, simply because the democratic government is the result of long-standing civil society interaction with the past illegitimate apartheid regime. In the example from the Philippines, Government and Civil Society has found a meeting place for communication, but the lack of a strong and transparent policy contributes to a situation where significant interest groups in society, with contradictory development objectives, are jeopardising progress. The examples from India and Pakistan illuminate that the lack of Government recognition and support is a vital drawback for modification of prevailing policy. It was also shown that contradiction of interest between, for instance, public health concerns and a concern for job opportunities is a very noticeable challenge for an effective water and environmental policy.

The most obvious shortcoming refers to a poor coverage. Large segments of the urban population are without easy access to water and also without sanitary arrangements. For roughly half of the urban population who reside in the unauthorised settlements, livelihood conditions are particularly tough and uncertain. Due to prevailing official policies and as a result of a centralised management system, public services are not provided to the informal areas. Administrative harassment is, moreover, often hindering civil society initiatives and other providers to organise water and sanitary services to the people living in these areas.

A disregard for the most basic needs of a large part of the urban area and the urban population is counterproductive to the objective of a stable urban system as a whole. The strength of the urban economy largely depends on a healthy population and the overall environmental situation in the urban system will not improve unless the problems in the unauthorised areas are tackled. A worsening of the environmental situation in one part of the urban area is not confined to that particular area but will affect a wider part of the urban system and the basin where urban areas are located. The current focus on poverty alleviation in the international discussion on development strategies in bi-lateral and multi-lateral policy formulation presumes a comprehensive and integrated approach for water and sanitation improvement for the entire urban system.

Civil society initiatives are of vital importance in this context. Bureaucracies who traditionally manage water and pollution abatement systems must learn that pollution management is too important a business not to involve local people. A successful involvement of local people will not be operational until serious effort is made to institutionalise the involvement of people in the management of their water and waste.

The developing world has, by and large, neglected to strengthen its local democratic base. Mutual mistrust between officials representing the government system and the civil society organisations represents a major drawback for the evolution of proper organisation of urban water management. The urban municipal authorities remain within a representative system of democracy and in most cases are asked to govern areas that are just too large and diverse for management while at the same time, they are refrained from working in other parts of the urban area. Responsibility will need to be devolved to a level that is easier to manage and more importantly, the form of governance of these institutions will need to be participatory and not representative in a formal sense. Decentralisation of responsibility can, of course, not take place unless the Government or municipality formally recognises the existence of civic organisations and provide strategic support.

As emphasised in the body of the chapter, water quality is a particularly urgent issue. The availability of information about water quality is of significant importance to local communities. Transparency in this regard and in terms of Government environmental policies would be vital for people to participate in the management of the resource. Representatives of civil society must be regarded as partners in a proper water pollution abatement strategy. The regulation of pollution demands strong, capable institutions and a culture of transparency and accountability. It also demands a high order of societal discipline. Currently regulation of pollution control is very weak. Institutions needed to manage pollution control at the local level properly do not exist. Pollution control is also about strengthening democracy, and to make it both local and participatory. Government agencies must stimulate relevant and necessary changes in this regard.

Given the size and rapid growth of urban systems, special consideration must be given to the fact that both the supply of water to the town or city and the disposal of wastewater and effluents, must be arranged within the context of the basin in which they are located. For the big cities, the technical as well as the institutional arrangements are increasingly complex and the option to divide different management tasks among various management organisations should be contemplated. For the bulk supply of water to cities, the municipality or other designated Government agencies play a major role. The distribution of water within the city could, however, be organised by civil society organisations or other concessionaires. Services offered by civic organisations and independent providers seem particularly suited to cater for the needs of the poor segments of the urban population and to people in areas that are not yet regulated.

Management of wastewater, sludge and pollution abatement in general, is similarly dependent on a commitment of the Government in terms of creating operational procedures for compliance with water quality standards, implementation of incentives and sanctions for waste handling, provision of sites for sludge disposal, etc. But an effective policy for the abatement of water pollution must involve management arrangements at the level of the water user, for instance, in terms of changes in production technologies and modification of water use in society. It is futile to expect that the serious threats against environmental sustainability, which currently are so obvious in most urban areas in the South, will be averted only through a policy that favours the building of end-of-pipe solutions, which is the typical public sector approach.

Ultimately, it is the actual use of water, both at the household level and in industry and other sectors of the economy that determines how much water is required and what happens to water when it passes through society. We have emphasised that the deplorable state of water and environmental quality in many cities and the ongoing deterioration in this regard, must be effectively dealt with and this can only be accomplished by focusing on the actual use of water. Since decisions about actual water use are not primarily an issue within the public sector domain, it is logical that those who are the main stakeholders in this regard be actively involved in policy and management. This does, in no way, reduce the role and responsibility of the government system. But it highlights the need for a proper partnership between the formal and the informal management institutions and organisations.

8.9 ACKNOWLEDGEMENT

Valuable comments to earlier versions of the manuscript have been given by Malin Falkenmark, Jan-Olof Drangert, Karin Kemper and Leif Ohlsson.

8.10 REFERENCES

Abrams, L.J. (1996) Water Policy Development in South Africa. *Cranfield International Water Policy Conference, Cranfield University, Bedford UK,* September 1996. Available from Website http://www.africanwater.org

Agarwal, A. & Narain, S. (1999) Making Water Management Everybody's Business: Water Harvesting and Rural Development in India. *Gatekeeper Series No. 87.* International Institute for Environment and Development. London.

Agarwal, A., Narain, S. and Sen, S. (1999) The Citizens' Fifth Report. Part I: National Overview and Part II: Statistical Tables. Centre for Science and Environment, New Delhi. India.

Agarwal, A. (2000) Crosscurrents. *Down to Earth*; **9**(2). (http:/www.oneworld.org/ cse/html/ dte/dte20000615/dte_cross.htm)

Agence de l'Eau. (2000) *Social Charter for Water: A New Approach to Water Management in the 21st Century.* Paris: Réalisation les éditions Textuel (ISBN 2-909317-50-1).

Ahmed, K., Khanduja, C. and Sethi, N. (2000) Master Plan for Anarchy. *Down to Earth* **9**(15), 22-30. December, Available at:
Http://www.cseindia.org/html/dte/dte20001231/dte_srep1.htm.

Arrow, K., Bolin, B., Costanza, R., Dasgupta, P., Folke, C., Holling, C.S., Jansson, B-O., Levin, S., Mäler, C-G., Perrings, C. & Pimentel, D. (1995) Economic Growth, Carrying Capacity, and the Environment. *Science.* **268**, 520–521.

Bath, V. (1999) Johannesburg 'Corporatises' to Finance City's Water Supply and Sanitation Needs. In *World of Water 2000: The Past, Present and Future.* Water and Wastewater International: Penwell Corporation

Berkes, & Folke, C. (eds.) (1998) *Linking Social and Ecological Systems. Management Practices and Social Mechanisms for Building Resilience.* Cambridge University Press. Cambridge.

Blomqvist, A. (1996) Food and Fashion: Water Management and Collective Action Among Irrigation Farmers and Textile Industrialists in South India. Linköping Studies in Arts and Science, no. 148. Motala, Sweden

Collignon, B. and Vézina, M. (2000) Independent Water and Sanitation Providers in African Cities. Full Report of a Ten-Country Study. World Bank. (http://www.wsp.org/English/pdf/providers.pdf).

Constitution of the Republic of South Africa. (1996) Act 108 of 1996. Pretoria: Government Printer (ISBN 0-620-20214-9)

Cosgrove, W.J. and Rijsberman, F.R. (2000) *World Water Vision: Making Water Everybody's Business.* London: Earthscan Publications Ltd.

Cronström, A. (1986) Vattenförsörjning och avlopp. (Water Supply and Sewerage). Stockholmsmonografier. Uppsala.

Drangert, J.O. (2000) On-site sanitation and small-scale water supply in the developing world - International experiences. Background paper to Sida's Water and Sanitation Strategy. Swedish International Development Cooperation Agency. Stockholm. (mimeo)

Drangert, P. (1997) The Aguaterias in Paraguay. Report to the World Bank (mimeo).

Enviroment News Service (ENS) (2000)

- Delhi Allows Polluting Industries to Reopen, January 17;
- India's Supreme Court Bans Delhi Industrial Water Pollution, January 27
- Three Dead in Delhi Protesting Closure of Polluting Industries, Nov. 22.

Falkenmark, M. (1997) Coping with Water Scarcity under Rapid Population Growth. In: *Proceedings from SADEC-EU Conference on Management of Shared River Basins. Maseru*, 20-21 May.

FAO. (2000) New Dimensions in Water Security. Authors: Lundqvist, J., Falkenmark, M., Gordon, L., Folke, C. and Ohlsson L. AGL/MISC/25/2000. The report is available on: ftp://ftp.fao.org/agl/aglw/docs/misc.25.pdf

Foster, S., Lawrence, A. and Morris, B. (1998) Groundwater in Urban Development. Assessing Management Needs and Formulating Policy Strategies. World Bank Technical Paper no. 390. World Bank, Washington D.C.

Foster, S. (1999) The interdependence of groundwater and urbanisation. In *Proc. of the 9th Stockholm Water Symposium.* "Urban Stability Through Integrated Water-Related Management", August 9-12, Stockholm. Pp. 125–138.

Gleick, P. H. (1998) *The World's Water 1998-1999*. Island Press, Washington, DC.

Gleick, P. H. (1999) The Human Right to Water. *Water Policy* **1**, 487–503.

Government of India (GoI). (1997) White Paper on Pollution in Delhi With an Action Plan. Delhi, August 8. Available on http://envfor.nic.in/divisions/cpoll/delpolln/html.

Hasan, A. (2000) Scaling up of the Orangi Pilot Project Programmes: Success, Failures and Potential. *Asian Mayors' Forum on "Fighting Urban Poverty"*. Shanghai, 26–29 June.

Hertsgaard, M. (1998*) Earth Odyssey. Around the world in search of environmental future* Broadway Books. New York.

Homer-Dixon, T.F. (1994) Environmental Scarcities and Violent Conflict: Evidence from Cases, in *International Security*, **19**(1), 5–40.

Jonsson, A. (2000) The Trace of Metals. Use, Emissions and Sediment Load of Urban Heavy Metals. Linköping Studies in Arts and Science, 221. Linköping, Sweden.

Khan, A.H. (1996) *Orangi Pilot Project: Reminiscences and Reflections* Oxford University Press. Karachi.

Kemper, K. (1996) The Cost of Free Water. Water Resources Allocation and Use in the Curu Valley, Ceará, Northeast Brazil. Linköping Studies in Arts and Science, No. 137. Motala.

Kemper, K. (2000) The Role of Institutional Arrangements for More Efficient Water Resources Use and Allocation. In: Water Security for the 21st Century - Innovative Approaches, Proceedings from the 10th Stockholm Water Symposium. *Water Science and Technology,*

Labunska, I., Stephenson, A., Brigden, K., Santillo, D., Stringer, R., Johnston, P.A. & Ashton, J.M. (1999) Organic and heavy metal contamination in samples taken at three industrial estates in Gujarat, India. Greenpeace Research

Laboratories, Exeter, UK. Technical Note 05/99. (http: //www.greenpeace.org/%7Etoxics/toxfreeasia/gujarat.pdf).

Lundqvist, J. (1998) Avert Looming Hydrocide. *Ambio* **27** (6), 428–433.

Lundqvist, J. (2000) Rules and Roles in Water Policy and Management. Need for Clarification of Rights and Obligations. *Wat Intl*, **25** (2), 194–201.

Macy, P. (1999) Urban Water Demand Management in Southern Africa: The Conservation Potential. Publications on Water Resources, No. 13. Sida. Stockholm.

Matsui, S. (1993) Industrial Pollution Control in Japan: A historiacal Perspective. In: *Environmental Pollution Control. The Japanese Experience.* Papers presented at the UNU International Symposium on Eco-Restructuring, 5–7 July 1993. United Nations University, Tokyo.

Nacario-Castro, E.N. (1997) When the Well Runs Dry: A civil initiative in watershed planning and management in the Philippines. Report edited by Jefferson Fox, Larry Fisher and Carolyn Cook and published by Ramon Aboitiz Foundation Incorporated and Cebu Uniting for Sustainable Water Foundation. Cebu City, The Philippines.

Narain, S. (1999) We all live downstream: Urban industrial growth and its impact on water systems. In *Proc. of the 9th Stockholm Water Symposium.* "Urban Stability Through Integrated Water-Related Management", August 9-12, Stockholm. pp. 75–83.

Narain, S. (2000) The Future of India's Urban Environment. In *Proc. from: A Swedish-Asian Forum on the Future of Asia's Urban Environment.* May 15–17. Ministry of Foreign Affairs. Stockholm.

Neto, F. and Tropp, H. (2000) Water supply and sanitation services for all: global progress during the 1990s. *National Resources Forum* **24**, 225–235.

National Water Act. (1998) Act 36 of 1998. Pretoria: Government Gazette.

Newman, P. (2001) Sustainable Urban Water Systems in Rich and Poor Countries. Steps towards a new approach. In: Water Security for the 21st Century - Innovative Approaches, Proceedings from the 10th Stockholm Water Symposium. *Water Science and Technology* **43** (4), 93-100.

OECD. (2000) Increasing Resource Efficiency. Available at: http:// www.oecd.org/env/efficiency/wastemini.htm.

Ohlsson, L. and Lundqvist, J. (2000) The Turn of the Screw. In: FAO 2000, pp. 4–76. Available at: ftp://ftp.fao.org/agl/aglw/docs/misc.25.pdf

Peters, N.E., and Meybeck, M. (2000) Water Quality Degradation Effects on Freshwater Availability: Impacts to Human Activities. *Water International*, **25** (2), 185–193.

Platt, R. (1999) Presentation given at Oxford University to the 4th Meeting of METRON Project Partners on 7th and 8th June 1999, hosted at the Oxford School of Geography.

Reisner, M. (1993) *Cadillac Desert: The American West and its Disappearing Water.* Revised Edition. New York: Penguin

Sampat, P. (2000) In Deep Trouble. The Hidden Threat of Groundwater Pollution. *Worldwatch Papers 154.* Washington DC.

Schreiner, B., and Naidoo, D. (1999) Water as an Instrument for Social Development in South Africa. In *Proc. of the 9th Stockholm Water Symposium,* Stockholm, "Urban Stability Through Integrated Water-Related Management", August 9–12, Stockholm. Pp. 253-266.

Serageldin, I. (1994) Water Supply, Sanitation, and Environmental Sustainability. The Financing Challenge. The World Bank, Washington D.C.

Shah, A.A. (2000) Asia's Urban Environment – The Challenges Ahead. In *Proc. from: A Swedish-Asian Forum on the Future of Asia's Urban Environment.* May 15–17, 2000. Ministry of Foreign Affairs. Stockholm.

Stanners, D. and Bourdeau, P. (eds.) (1995*) Europé's Environment. The Dobris Assessment* European Environment Agency. Copenhagen, Denmark.

Swyngedouw, E. (1998) *Modernity & Hybridity – The Production of Nature: Water and Modernization in Spain.* Working Papers in Geography 97-12, School of Geography, Oxford University.

Tiwari, M. and Mahapatra, R. (1999) What goes down must come up. *Down To Earth.* **8** (7).

Tortajada, C. (1998) Water Supply and Wastewater Management in Mexico: An Analysis of the Environmental Policies. *Int. Journal of Water Resources Development* **14** (3), 327–337.

Tropp, H. (1998) Patronage, Politics and Pollution. Precarious NGO-State Relationships: Urban Environmental Issues in South India. Linköping Studies in Arts and Science, No. 182. Linköping, Sweden

Turton. A.R. and Ohlsson, L. (1999) *Water Scarcity and Social Adaptive Capacity: Towards an Understanding of the Social Dynamics of Managing Water Scarcity in Developing Countries.* Paper presented in Workshop No. 4: Water and Social Stability at the 9th Stockholm Water Symposium "Urban Stability through Integrated Water-Related Management", hosted on 9–12 August by the Stockholm International Water Institute (SIWI). Also available as MEWREW Occasional Paper No. 18 from Website http://soas.ac.uk/Geography/WaterIssues/OccasionalPapers/home.html

Turton, A.R. (2000) Water and Sustainable Development: A Southern Perspective. Forthcoming in *Encyclopedia of Life Support Systems (EOLSS).* UNESCO: EOLSS Publishers Ltd. Also available in draft form on Website http://www.up.ac.za/academic/libarts/polsci/awiru

Turton, A.R. and Meissner, R. (2000) *The Hydro-Social Contract and its Manifestation in Society.* AWIRU Occasional Paper. Available on Website http://www.up.ac.za/academic/libarts/polsci/awiru

UN (2000) Progress made in providing safe water supply and sanitation for all during the 1990s. Report of the Secretary-General (E/CN.17/2000/13). United Nations, New York.

Water Services Act. (1997) Act 108 of 1997. Pretoria: Government Gazette.

Water Supply and Sanitation Collaborative Council (WSSCC) (2000) Vision 21: A Shared Vision for Hygiene, Sanitation and Water Supply and a Framework for Action. Geneva, Switzerland.

World Bank. (1990) *Structural Adjustment and Sustainable Growth: The urban agenda for the 1990s.* World Bank, Washington DC.

World Resources Institute (WRI). (2000) World Resources 2000-2001. People and Ecosystems. The Frying Web of Life. The World Resources Report is produced as a collaborative effort by UNDP, UNEP, IBRD and WRI. Oxford, UK.

World Water Commission (WWC). (2000) *World Water Vision: Commission Report – A Water Secure World. Vision for Water, Life, and the Environment.* London: Thanet Press

Personal communication:

Lundberg, Hans, 2000. Research Director, Swedish Environmental Research Institute, Hälsingegatan 43, S-100 31 Stockholm.

Mujwahuzi, Mark, 2000. Professor, University of Dar es Salaam, Tanzania

Mutale, Michael. 2000. Executive Secretary, African Water Resources Management Forum (AWRMF) Secretariat, c/o Water and Sanitation Programme - Africa (WSP-AF), Nairobi.

Mwanza, Dennis D. 2000. Director, Water Utility Partnership for Capacity Building in Africa, WUP, P.O. Box 2642, 05 Abidjan, Ivory Coast.

Reinius, Lars-Gunnar. 2000. Stockholm Water Company, Maintenance and Operation Department, Stockholm.

9

Outlook for the 21st Century

José Alberto Tejada-Guibert and Čedo Maksimović[1]

Water is a most precious resource, not only because it is essential for life, but also because it is subjected to a mounting demand, while its availability in the form suitable for human use is being reduced significantly through pollution and mismanagement. True to its name, this book tackles the question whether the problems concerning the urban dimension give rise to a *deadlock* or whether hope for their resolution is maintained? Hope not only for the people who live in cities or who are on a gradual – or sudden – move to urbanised areas, but in general for society at large, because the urban pressure for water and for the disposal of wastewater reaches much farther than the geographical confines of the cities. The collective answer of the authors to the question posed has been unequivocal: yes, there is hope.

[1] *With contributions from the principal authors of the chapters.*

Edited by Čedo Maksimović and José Alberto Tejada-Guibert. ISBN: 1 900222 76 0

9.1 INTRODUCTION

The reader has been led through diverse subjects in this book – from the technical principles and practices to relevant, economic, social and institutional aspects in urban water systems – all having the underlying thread of examining the current situation and providing an outlook for the future.

This chapter endeavours in distilling the views and propositions made throughout the book and in providing a synthesis on promising approaches in securing safe rehabilitation of the outdated urban systems and in planning, developing and managing sustainable systems for the future generations.

One of the central concepts that we encounter time and again in the book is that new approaches, even a new mentality, are required with a sense of urgency to face successfully the current urban water challenges, especially in the developing countries. To be sure, this is not a revelation that has suddenly come about. A process of growing awareness has been evolving over the past few years and becoming a formal set of propositions. The Paris Statement (UNESCO & Académie de l'Eau, 1998), issued on the occasion of the symposium on Water, the City and Urban Planning (Paris, April 1997), recognised the growing acuteness of the urban water problems and, under an integrated approach to environmental and water management, called for action on:

- The implementation of demand management measures, recognising the economic value of water but having social sensitivity;
- The alleviation of competing needs for water of rural and urban areas, considering, among others, the safe supply of urban stormwater and wastewater for agriculture;
- The integrated management of surface and ground water in urban areas.
- The timely consideration of environmentally sound projects that will increase the availability of water where it is needed;
- The development of appropriate approaches for urban drainage according to climatic differences, considering the special problems of developing countries;
- The investigation of novel approaches, citing, among many others, dry sanitation as an alternative to traditional water-borne sanitation; consideration of stormwater as a resource for less demanding uses, local treatment of stormwater using biological systems, and development of technologies to recycle nutrients from wastewater;
- The adoption of non-structural means for urban flood mitigation, as structural solutions have proved to be at best partial solutions;

- The active involvement and participation of all stakeholders and a willingness to invert the traditional top-down decision;
- The specific and urgent problems of the peri-urban, less privileged areas in the outskirts of cities in developing countries; and
- Conception and application of new solutions and systems when existing ones have proved inadequate or may be infeasible, considering the unique character and circumstances of each city.

It may be added that other approaches recently developed, such as the Household Centred Environmental Sanitation (Kalbermatten et al., 1999) contain similar propositions. The elements of the Paris Statement, enunciated as guiding concepts, have found in large measure substantiation and elaboration in this book.

9.2 LEGACY FROM THE PAST

The water-related environmental problems that cities are facing can be grouped broadly into four overlapping and interacting categories: (i) problems of access to water and sanitation infrastructure and services; (ii) problems of pollution from urban wastewater; (iii) problems of infrastructure ageing, resource degradation; and (iv) problems of water related hazards. In developed countries categories (i) and (iv) are either under control or less prevalent. Cities in developing countries and lower-income urban areas often face all four sets of problems (intensified in many cases, as they tend to occur simultaneously and with higher intensity over longer periods of time).

It is predicted that almost all of the world's urban population growth will occur in urban areas of less developed countries (as illustrated in the left part of the Figure 9.1). These areas are already overloaded by their own waste, suffer from inadequate sewer systems, and experience enormous difficulties in supplying running water to sprawling peri-urban locations, illegal settlements, and decaying city centres, placing additional pressure on the economies of these countries. About 50% of the urban population in the developing countries reside in areas where government representatives are unable or reluctant to work due to formal and circumstantial reasons. Cities in more developed countries (right side of the Figure 9.1) will experience two types of ageing: ageing of the infrastructure calling for its overall overhaul and modernisation and ageing of its population calling for higher life quality and shifting emphasis on the types and levels of services and facilities required.

Developed countries have gone through a period of industrial growth that more often than not left a persistent imprint in the form of environmental damage, such as shown in the Figure 9.1 (left). The balance between the pollution and self-purifying mechanism of nature was many times thrown into disarray. Environment may have been degraded such a degree that in many locations it is assessed that it might take another 50–100 years for the environmental mitigation to succeed at a high cost. It is argued that the developing countries have a chance not to repeat the same mistakes and can avoid future costs (Figure 9.1 right). The cities of developed countries also have the massive legacy of existing infrastructure built on the basis of what was an acceptable technology when these works were implemented – this can be a very heavy load when age, technological obsolescence and new conditions point to a change.

Figure 9.1 Urban contrasts: Both settlers have a "complete urban water infrastructure": running water, sewerage and solid waste disposal facilities and an urban stream to carry away the storm runoff (Photographed by Č. Maksimović).

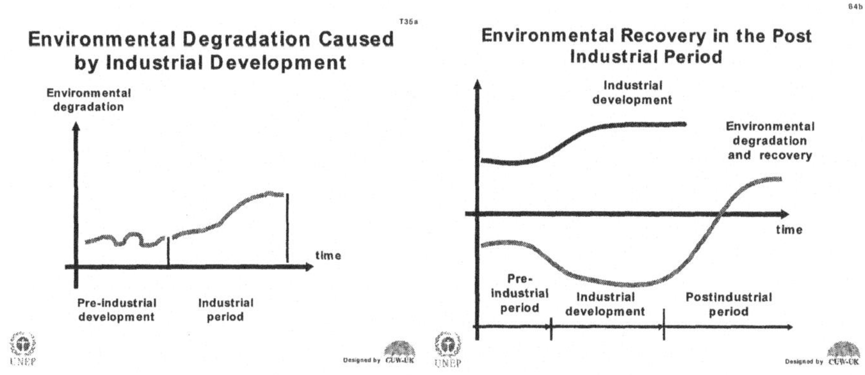

Figure 9.2 Environmental degradation caused by industrial revolution (left) and costly recovery in the post-industrial period (right). Do developing countries have to make the same mistakes in the future as the developed ones did in the past?

Urban water management facilitates the sustainable provision of specific services, including water supply, flood protection and drainage, wastewater treatment and disposal, and maintenance of a general amenity function of water resources. The currently pressing tasks of water management tend to differ in developed and developing countries. While in developing countries there is a need for developing new supply capacity, in developed countries, the focus seems to be on reducing further growth of withdrawals by demand management, water saving, and wastewater reuse and recycling. Urban water management has to "accommodate" the relatively new function of protecting receiving waters as a recreational, aesthetic, environmental and ecological amenity.

Flood protection is practised widely, with emphasis on 'living with floods', in many cases based on structural flood protection measures (major civil works), increasingly complemented now by non-structural measures. The degree of protection provided largely reflects the socio-economic development of the society. Flood control schemes and surface water quality protection are supported by stormwater management in urban areas, with control of runoff generation and enhancement of its quality. In developing countries, the drainage facilities that would serve to perform flood control are often not properly maintained, accentuating the impact of the flood. Despite the general acceptance of the concept of sustainable urban drainage systems based on broader application of control near the source and resources recycling, the appropriate solutions are still to be sought for both developed and developing countries.

Cities and societies in developing countries require not only avoiding the mistakes that industrialised societies made in the past, but also learning to recognise what approaches, even if successful in urban areas of developed countries, might not work so well or not at all in their cities. The conditions can be quite different, many of which are quite obvious: usually less human and financial resources, an incipient or inappropriate institutional structure and greater political instability. Some are subtler: different dynamics (cities in developed countries and their current systems developed over many decades or even over one or two centuries), cultural receptiveness to certain solutions, climatic differences (technologically appropriate solutions for temperate climates may not work in the humid tropics or arid climates). Other variables are just beginning to be evaluated, such as the effect of the level of corruption on development.

The rapidly expanding urbanisation in developing countries affects dramatically the water resources in the areas of concern because of physical, chemical, biological, and combined impacts. The need for integrated management becomes particularly evident as these water resources become depleted or degraded. Beyond the physical and environmental aspects, the whole framework of society is involved, directly or indirectly in the quest of acceptable and lasting solutions: governance and policy issues, interaction with other competing sectors, responsive institutional arrangements, involvement of civil society in the decision making process, water rates (are they economically and socially effective?), and a host of other considerations. Shortcomings exist in developed and developing countries, but they are indeed deeper, wider and more entrenched in the developing world.

9.3 OUTLOOK FOR THE FUTURE

9.3.1 Managing urban water in a time of change

Integrated water management can be understood as the application of a long-term policy for developing water resource uses in a sustainable and socially balanced manner. An implementation strategy needs to be developed, recognising the existing legislative, administrative and other constraints and conditioning factors. Urban areas with their specific management problems and solutions represent only sub-elements of river catchments, and their management has to support the overall catchment planning.

The planning and implementation of integrated urban water management require a full spectrum of tools and procedures, ranging from communication with, and empowerment of, the public to sophisticated computer tools and techniques. It is increasingly recognised that successful management of urban waters cannot be accomplished with the conventional prescriptive water management. Adaptive management, that is, proceeding with solutions along the envisaged optimal path, while collecting additional data and monitoring the system performance and making further adjustments as required, has a better chance of success. This requires much better coordination and open communication channels between the relevant players. It entails not only the dialogue between authorities and management bodies, but also a better understanding and handling of damaged "interfaces", resulting from interactions between different subsystems of urban water system. Furthermore, many aspects of urban water management (land use, industrial development, environmental laws) and the associated socio-economic issues extend outside of the authority of urban water managers and, consequently, the carrying out of the protection, restoration and management of urban waters will require co-operation of the entire society.

However, conditions do not remain stationary, hence best solutions do not necessarily remain constant with time. This is particularly true in a world of increasing globalisation, which is manifesting itself through changes in patterns of consumption, production, distribution and investment. Such changes can lead to greater economic efficiency – but likewise contribute to increased pressure on resources. Thus, globalisation goes beyond economics – it is a social, cultural and technological phenomenon with evident environmental implications.

Four broad areas of on-going transformations in water resources have been cited in the text: (i) *conceptual breakthroughs,* including shifting paradigms in terms of ecosystems, sustainability, non-linear and non-hierarchical systems, complexity, uncertainty, and interdependence of surrounding environments; (ii) *methodological advances,* especially multi-purpose/multi-objective approaches, Decision Support Systems, Risk Analysis, and the implications of rapidly expanding computational power and access and of new information and communication technologies; (iii) *organisational mobilisation*, in terms of new administrative mechanisms, institutional arrangements, participatory processes, renewed interest in river basins contingency planning, etc.; and, (iv) *contextual changes*, signifying the full range of on-going and future quantity and quality problems, new areas of concern, shifting priorities, potential socio-political intervention mechanisms and comprehensive resource policies.

The strategy of change outlined here builds on that first outlined by Butler & Maksimovic (1999) and emphasises the seven Is: *Integration* (the development of coordinated and integrated urban water models); *Interaction* (or the search for the additive, cumulative synergistic effects of complex urban systems); *Interfacing* (especially with the public and the environment); *Instrumentation* (in terms of real time control, sensors and noninvasive techniques); *Intelligence* (and the expansion of data, information, and knowledge through GIS [Geographic Information Systems], and hydroinformatics); *Interpretation* (the complementarity of data and judgement, and a wise combination of structured reasoning and disciplined imagination); and, finally *Implementation* (the capacity for true and concerted action and the transformation of policies into reasonable practice that combines corrective and proactive strategies and tactics).

9.3.2 A new paradigm at work: emerging technologies vs. conventional practices

The ongoing commonplace and deplorable degradation of the urban environment is a threat to the sustainability of the entire urban system. The goal of sustainability is well supported by the principle of total management of the water cycle, including, first of all, provision of the basic healthy drinking water and elementary sanitation to all. However, to ensure meeting these needs of society in the long run new approaches will have to be developed, tested and applied:

(a) the appropriate techniques for re-use of treated wastewater (for disposing pollutants and substituting for sub-potable water supply);
(b) integrated stormwater, groundwater, water supply and wastewater based management; and
(c) water conservation based approaches including more efficient use of water, substitute landscape forms and substitute industrial processes.

While in different cases one may aim at the same goal, the methods will differ. Even though an ever-increasing fraction of the world's population is served by proper sanitation (or in WHO parlance "improved sanitation") and a high percentage of population in developed countries is connected to wastewater treatment facilities, there is a large population in developing countries, which is not served with proper sanitation. Thus, the emphasis in developing countries would normally be placed in increasing the availability of proper sanitation, and in developed countries, the main goals regarding sanitation would be to improve the efficiency and sustainability of wastewater collection and treatment systems, by both changes in environmental behaviour and technological advances.

A new urban water paradigm to handle the types of problems described above is emerging. It calls for the implementation of demand management practices, resources recycling, and environment-friendly sustainable solutions in planning, development and management of urban water. The concept of demand management implies encouraging a change in life styles with emphasis on recycling, waste minimisation (leakage reduction), reuse and reclamation, which ultimately may be conducive to diminishing demands for water rather than continuously increasing supply. This change would go coupled with the use of potential technological breakthroughs associated with the "post-industrial city" that can alter the form, structure and operation of urban aggregations and, thus, the shape of solutions to urban water problems. This idea is applicable in both developing and developed countries.

The solutions analysed and advocated in chapters 3 through 6 call for the implementation the technologies already developed and proven to work in potable, wastewater and stormwater systems in developed countries. These technologies, making use of informatics technology up to a moderate scale, should be further tested, implemented on broader scale and made more sustainable. These include: advanced treatment; more efficient demand driven systems; "customer driven" corporate information systems; reduced cost and increased reliability data acquisition and processing systems; methodology for fault detection, state description and optimisation of systems' rehabilitation scenarios; improvement of water quality in the distribution systems; and outputs from wastewater treatment plants and from stormwater disposal facilities. Special attention is being paid to the innovative methods of drinking water treatment that will significantly reduce the risk of cryptosporidium breakthough, reduce the risk of presence of potentially harmful organic micropollutants in drinking water, especially those arising as byproduct compounds from disinfection. In the case of developing countries a proper balance between affordability, technical competence of the society and sustainability improvement should be sought.

It must be recognised that not all of the high technology solutions being developed and applied in the developed world will be immediately "exportable" to developing countries. As a matter of fact, it might not be desirable to aim indiscriminately for this. There are documented cases where the flow in technology has reversed directions – new approaches developed in the South for local conditions that result applicable in the North.

9.3.3 Social, institutional and financial aspects

The major elements that must be contemplated for the implementation of the concept of sustainable urban water development in the 21st century, as described in the book are:

- establishing efficient and just, institutional and legal arrangements for water management appropriate to the country's degree of social and cultural conditions and economic development;
- developing an appropriate institutional framework that integrates the transportation, housing and land use sectors into the water resources sector, capable of involving the stakeholders in the decision-making and management processes, having the requisite dynamism based on adequate financing and clearly delineated responsibilities;
- finding the appropriate financing for the heavy infrastructure investments required in the developing world and for the rehabilitation/upgrading of existing urban water infrastructure in the developed world;
- adopting rates for water services that, as far as possible, reflect the total cost of provision so as to meet the costs of operations, debt service and capital expansion, that that encourages efficiency in the use of resources, making the users and the polluters pay adequately and justly, and that seeks to make services affordable to all, including the poor.;
- encouraging the effective involvement of the civil organised society and water users, including groups from different backgrounds and on totally different issues, in the early stages of the planning process and in decision-making; and
- implementing innovative means of education, based on the updated concepts of environmental friendliness - sustainability increasing social awareness of population towards the caring for the urban water resources, and balancing of professional analytical studies and expert opinion against the concerns of various affected constituencies.

9.4 CONCLUDING REMARKS

It is appropriate here to quote from the message of the Secretary-General of the United Nations on the occasion of World Water Day 2001 (United Nations, 2001):

"Access to safe water is a fundamental human need and, therefore, a basic human right. Contaminated water jeopardizes both the physical and social health of all people. It is an affront to human dignity…

In this new century, water, its sanitation, and its equitable distribution pose great social challenges for our world. We need to safeguard the global supply of healthy water and to ensure that everyone has access to it".

This is a tall order – but society is morally bound to do its best to come through. The urban environment, home to an ever-growing majority of mankind, will provide the most severe testing ground to society's resolve. The rapidly evolving conditions compound the challenge. This book shows that approaches and solutions exist, but they do call for new attitudes and for a joint effort going from the highest spheres of government to the community based stakeholders. Technology, while being a key component to success, must remain subservient to the goals of sustainability and social equity.

9.5 REFERENCES

Butler, D. & Maksimović, Č. (1999). Urban water management – Challenges for the third millennium, *Progress in Environmental Science*, **1** (3), 213–235.

UNESCO & Académie de l'Eau (1998). *Proceedings Symposium Water, the City and Urban Planning, Paris 10-11 April 1997.* UNESCO Paris, France.

Kalbermatten, J.M., Middleton, R. & Schertenleib, R. (1999). Household-Centred Environmental Sanitation. EAW-SANI, Swiss Federal Institute for Environmental Science and Technology.

United Nations (2001). Press release SG/SM/7738 OBV/200. New York.

Index

IWA Publishing's authorised EU representative for General Product Safety Regulations is Diane D'Arras, 15 rue Duret, 75116 Paris, France, e-mail: safety@iwap.co.uk.

Printed and bound by CPI Group (UK) Ltd, Croydon, CR0 4YY
12/05/2026
02108864-0009